T0180589

Advances in Intelligent Systems and Computing

Volume 503

Series editor

Janusz Kacprzyk, Polish Academy of Sciences, Warsaw, Poland
e-mail: kacprzyk@ibspan.waw.pl

About this Series

The series "Advances in Intelligent Systems and Computing" contains publications on theory, applications, and design methods of Intelligent Systems and Intelligent Computing. Virtually all disciplines such as engineering, natural sciences, computer and information science, ICT, economics, business, e-commerce, environment, healthcare, life science are covered. The list of topics spans all the areas of modern intelligent systems and computing.

The publications within "Advances in Intelligent Systems and Computing" are primarily textbooks and proceedings of important conferences, symposia and congresses. They cover significant recent developments in the field, both of a foundational and applicable character. An important characteristic feature of the series is the short publication time and world-wide distribution. This permits a rapid and broad dissemination of research results.

Advisory Board

Chairman

Nikhil R. Pal, Indian Statistical Institute, Kolkata, India
e-mail: nikhil@isical.ac.in

Members

Rafael Bello, Universidad Central "Marta Abreu" de Las Villas, Santa Clara, Cuba
e-mail: rbellop@uclv.edu.cu

Emilio S. Corchado, University of Salamanca, Salamanca, Spain
e-mail: escorchado@usal.es

Hani Hagras, University of Essex, Colchester, UK
e-mail: hani@essex.ac.uk

László T. Kóczy, Széchenyi István University, Győr, Hungary
e-mail: koczy@sze.hu

Vladik Kreinovich, University of Texas at El Paso, El Paso, USA
e-mail: vladik@utep.edu

Chin-Teng Lin, National Chiao Tung University, Hsinchu, Taiwan
e-mail: ctlin@mail.nctu.edu.tw

Jie Lu, University of Technology, Sydney, Australia
e-mail: Jie.Lu@uts.edu.au

Patricia Melin, Tijuana Institute of Technology, Tijuana, Mexico
e-mail: epmelin@hafsamx.org

Nadia Nedjah, State University of Rio de Janeiro, Rio de Janeiro, Brazil
e-mail: nadia@eng.uerj.br

Ngoc Thanh Nguyen, Wroclaw University of Technology, Wroclaw, Poland
e-mail: Ngoc-Thanh.Nguyen@pwr.edu.pl

Jun Wang, The Chinese University of Hong Kong, Shatin, Hong Kong
e-mail: jwang@mae.cuhk.edu.hk

More information about this series at http://www.springer.com/series/11156

Francisco Campos Freire · Xosé Rúas Araújo
Valentín Alejandro Martínez Fernández
Xosé López García
Editors

Media and Metamedia Management

Springer

Editors

Francisco Campos Freire
Faculty of Communication Sciences
University of Santiago de Compostela
Santiago de Compostela
Spain

Xosé Rúas Araújo
Faculty of Social and Communication
  Sciences
Universidad de Vigo
Pontevedra
Spain

Valentín Alejandro Martínez Fernández
Faculty of Communication Science
University of A Coruña
A Coruña
Spain

Xosé López García
Faculty of Communication Sciences
University of Santiago de Compostela
Santiago de Compostela
Spain

ISSN 2194-5357   ISSN 2194-5365   (electronic)
Advances in Intelligent Systems and Computing
ISBN 978-3-319-46066-6   ISBN 978-3-319-46068-0   (eBook)
DOI 10.1007/978-3-319-46068-0

Library of Congress Control Number: 2016951706

This Springer imprint is published by Springer Nature
The registered company is Springer International Publishing AG
The registered company address is: Gewerbestrasse 11, 6330 Cham, Switzerland

Editorial Project Developed by XESCOM

This editorial project was developed by the International Research Network on Communication Management (R2014/026 XESCOM) and supported by the Regional Ministry of Culture, Education, and Organization from the Xunta de Galicia (Spain). It is made up of the research groups Novos Medios (www. novosmedios.gal), from the University of Santiago de Compostela; iMarka, from the University of A Coruña (www.imarka.es); and NECOM, from the University of Vigo (necom.uvigo.es), and the collaboration of researchers from communication groups at the Technical University of Loja and the Pontifical Catholic University of Ecuador—barra headquarters—as well as scholars from Portugal, Brazil, and Mexico.

Editorial coordination: Xosé Rúas Araújo and Iván Puentes-Rivera
Editorial review: Andrea Valencia-Bermúdez
Book layout and formatting review: Sabela Direito-Rebollal

Foreword

A total of 118 authors from 31 Spanish, Portuguese, English, and Ecuadorian universities have contributed to this book, and it was edited, coordinated, and introduced by professors Francisco Campos Freire and Xosé López García, from the University of Santiago de Compostela, Xosé Rúas Araújo, from the University of Vigo, and Valentín Alejandro Martínez Fernández, from the University of A Coruña.

Media and Metamedia Management has contributions from seven prestigious experts, who offer their expertise and the view from their vantage point on communication, journalism, advertising, and audio-visual, corporate, political, and digital communication, paying special attention to the role of new technologies, the Internet, and social networks, also from an ethics and legal dimension.

Readers can also enjoy 66 articles, grouped into diverse chapters, on Journalism and cyberjournalism, audio-visual sector and media economy, corporate and institutional communication, and new media and metamedia. All of them were subjected to double-blind peer review.

Contents

Editors and Contributors

About the Editors

Francisco Campos Freire is the senior professor of journalism at the Faculty of Communication Sciences of the University of Santiago de Compostela (Spain). He has a BA in journalism from the Faculty of Information Sciences in the Complutense University of Madrid, a MBA degree in industrial and service business management, and a Ph.D. in communication and contemporary history in the University of Santiago de Compostela. His research lines are focused on the study of Cultural and Communication Industries, management of informative and audio-visual business, communication policies, family business and research groups, analysis and accounting of quality in communication organizations, and corporate social responsibility.

Xosé Rúas Araújo is the tenured lecturer of Techniques of Electoral and Institutional Communication at the Faculty of Social Sciences from the University of Vigo. He coordinates the research group Necom: Neurocommunication, Advertising and Politics (www.necom.uvigo.es), Pontevedra, Spain.

Valentín Alejandro Martínez Fernández is the tenured professor of commercialization and market research at the University of A Coruña (Spain). He has a Ph.D. in information sciences (Complutense University of Madrid); has a MBA from the University of A Coruña; and has a degree in information sciences from the Complutense University of Madrid. He is the member of the Prometeo Project, which belongs to the National Secretary of High Education, Science, Technology, and Innovation (SENESCYT) of the Ecuadorian government.

Xosé López García is the professor of journalism at the communication sciences, Department of the University of Santiago de Compostela. For twenty years, he has worked as a journalist in Galicia and is a member of the Galician Culture Council and head of the New Media Research Group Santiago de Compostela, Spain.

Contributors

Silvia Alende Castro, Ph.D., from the University of Vigo (Facultade de Ciencias Sociais e da Comunicación, Pontevedra). Her doctoral thesis *Periodismo de Prevención en Galicia. El concepto de comunicación útil en la prensa diaria* earned an outstanding grade plus Cum Laude. orcid.org/0000-0002-4877-262X, Pontevedra, Spain.

Jaime Álvarez de la Torre is a Ph.D. student of planification and organization of tourism at the University of A Coruña, Spain, and the member of the Investigation Group in Applied Marketing in Faculty of Economics and Business at University of A Coruña, A Coruña, Spain.

Sergio Álvarez Sánchez He is a graduate on advertisement and PR (with his results granting him the End of Studies Prize) and master in communication of the organizations, having completed both of those studies at the Information Sciences Faculty of Universidad Complutense de Madrid (UCM). He is currently working on his Ph.D., based on the cited conclusions for this exploratory study, as a member of the UCM audio-visual communication, advertisement, and PR programme. He has contributed to different activities carried over by the departments for audio-visual communication and advertisement I and applied economy IV at the same university, Madrid, Spain.

Lorena Arévalo Iglesias She has a degree in communication sciences. She is specialized in creativity and contemporary art and is developing her career as researcher at the University of Vigo. She is currently working on her Ph.D. on the potential of urban art as an element of cohesion in small communities, Vigo, Spain.

Verónica-Alexandra Armijos-Buitrón has master's degree in management and business administration from the Instituto de Estudios Bursátiles, IE., is a Ph.D. candidate in business administration from Universidad Nacional de Rosario-Argentina Professor at UTPL, and is the member of the research group "Innovation and New Business", Madrid, Spain.

Ramiro Armijos-Valdivieso has master in applied economics, Universidad Autónoma de México and certification programme in business performance andis an economist at the Universidad Técnica Particular de Loja and permanent professor and Researcher and has research interest in the area of economics, banking and finances, hospitality and tourism, accounting, and business administration. Universidad Técnica Particular de Loja. Financial and Administrative Director, Loja, Ecuador.

Patricio-Mauricio Artieda-Ponce is a Master Haute Cuisine from the French school "Le Cordon Blue". He has a degree in tourism management and planning from the UTPL and in management of Hotel Companies from the Pontifical Catholic University of Ecuador. He received National Chef Instructor issued by WACS y ACE; International Chef Instructor from the French Culinary Academy;

Coordinator of the Degree in Gastronomy; and Professor and Executive Chef for 8 years in the UTPL. At present, he is studying a master in tourism and health, Loja, Ecuador.

María Barreiro-Gen She is an assistant professor at the Faculty of Economics and Business (University of A Coruna) and lecturer at the National University of Distance Education (UNED). Her research is focused on statistical analysis, labour market, and social exclusion. She is a member of the Economic Development and Social Sustainability Unit (EDaSS), A Coruña, Spain.

Ana Bellón Rodríguez She has a degree in journalism, MA in news edition, and Ph.D. in journalism. She is in charge of communication at the Consejo Superior de Investigaciones Científicas (CSIC) in Galicia, and she is a journalism-associated professor in the USC. Her lines of research include Free Press, Composition Models, and Scientific Journalism, Santiago de Compostela, Spain.

Jaime Cabeza Pereiro He is a professor of labour and social security law at the University of Vigo, project manager of several European, Spanish, and Galician research projects, author, coauthor, and editor of several monographs, chapters in books, and more than a hundred of academic papers in labour and social security law, speaker at national and international conferences and courses in labour and social security law and community social law, and arbitrer at the trade union elections of the Province of Pontevedera since 2001, Vigo, Spain.

David Caldevilla Domínguez He has a Ph.B. and a Ph.D. degree in information science from the Complutense University of Madrid and a diploma course in teaching from the University of Zaragoza. He is a professor of audio-visual communication, journalism, and public relations at the Faculty of Information Sciences from the Complutense University. He was professor at the European University of Madrid and the School of Public Affairs Studies (ESERP), in Spain, and the Portuguese Institute of Marketing Admimistração—IPAM—in Oporto (Portugal). He was the author of the books *The Spielberg seal, Manual of Public Relations, Public Relations and their foundation, The inside of corporate communication, Public Relations and culture*, and *Brief Dictionary of Mexicanisms-Gachupinisms* and *Gachupinisms-Mexicanisms.*, author of more than 30 book chapters and more than 70 scientific articles, and lecturer and speaker at over 70 international conferences.

María José Cantalapiedra She received her Ph.D. from the University of the Basque Country and is a professor at the Faculty of Social Science and Communication. Among her lines of research are the evolution of journalistic genres and their transposition to the Web. Author of more than twenty publications, she currently directs a research project funded by the Spanish government, Barakaldo, Spain.

Andreu Casero-Ripollés He is an associate professor in the Department of Communication Science and the director of the journalism degree at the Universidad Jaume I de Castelló (Spain). He received his bachelor's degree from the Universitat Autònoma de Barcelona and his doctorate from the Universitat Pompeu Fabra. He has also been a research fellow at Columbia University and the University of Westminster, among others. He is the author of various books and articles on political communication and transformations in digital journalism, Castellón, Spain.

Patricia-Marisol Chango-Cañaveral has MBA in Quality and Productivity Management from the University of Armed Forces ESPE; Food Engineer; Trainer Ministry of Tourism; Quality Control Supervisor at the Group Hanaska Catering Service; and Teacher at the Referential Technical College Luis Fernando Ruiz. At present, she teaches in the degree of gastronomy at the Department of Entrepreneurial Sciences, Hotel, and Catering Department of the UTPL, Loja, Ecuador.

Lluís Codina He is a lecturer in communication science at the Pompeu Fabra University in Barcelona (Spain). In addition to teaching undergraduate courses in journalism and audio-visual communication, he coordinates the school's master's degree in social communication (MUCS), Barcelona, Spain.

Raphael Cohen-Almagor is chair in politics at the Faculty of Arts and Social Sciences of the University of Hull (United Kingdom). He received his Ph.D. in political theory from Oxford University and has published numerous articles and book chapters in the fields of political science, law, Israel studies, philosophy, media ethics, medical ethics, education, sociology, and history. His latest book (2015) is confronting the Internet's Dark Side: Moral and Social Responsibility on the Free Highway (Cambridge University Press).

Jesús Ángel Coronado Martín He is architect, has Ph.D. in construction and architectural technologies, and has master in integrated management of projects, at the Polytechnic University of Madrid. Research work reflected in four articles in magazines JCR and three symposiums at international congresses. He is a professor at Técnica del Norte University, PUCESI (Ibarra, Ecuador) and Antonio de Nebrija University (Madrid). Professionally, he has worked in companies of the Spanish Real State Sector (Axxo, Tasa Group, etc), Ibarra, Ecuador.

Mónica-Patricia Costa-Ruiz Has master in business management and economics from the Universidad Técnica Particular de Loja (UTPL). She coordinates the master of management of corporate social responsibility at the UTPL. She is the member of the research group "Innovation and New Business" and professor at the UTPL, Loja, Ecuador.

Carmen Costa-Sánchez has Ph.D. in communication at the University of Santiago de Compostela, received Extraordinary Doctorate Prize awarded by University of Santiago de Compostela Communication Science Faculty and Latina de

Comunicación Award for the second most cited article between 2008 and 2014 and is a corporate communication professor at University of A Coruña. *Corporate Communication: Keys and Scenarios* editor (*Comunicación Corporativa: Claves y escenarios*, 2014, Ed. UOC), A Coruña, Spain.

Verónica Crespo-Pereira Has a degree in advertising and public relations from the University of Vigo and Master in management and audio-visual production at the University of A Coruña. She is a Ph.D. candidate at the University of Vigo and works as a researcher at the mentioned institution. Her professional career has been developed in the audio-visual sector, Pontevedra, Spain.

Alberto Dafonte-Gómez Has Ph.D. in advertising and public relations and was a lecturer of "Theory and Practice of TV Communication" at the Faculty of Social Sciences and Communication (University of Vigo). His research focuses on TV formats, audio-visual content on Internet, and advertising, Vigo, Spain.

Miquel de Moragas Spà He has Ph.D. in philosophy, is a researcher at the Institute of Communication (InCom-UAB), and is the member of the Institut for Catalan Studies (ICE) at the Faculty of Communication Sciences from the Autonomous University of Barcelona (UAB). He works primarily on communication theories and policies and studies on sport from a cultural point of view, Barcelona, Spain.

María del Mar Lozano Cortés has BA in audio-visual communication at the University of Extremadura (2015). She is currently doing postgraduate studies related to marketing and consumer behaviour at the University of Granada.

María del Mar Rodríguez She received her Ph.D. from the University of the Basque Country (UPV/EHU). She is currently in charge of content for the Eroski Consumer project and its Website www.consumer.es. With her specialization in crisis communication and digital journalism, she has published around ten academic articles. She has conducted research at the Catholic University of Chile (2003) and the Catholic University of Argentina (2004), San Sebastian, Spain.

Sabela Direito-Reboliai is a Ph.D. candidate in communication and contemporary information and researcher at Novos Medios (USC) and has degree in communication from the University of Santiago de Compostela (USC), master in communication and creative industries (USC), and degree in Movie and TV Scripts, Santiago de Compostela, Spain.

Yolanda Dominguez-Lopez has bachelor in tourism at the Business Administration and Tourism School, University of Vigo, and master in Inland and Health Tourism Management and Planning from the University of Vigo, Vigo, Spain.

Viviana Espinoza-Loaiza is a titular professor at the Universidad Técnica Particular de Loja, master of bank management and finance, master in agro-food economics and the environment by the Universidad Politécnica de Valencia, and engineer in banking and finance. Their lines of research are closely linked to the social economy in a special way finances popular and solidarity, the development of local capacities, and social capital, Valencia, Spain.

Tania Fernández Lombao is an assistant professor from the University of Santiago de Compostela and has Ph.D.—"Corporate Social Responsibility in European Public Broadcasters"—in Communication and Creative Industries from the same university. She has a degree on journalism from the USC, and at present, she combines conference's participation with her journalism practice at the radio station Cadena Ser. Also, she works as eventual correspondent at Agencia EFE and Europa Press, Lugo, Spain.

Ana Belén Fernández Souto is a professor at the University of Vigo. She has a Ph. D. in advertisement and public relations. she is a lecturer at the Universities of Uruguay, Brazil, México, Costa Rica, Chile, Italy, and Portugal. She lectures on public relations, protocol, crisis communication, and events.

Jessica Fernández Vázquez has Ph.D. from the University of Vigo. She has a degree in advertising and public relations and has master's degree in communication research and the title Specialist in Advertising, Marketing, and Consumption. Actually, she is developing her career as a researcher at the University of Vigo, currently working on a doctoral thesis about communication and terrorism, but she has some publications about public relations and corporate communications, Vigo, Spain.

Darío Flores Medina Born in Teide, Gran Canaria, Canary Islands, Spain. Finishing the architect major in the ULPGC, later he moved to Madrid where he finished his doctorate studies in the Escuela Superior de Arquitectura at the Polytechnic University of Madrid. His career has been focused on investigation and architecture, collaborating in different professional studies and universities in Spain and Ecuador.

Joan Francesc Fondevila Gascón is a doctor in journalism and communication sciences (Autonomous University of Barcelona), professor of universities (Pompeu Fabra University, Ramon Llull Blanquerna-University, Mediterrani University School of the University of Girona, Open University of Catalonia, University of Barcelona, Autonomous University of Barcelona, and Camilo José Cela University), CECABLE (Cable Studies Center) director, and principal researcher of the Research Group on Digital Journalism and Broadband and the Research Group on Innovative Monetization Systems on Digital Journalism, Marketing, and Tourism (SIMPED). He has published numerous scientific articles indexed on journalism, telecommunications, and social sciences. He has received numerous awards for his scientific, teaching, and management activity, Terrassa, Spain.

Bárbara Fontela Baró is a BA communication student at the University of A Coruña and IE Business School Master in Management February 2017 candidate. He received grants from Amancio Ortega Foundation and Pedro Barrié de la Maza

Foundation for studying abroad in Seattle, USA. Corporate Communication Department Intern at Gas Natural Fenosa and Research Intern at University of A Coruña, A Coruña, Spain.

Julia Fontenla Pedreira She has bachelor's degree in journalism at the University of Santiago de Compostela and master in language and communication in business at the University of Vigo. Currently, she is a Ph.D. candidate in communication and contemporary information and specialist in organizational communication and corporate, institutional press offices and private sector, social networks, and electronic journalism. Her professional experience is linked to different means of communication and university professor at Pontifical—Catholic University of Ecuador, Ibarra, Ecuador.

Manuel Gago Mariño He is an associate professor in the Faculty of Sciences of Communication of the University of Santiago de Compostela and director of culturagalega.gal, the digital platform of the Galician Culture Council. He has Ph.D. in journalism, researching on digital discourse, information architecture, and social media.

Fabián-Mauricio Gaibor-Monar has master in food processing; has degree in gastronomic management; is an executive chef of the del Catering Service/Restaurant"El Maizal; is planner, trainer, and advisor for the implementation of the Restaurant "Dulce Carbón"; and is a teacher at the Occupational Training Centre Canadian School. He also teaches at the UEB, ESPOCH and UTPL.

Fanny Galarza She has a Ph.D. in communication and journalism from the University of Santiago de Compostela (Spain). She has published papers in journals indexed in Scopus and Latindex and in conference proceedings. She is the editor of the book "Comunicación estratégica en las organizaciones" (Revista Latina). At present, she coordinates the degree in public relations and directs the project "Strategic Communication Observatory" in the UTPL (Ecuador). Her lines of research are focused on branding, social networks management, and traditional and digital corporate and strategic communication traditional and digital, Loja, Ecuador.

María García García has Ph.D. in audio-visual communication at the University of Extremadura (2012). She is a lecturer in the Department of Information and Communication at the University of Extremadura. She has participated in several publicly funded research projects, both regional and national, and made various predoctoral teaching, and research stays in countries such as Portugal and France; other postdoctoral stays in the UK and Ecuador. Her research focuses on business communication, especially digital communication in SMEs. http://orcid.org/0000-0002-7262-1602.

Aurora García González has Ph.D. in public communication. She is a journalism professor at the University of Vigo and the main responsible of Radio communication subject of the audio-visual communication degree and radio models:

innovations and social uses subject of master communication research. She has published several research papers in the field of communication, as "Radiomorfose em contexto transmedia" (ISSN: 978-958-738-286-0), Vigo, Spain.

José Vicente García Santamaría He holds a degree in journalism and audio-visual communication from the UCM and a doctorate in communication science from the Universidad Rey Juan Carlos. His extensive professional experience includes communication and marketing consulting work for Spanish and foreign corporations and communications positions in ICO and BBVA. He is currently a professor in the Department of Journalism of the Universidad Carlos III and head of the research group Instituto para la *Innovación Periodística.*

Berta García Orosa She is the professor of journalism at the University of Santiago de Compostela in Spain, with 12 years of research experience. She holds a BA in communication sciences, political sciences, and administration, as well as a Ph.D. in communication. Her lines of research include digital and print media, organizational communication, cultural industries, and media literacy, Santiago de Compostela, Spain.

Aingeru Genaut Arratibel He lectures at the Faculty of Social Science and Communication at the University of the Basque Country (UPV/EHU) since 2001. He received his Ph.D. from the UPV/EHU and has published more than a dozen articles in national and international magazines. His principal area of research is focused on new technologies and digital journalism.

Monike Gezuraga Amundarain She is a pedagogue, obtained MA in special education and Ph.D. in education and is a lecturer at the Department of Didactics and School Organization of the University of the Basque Country UPV/EHU, within the degree of social education. Her research is focused on Service-Learning (S-L) and the link between University-Society. She is the member of the S-L promotion team at UPV/EHU, the Zerbikas Foundation, and the Spanish Service-Learning University Network (ApS-U), Bilbao, Spain.

Flávia Gomes-Franco e Silva She has degree in journalism from the Universidade Federal de Goiás (UFG) and Ph.D. in Communication from Universidad Rey Juan Carlos (URJC). She has taught courses at different degrees of URJC and currently works as a lecturer in various graduate programmes at this university. She is a researcher of the National R&D Project "The Research System on Social Practices in Communication: Map of Projects, Groups, Lines, Objects of Study and Methods" (CSO2013-47933-C4-1-P), Madrid, Spain.

Cristina González-Díaz She holds an undergraduate degree in advertising and public relations. She is a faculty member at the University of Alicante's Department of Communication and Social Psychology, where she teaches undergraduate courses in advertising and public relations. She is also the director of research project "A Study of the Presence of Health Claims in Food Advertising", funded by the University of Alicante, Alicante, Spain.

Juan Enrique Gonzálvez Vallés He has Ph.D. in Information Sciences from the Complutense University of Madrid and degree in journalism. He currently belongs to the Department of Audiovisual Communication and Advertising at San Pablo CEU University, where he teaches Fundamentals of Advertising, Market Research, Radio and Radio Production and Direction. He has taught Marketing and Commercial Management in Camilo José Cela University in Madrid. He is also a part of Concilium Research Group and of the entities Spanish Society for the Study of Iberoamerican Communication and Forum XXI. Professionally, he has served on several media and coordinated the project of the area of and Communication and Sponsorship of BBVA: "I go with Carlos Soria", Madrid, Spain.

Manuel Goyanes He has a Ph.D. in communication science from the University of Santiago de Compostela and master's degree in statistics. Currently, he is an assistant professor in media management at Carlos III University of Madrid.

Carlos Granda Tandazo He is a Ph.D. candidate in communication and journalism from the University of Santiago de Compostela (Spain). He has published articles in journals indexed in Scopus and Latindex and in conference proceedings. He is the member of the "Strategic Communication Observatory" at the UTPL (Ecuador). He heads various institutional projects of the UTPL. His lines of research are focused on branding, city brand, digital brand, social networks management, and traditional and digital corporate and strategic communication, Loja, Ecuador.

Estrella Gualda She is accredited as full professor by the ANECA and holds a Ph. D. in sociology and a MA in politics and sociology sciences from the Complutense University of Madrid. She is a professor of sociology at the University of Huelva and director of the social studies and Social Intervention Research Centre. She has been visiting professor at the Princeton University (USA), Arizona State University (USA), and Universität Augsburg (Germany), directed many projects and published several books, book chapters, and scientific articles in the fields of sociology, social exclusion, migrations and minorities, cross-border issues, and social networks, Huelva, Spain.

Javier Guallar He is a lecturer at the Faculty of Library and Information Science of the University of Barcelona and at the Blanquerna School of Communication Sciences of Ramon Llull University. He is also subeditor of the journal "El profesional de la información" (EPI) and content curator of loscontentcurators.com.

Mar Iglesias-García She is journalist and lecturer at the Department of Communication and Social Psychology from the University of Alicante (Spain), where she teaches undergraduate courses in advertising, public relations, and tourism. She is the director of digital journal ComunicandoUA.com since 2010, Alicante, Spain.

Andoni Iturbe Tolosa, Journalist He has degree in journalism and Ph.D. in audio-visual communication and is a lecturer in the journalism II Department of the University of the Basque Country UPV/EHU in 2015 and in audio-visual communication of the University of Mondragon/Mondragon Unibertsitatea (2010–2012). He is the president of the Cinematographic Section of the Society of Basque Studies (Eusko Ikaskuntza/Sociedad de Estudios Vascos) and member of AEHC. His research focuses on communication and the dialectics between cinema and television, Legorreta, Spain.

Leire Iturregui Mardaras She is a doctor of journalism from the University of the Basque Country (UPV/EHU); she is a graduate in journalism and political science and public administration, specialized in International Relations, also from UPV/EHU. Since 2009, she has held a post as lecturer in the Department of Journalism II in the Faculty of Social Science and Communication at UPV/EHU. Her lines of investigation focus on war reporting, digital journalism, and institutional communication. She has worked at local broadcasters of the radio station Cadena SER and has collaborated with the newspaper *El Correo*. From 2008 to 2011, she was communications officer at the association of Basque Municipalities (EUDEL).

Óscar Juanatey-Boga He has a Ph.D. in economics and business from the University of A Coruña (UDC), master MBA in Management and Business Administration, master in Business Management and Marketing, and master in Business Communication by the UDC. Professor of marketing and market research at the Faculty of Economics and Business at the same university is the author of articles on media and communication strategies, A Coruña, Spain.

Iván Lacasa-Mas He is an associate professor, School of Communication Sciences, Universitat Internacional de Catalunya (UIC Barcelona, Spain). He is a researcher at the Labcom group, currently working on "Innovation and development of online media in Spain. Applications and technologies for the production, distribution, and consumption of information" project (reference CSO2012-38467-C03), which is financed by the Ministry of the Economy and Competitiveness, Barcelona, Spain.

Diana Lago Vázquez He was from Santiago de Compostela, Spain, and has graduated in journalism by the University of Santiago de Compostela with a master's degree in communication and creative industries. She is a Ph.D. student in the area of communication and contemporary information from the same university, and she has focused on the study of social networks and their relationship with the media, especially television, Santiago de Compostela, Spain.

Moisés Limia Fernández He has Ph.D. in journalism and is currently working as a postdoctoral fellow at the University of Minho (Portugal). He has two lines of research. Firstly, his personal research revolves around the relationships between journalism and literature. Secondly, as part of the research group *Novos Medios*, his research focuses on social networks and the diverse narratives and ways of participating in journalism that have emerged since the Web 3.0.

Ana María López Cepeda She has a degree and a Ph.D. in communication science from the University of Santiago de Compostela (USC) and a degree in law from the National University of Distance Education (UNED). At present, she is a professor at the University of Castile-La Mancha. Her areas of expertise are Communication Policies, Media Structure, Transparency, and Governance.

Mónica López-Golán She holds a Ph.D. from the University of Santiago de Compostela (USC). For 15 years, she worked as a television producer at Gestmusic Endemol SAU and currently is a professor and researcher at the Catholic University of Ecuador, Ibarra, Ecuador.

Carla López Rodríguez She has Ph.D. graduate from the University of Vigo. She has a degree in advertising and public relations and production in audio-visual communication and a master in strategic management and innovation in communication. Her research is focused on communication strategies in order to develop an efficient and effective communication in different situations using new technologies (web 2.0), Vigo, Spain.

Alex-Paul Ludeña-Reyes is a Ph.D. candidate in Sustainable Development and Innovation of Tourist Destinations from the University of Las Palmas de Gran Canaria—Spain. He has a master in sustainable development of tourist destinations from the same university. He has a degree in management of tourism and hotel companies from the Technical Particular University of Loja, Loja, Ecuador.

María Del Rosario Luna After attending the School of Philosophy and Literature (University of Buenos Aires), she received her doctorate degree from the University of Salamanca. Specializing in education and audio-visual communication, she implements Community Communication projects targeted at young adults and teachers. She has been a professor in several Ibero-American universities: Buenos Aires (Argentina), Salamanca (Spain), Santiago de Compostela (Spain), Nacional Este (Paraguay), and the University of Extremadura, where she currently works, Badajoz, Spain.

Lidia Maestro Espínola She received a doctoral degree in audio-visual communication and advertising from the Universidad Rey Juan Carlos. She also holds a licentiate degree in advertising and public relations from the UCM and an MBA from the EOI. She teaches undergraduate- and graduate-level courses in communication and marketing at the UNIR, where she serves as director of the undergraduate Marketing and International Business programme and academic coordinator of the master's degree programme Integrated Advertising: Creation and Strategy.

María-Dolores Mahauad-Burneo is a master in economics to environment and agro-food industry, Universidad Politécnica de Valencia; business administration engineer, Universidad Técnica Particular de Loja; and permanent professor, Business Administration School. He is the professor of the Subject Entrepreneurship and Strategic Management, head of Department of Enterprises Organization, and member of the Research team for Innovation and New Business, Loja, Spain.

Iñigo Marauri Castillo He lectures at the Faculty of Social Science and Communication at the University of the Basque Country (UPV/EHU). He received his Ph.D. in 2008, has carried out research in Chile and Argentina, and has published around ten articles in national and international magazines. He has worked at the newspapers El Correo and El País, the magazine Eroski Consumer, and its Website www.consumer.es, and Basque public radio and television (EITB), Vitoria, Spain.

Israel Márquez He has Ph.D. in information sciences and journalism studies from the Complutense University of Madrid (Extraordinary Ph.D. Award 2012/2013) and has master's degree in information and knowledge society from the Open University of Catalonia. He has been visiting researcher at the School of Literature, Media, and Communication (Georgia Institute of Technology) and at the IT University of Copenhagen. He is now a postdoctoral researcher at the Open University of Catalonia (Juan de la Cierva Research Grant Program).

Rosa Martín Sabarís She has Ph.D. in information science from the UPV/EHU and is a professor of the Department of Audiovisual Communication and Advertising at the Faculty of Social Science and Communication from the UPV/EHU. She teaches on the UPV/EHU's master's in social communication and the master's in corporate communication. Her research focuses on news production processes, professional sociology among journalists, electoral communication, and the image of immigration. She coauthored the EITB Style Guide, together with María José Cantalapiedra and Begoña Zalbidea. She has occupied several management posts and been editor of the communication studies magazine *ZER*.

Guadalupe Meléndez González-Haba She is a graduate in audio-visual communication (University of Extremadura) and master in communication management (UCAM). She was awarded a predoctoral research scholarship (PRS) to study TV advertising as an influencing factor on eating disorders. Later, she developed her teaching and research in the field of audio-visual communication and advertising at the Universities of Extremadura and Cádiz (Spain) and Veracruz (Mexico). Currently, she is a professor at the University of Cádiz, Jerez de la Frontera, Spain.

María Isabel Míguez-González She has Ph.D. in communication, advertising, and public relations and is a lecturer of "Theory and practice of public relations" and "Communication Management" at the Faculty of Social Sciences and Communication (University of Vigo). Her research focuses on public relations, communication management, and tourism. Since 2013, she is a part of the research group on "Neurocommunication" (NECOM) at the University of Vigo.

Pedro Mir Bernal is a doctor in marketing (CEU), professor at the University of Navarra and Pompeu Fabra University, and member of CECABLE (Cable Studies Center), the Research Group on Digital Journalism and Broadband and the Research Group on Innovative Monetization Systems on Digital Journalism, Marketing, and Tourism, Barcelona, Spain.

Verónica Mora-Jácome She is an assistant professor at the Department of Entrepreneurial Sciences, Universidad Técnica Particular de Loja. She is studying a master in domestic tourism and health from the University of Vigo. She has a degree in strategic market management from the Technical Particular University of Loja and a degree in management of tourism and hotel companies from the same university, Loja, Ecuador.

María Cruz Negreira Rey She is a journalism graduate and Ph.D. student at the University of Santiago de Compostela, in the Contemporary Communication and Information Programme, whose research centres on hyperlocal media and innovation, Santiago de Compostela, Spain.

Isabel Novo-Corti She is a professor at the Faculty of Economics and Business (University of A Coruna), lecturer at the National University of Distance Education (UNED), and head of the department of Economic Analysis and Business Administration. Her research is focused on groups at social exclusion risk: disability, gender, inmates, etc. She is the head of the Economic Development and Social Sustainability Unit (EDaSS), A Coruña, Spain.

Adriana Paíno Ambrosio She has graduated in journalism from the University of Valladolid; she then graduated from a master's degree in communication and creative industries from the University of Santiago de Compostela. She is currently working on her doctoral dissertation in the Department of Sociology and Communication in the University of Salamanca. Her research focuses on the study of new digital formats in journalism and transmedia storytelling, Valladolid, Spain.

Jhoana-Elizabeth Paladines-Benítez is a Ph.D. candidate in business administration from the Universidad Nacional de Rosario—Argentina—and is a professor at UTPL and part of the research group "Innovation and New Business".

Sara Pérez Seijo is a researcher at the Faculty of Journalism of the University of Santiago de Compostela (Spain), currently working with the research group Red XESCOM.

María José Pérez Serrano She holds a degree in journalism from the UCM and an MBA with a concentration in financial management. Her thesis for a doctoral degree obtained from the UCM was honoured with an award of special distinction

2006–2007. She is a member of the faculty of the Universidad Complutense de Madrid's Department of Journalism IV (Media Business). Her teaching and research are centred on media companies with a special focus on concentration within the sector and its impact on media pluralism.

Blanca Piñeiro Torres is a Ph.D. candidate in communication sciences at the University of Vigo and has BA degree in journalism (2002) from the University of Santiago de Compostela in advertising and public relations (2005) and audio-visual communication (2010), communication research masters (2011) from the University of Vigo, and professional experience in Advertising, Public Relations, and Journalism at the Congress of Deputies of Spain, Parliament of Galicia and Pazo da Cultura of Pontevedra.

Esperanza Pouso Torres She has a degree in journalism from the Complutense University of Madrid. She has a master's in communication research from the University of Vigo. At present, she is a Ph.D. student at the Department of Psychology and Communication at the University of Vigo, Pontevedra, Spain.

Iván Puentes-Rivera has degree in advertising and public relations and post-graduate course in research on communication and is the member of the research group "Necom: Neurocomunicación, Publicidad y Política" ("Necom: Neurocommunication, Advertising and Politics") of the UVigo. He has lectured at the graduate courses in advertising and PR and audio-visual communication. He has enjoyed research stays and teaching periods at the Universidade do Minho and the Universidade dos Açores. His research focuses on political communication, public relations, and social networks.

Rosario Puertas Hidalgo is a professor at the Universidad Técnica Particular de Loja, Ph.D. candidate in communication and creative industries from the University of Santiago de Compostela, and coordinator of digital communication in the direction of communication from the Universidad Técnica Particular de Loja. Her research interests are linked to the management of communication in social networks, advertising, and press.

María-Magdalena Rodríguez-Fernández has Ph.D. in economic and business sciences from the University of A Coruña. She is a professor at the Commercialization and Market Research Department of the same university. She is author and coauthor of various articles in journals and books. Her lines of research are related to communication, marketing, new technologies, and tourism, A Coruña, Spain.

Mª Isabel Rodríguez Fidalgo She has a Ph.D. from the University of Salamanca, where she graduated in audio-visual communication and in social work. She is professor at the University of Salamanca, within the areas of audio-visual communication and publicity. She conducts research focusing on new multi-screen media, new hypermedia–transmedia storytelling (fiction and interactive non-fictional documentaries), and cyberjournalism, about which she has published several articles in indexed journals, as well as books, Salamanca, Spain.

Emma Rodríguez Rodríguez is an associate professor of labour and social security law, Universidade de Vigo (Spain), researcher at several European, Spanish, and Galician research projects, author, coauthor, and editor of five monographs, chapters in books, and more than a eighty of academic papers in labour and social and security law, speaker at national and international conferences and courses in labour and social security law and community social law, and academic coordinator of a master's degree in labour management at the University of Vigo since 2011.

José Rodríguez Terceño has doctor's degree in audio-visual communication and information sciences from the Complutense University of Madrid and is the member of complutense Concilium validated research team. He has published more than 20 chapters in books and authored more than 15 articles in refereed journals. His research interests are Cinema and persuasive communication.

Diego Rodríguez-Toubes has bachelor's degree in economics and business administration from the Basque Country University and Ph.D. with distinction from the University of Vigo andis an assistant professor at the Department of Business Administration and Marketing from the University of Vigo, campus Ourense, Spain.

Ana Isabel Rodríguez Vázquez has Ph.D. in communication from the University of Santiago de Compostela (USC). She has a degree in information sciences from the Complutense University of Madrid (UCM). She worked in TV, press, and the Internet for more than a decade. At present, she is lecturer at the USC. Among her main lines of work are information, programming, audiences, and TV production. She is a member of the Research Group of Audiovisual Studies (GEA) from the USC. http://orcid.org/0000-0001-7975-1402 (Santiago de Compostela, Spain).

Clide Rodríguez-Vázquez has Ph.D., University of A Coruña, master in tourism management and planning, University of A Coruña, and certification programme in tourism, University of A Coruña and is a professor, Area of Marketing and Market Research, University of A Coruña, professor at Economics and Business School, and Professor of the Subjects of Marketing and Commercial Distribution, A Coruña, Spain.

Josep Rom Rodríguez, Doctor in Advertising and Public Relations (URL) is an assistant dean and director of the Research Group on Strategy and Creativity in Advertising and Public Relations (GRECPR) of the Faculty of Communication and International Relations of the Ramon Llull Blanquerna-University, Barcelona, Spain.

María Ruiz Aranguren has Ph.D. in journalism from the University of the Basque Country (UPV/EHU, 2014). She currently works there as postdoctoral researcher. She worked for more than five years as a journalist in print and radio (*Diario de Noticias de Navarra, RNE Madrid, El Correo, Deia, Onda Vasca*), specializing in cultural and social information. Her lines of research focus on strategic communication, processes of discursive construction, and international migration. She collaborates on the cultural supplement *La Pérgola*.

Eulalia-Elizabeth Salas-Tenesaca is a professor at the Universidad Técnica Particular de Loja. He has master's degree in educational leadership and management, is an engineer in management in banking and finance, is an engineer in computer systems, and is candidate for doctor of administrative sciences by the University Mayor of San Marcos of Lima—Peru. His research interests are linked to the social economy, finances popular and solidarity, social development, and public policy in the popular sector analysis.

Aurora Samaniego-Namicela is a professor at the Universidad Técnica Particular de Loja, has master of management in banking and finance, and is a commercial engineer. Responsible for bonding with the community projects, its research lines are linked to social economy, the study of gender in the popular and solidary sector.

Eva Sánchez-Amboage has Ph.D. in tourism management and planning from the University of A Coruña. She has a master in tourism management and planning, a master in teaching BAC, FP, and foreign languages, and a degree in tourism from the University of A Coruña. At present, she works at the UTPL, A Coruña, Spain.

Estefanía Sánchez-Cevallos She is an assistant professor at the Department of Entrepreneurial Sciences, Universidad Técnica Particular de Loja and a Ph.D. candidate in comprehensive development and innovation of tourist destinations from the University of Las Palmas—Spain. She has a master in comprehensive development of tourist destinations from the same university and a degree in management of tourist and hotel companies from Universidad Técnica Particular de Loja, Loja, Ecuador.

Eva Santana López is a doctor in advertising and public relations (UAB), associate professor, coordinator of the degree of advertising, public relations, and marketing, and member of the Research Group on Strategy and Creativity in Advertising and Public Relations (GRECPR) of the Faculty of Communication and International Relations of the Ramon Llull Blanquerna-University), Sant Cugat, Spain.

Javier Serrano-Puche He is a senior lecturer within the School of Communication and a research fellow for the Center for Internet Studies and Digital Life, both at the University of Navarra. He also collaborates with the Institute for Culture and Society. Professor Serrano-Puche has been a visiting fellow at the London School of Economics and has published several papers and book chapters about the social impact of digital technologies, Pamplona, Spain.

Carmen Silva Robles has Ph.D. in advertising and public relations and BA in journalism from the University of Seville. He is a professor and researcher at the Department of Marketing and Communication of the University of Cádiz and collaborator in the BA degree programme in Communication and the MA programme in Communication, Marketing, and ICT of the Open University of Catalonia, Cádiz, Spain.

Alba Silva Rodríguez She has Ph.D. in journalism at the University of Santiago de Compostela (Spain) and is the member of the research group "Novos Medios" (GI-1641), whose research lines are focused on the analysis of strategies, rhetoric, and technological formats for emerging markets in communication. She is the secretary of the journal RAEIC (Spanish Journal of Communication Research). As a researcher, her specialization in the analysis of digital communication focused on the study of media conversation in social media as well as the development of journalistic contents in mobile devices is particularly noteworthy, Santiago de Compostela, Spain.

José Sixto García He has Ph.D. in journalism and MBA and MA in formation in teaching. He is the director of Instituto de Medios Sociales. His lines of research deal with Social Media and Community Management. Between his publications, we highlight Las redes sociales como estrategia de marketing online and *Flujos comunicativos y búsqueda de talento para la marca personal*, Santiago de Compostela, Spain.

Xosé Soengas Pérez He is a professor in audio-visual communication and advertising at the University of Santiago de Compostela (USC). His research lines focuses on production and news programming in radio and television, and the quality of news contents in audio-visual media—control, censorship, and manipulation-. He is the member of the Research Group of Audiovisual Studies (GEA) from the USC. http://orcid.org/0000-0003-3246-0477, Santiago de Compostela, Spain.

Blas José Subiela Hernández has Ph.D. in communication and degree in advertising and public relations. He is the chief researcher in the group called "Advertising and Public Relations: Redefinitions, audiences and media" within the Department of Communication at Universidad Católica San Antonio de Murcia (Spain). His research interest is in the field of visual communication and rhetoric of the image, with special focus to typography and corporate visual identity.

Abel Suing has Ph.D. in communication (2012) from the University of Santiago de Compostela. He is the professor of the Department of Communication Sciences of the Technical University of Loja (UTPL) and researches on communication policies and television. He is the member of the Ecuadorian Society of Communication Researchers, SENESCYT alumnus. He directed between 2004 and 2010 the School of Communication at the UTPL. In 2011, he headed the project "Adaptation of the Model Plan of journalism studies of UNESCO to the curriculum of communication of the UTPL".

Raquel Tinoco-Egas is a full-time auxiliary professor at Technical University of Machala in Ecuador, and has Ph.D.(c) in economic analysis and entrepreneurial strategy at A Coruña University in Spain and M.Sc. in international business development at University of Neuchâtel in Switzerland. Languages: Spanish (native), English, French, and German.

Carlos Toural Bran He is a professional journalist and associate professor at the Department of Communication Sciences of the School of Communication Sciences from the University of Santiago de Compostela. He is author and coauthor of several published studies on information architecture and online media, Santiago de Compostela, Spain.

Ana Cecilia Vaca Tapia She has MBA from the University of Santiago de Compostela (USC) and currently is a professor and researcher at the Catholic University of Ecuador, Ibarra, Ecuador.

Andrea Valencia-Bermúdez She is Ph.D. candidate in communication and contemporary information from the University of Santiago de Compostela (USC) and student of Applied Languages and Translation at the University of Vic (Barcelona). Formerly, she obtained a degree in journalism and a master in communication and creative industries from the USC. At present, she works in the group Novos Medios from the USC. She has participated in many international and national conferences and published works indexed in Scopus and Thomson, Ourense, Spain.

Martín Vaz Álvarez is a journalism student at the University of Santiago de Compostela, Spain. The author's experience in the audio-visual area includes the production of institutional videos, music videos, short fiction films, and documentaries. For the past year, he has been experimenting with mobile devices as part of a study about the possibilities of new media. He has completed courses on mobile production in London and audio-visual communication in Berlin, where he is currently studying at the Universität der Künste.

Montse Vázquez-Gestal She is a professor at Universidade de Vigo (Spain) andhas Ph.D. in information sciences. She is a lecturer in some Italian universities, as Universities of México and Portugal. She teaches in creativity and advertisement. She is the main head of the investigation group CP2, in Vigo's University, Vigo, Spain.

Jorge Vázquez Herrero He is a Ph.D. student in communication and contemporary information at the University of Santiago de Compostela, and he develops research about interactive non-fiction digital narratives focusing on interactive documentary and journalism, A Coruña, Spain.

Pablo Vázquez-Sande He has Ph.D. in communication from the University of Santiago de Compostela and Reed Latino Prize for the best dissertation on political communication, with the first research linking storytelling and political communication. He is currently combining his work at a department of institutional communication with teaching, while continuing research on communication. He has been present in several national and international congresses and published in various prestigious journals, Lugo, Spain.

María Victoria-Mas She is an assistant professor, School of Communication Sciences, Universitat Internacional de Catalunya (UIC Barcelona, Spain). She is a researcher with the Labcom Group and with the "Innovation and development of online media in Spain. Applications and technologies for the production, distribution, and consumption of information" project (reference CSO2012-38467-C03), which is financed by the Ministry of the Economy and Competitiveness, Barcelona, Spain.

Christian Viñán-Merecí He is an assistant professor at the Department of Entrepreneurial Sciences, Technical Particular University of Loja and Ph.D. candidate in sustainable development and innovation of tourist destinations from the University of Las Palmas de Gran Canaria—Spain. He has a master in sustainable development of tourist destinations from the same university. He also has a master in business administration and management from the Technical Particular University of Loja, the Spanish Institute of Market Studies and the Complutense University of Madrid. He has graduated in market strategic management from the Technical Particular University of Loja, Loja, Ecuador.

Jenny J. Yaguache has Ph.D. in communication and journalism from the University of Santiago de Compostela. She is a professor of Organizational Communication Administration and Management of Information and Public Relations at the Technical University of Loja. Also, she is responsible for the organizational communication department. As a researcher, her areas of expertise are management and financing of media, organization and management of the news organization, cultural Industries, and corporate communications. She has some scientific publications and presentations at national and international conferences magazines, Loja, Ecuador.

Hernán Yaguana He has Ph.D. in communication and journalism from the University of Santiago de Compostela. His publications include radio, an evolving medium (2013), published in Social Communication and Communications issues—Spain, 85 years of Ecuadorian radio-broadcasting (2014), published in Ciespal—Ecuador, approaching terrestrial digital radio from the online radio (2013) Scientific Journal of communication Latindex, and Ecuadorian university radio, a new challenge for a new era in college radio in America and Europe (2014), published by Fragua—Spain.

Part I
Introduction

Chapter 1
Metamedia, Ecosystems and Value Chains

Francisco Campos Freire

Abstract The present chapter analyses the three emerging trends in the new organizational forms: deconstruction and fragmentation of value chains, positive and negative polarization based on the competitive advantages of economies of scale; and alter native models for exploiting the digital commons. It therefore analyses the link between metamedia and ecosystems to characterize some new forms of organization and valuation of communication networks and industries.

Keywords Metamedia · Media ecosystem · Communication · Organisational forms · ICT technologies · Media ownership

Metamedia and digital networks make up the new architecture of the communication society, in which million of citizens, organisations and other social actors establish relationships. This architecture becomes an ecosystem because of its complexity, interdependence, and diversity of species and self-organisation. The new techno-cultural setting disrupts the ways of interacting, but also the models of communication, production, commercialization, and ownership of the different forms of value of almost every industry, and especially those related with information, communication and culture.

This special issue addresses these new elements in communication and, in one way or another, most of the contributions belong to the 1st symposium "From media to metamedia", organized by the International Research Network on Communication Management (R2014/026 XESCOM). The network was launched in 2014 by Novos Medios Research Group, from the University of Santiago de Compostela; iMarka from the University of A Coruña; and NECOM from the

F. Campos Freire (✉)
International Research Network on Communication Management (XESCOM),
University of Santiago de Compostela (USC), Santiago, Spain
e-mail: francisco.campos.freire@gmail.com

F. Campos Freire
Prometeo from the Ecuadorian SENESCYT at the UTPL and PUCESI, Loja, Spain

© Springer International Publishing Switzerland 2017
F.C. Freire et al. (eds.), *Media and Metamedia Management*,
Advances in Intelligent Systems and Computing 503,
DOI 10.1007/978-3-319-46068-0_1

University of Vigo. Another international groups from Portugal, Brazil, Ecuador, Mexico and Colombia joined them.

In the past two years, XESCOM, which has more than 75 researchers, has published five research books; participates in five international and four national projects. Also, it conducted a survey including 5500 professors from the three Galician universities on the use of general and scientific social networks; launched the first Spanish Barometer on Communication Management, carrying out surveys to 350 media companies; published 65 scientific papers in conferences and indexed journals; organized three workshops in international conferences; and three training seminars on methodologies using digital research techniques.

The I International Symposium on Communication Management, which was held in November 2015 at the Faculty of Social and Communication Sciences (Pontevedra), received more than 100 papers, from which 87 were selected. Authors are from 25 universities from five countries (Spain, Portugal, Ecuador, Mexico and Brazil). After a peer-to-peer review, 50 works were selected. The result is the present volume.

This issue starts with the works of the coordinators from each of the research groups belonging to the network: Xosé López García–Novos Medios-; Xosé Rúas Araujo–Necom-; and Valentín Alejandro Martínez–iMarka-. The following chapters are the contributions from the first symposium on Communication Management, which develop some of the aspects involving the digital transition from the media to metamedia.

My contribution focuses on three emerging trends in the new organisational forms: deconstruction and fragmentation of value chains; positive and negative polarization based on the competitive advantages of economies of scale; and alternative models for exploiting the communication commons. It therefore analyses the link between metamedia and ecosystems to characterize some new forms of organisation and valuation of communication networks and industries.

The cultural convergence, caused by the new digital technologies and the network infrastructure, has triggered a long-lasting and deep transition process from traditional media to the new metamedia (Jenkins and Deuze 2008). This lead to an "hybrid ecology" (Benkler 2006), in which news contents are mixed with fiction, advertising, propaganda, non-profit information, politics and religion. And these contents are not only generated by media companies, but also by public and philanthropic institutions, amateurs, activists and robotic actors with artificial intelligence (Latour 2001, 2012).

The prolific use and development of the concepts of ecology and media environment, remediation, hybridization, and metamediation, is the result of the intertextual mosaic (Scolari 2010), produced by the classics by Marshal McLuhan (1964); Postman (2000, 2004) and the schools belonging to the media ecology from Toronto and New York (the media are species living within a communication ecosystem); Postman's moral reflections (an environment is a messaging complex system that requires humans to think, feel, and behave in a particular way); and the stories by Innis (1952) on the socio-economic evolution of the media.

Remediation (Bolter and Grusin 1999; Jenkins and Deuze 2008) represents the conceptualization on an evolutionary and convergent process of the new digital media, starting from the traditional media, adapting their existing features and others not yet developed. Remediation is the result of technological convergence, which produces vertical and horizontal innovation processes, linking together the media and their audience, from top to bottom, and from bottom to top. That produces contradictory changes in production, consumption and value relationships, which, in turn, provoke massive fragmentations and mega concentrations.

Convergence has changed the correlation of forces between the media and their publics, opening up more intense collaboration spaces (interaction, participation and co-creation) with users, but also new conflicts, since audience become partner producers and competitors of contents, giving rise to many ecosystems of micromedia, mesomedia and megamedia. Media and user-generated contents also create a global digital culture that Manovich, (Jenkins and Deuze 2008) refer to as remix and hybridization.

Hybridization is a complex concept, since it goes beyond technology and touches the socio-economic and sociological dimension, compromising not only the media experience, but also the "miscegenation in cultural consumption" of receivers, within a context of globalization, modernity, and new types of reception, enjoyment and ownership (Canclini 1989, 1999, 2007). But it has much significance because it connects the focus of reception with the techno-cultural convergence and the political economy in a context of a modern global society.

Metamedia emerge within that convergence culture. They are understood as digital media with new properties, produced by combining networks and software tools for creating, editing, disseminating and interacting between issuers and users of communication. The metamedia reinvent the way, use and access to the raw material of traditional media—information and entertainment-, using "existing or not yet invented" computer techniques, in a process of ongoing or groundbreaking innovation (Manovich 2005 and 2008).

The concept of metamedia was also anticipated by McLuhan (1964) and developed by Kay and Goldberg (1977), but Lev Manovich (2008: 288) was the one who studied it. His analysis proceed through four stages: (a) 1960–1970, experimentation and innovation, lining with the first steps of the Internet; (b) 1980–1990, commercialization and massive application of the Internet and the emergence of the web; (c) 1990–2003, development of the web 1.0; and (d) 2004–2015, convergence and hybridization through the explosion of the web 2.0 and digital social networks.

Digital ecosystems are organisational models adopted by computer technicians to represent the architecture of networked communication between servers, databases, and collaborative experiences of users, acting as participants of these operating systems (Norman and Draper 1986). They are decentralised and self-distributed models within an environment of complex interactions (Boley and Chang 2007). Also they require interaction, self-organisation, scalability and semantics (code) of relation for managing information's entropy.

The comparison between natural and digital ecosystems, which comes from the Macluhian and ecological influence, not only has organisational implications on the interaction of objects from the systemic communities, but also on the economic sustainability. In a natural ecosystem, biotic elements (live organisms) coexist with abiotic ones (inorganic, proteins, climate). The first ones may be autotrophic, capable of self-nourishing; and heterotrophic, which feed on the first ones. Therefore, if the ecological basis is the balance and sustainability between species, that philosophy might be applicable to the digital and innovative economic ecosystems in which the media are living.

An ecosystem is a utility paradigm for understanding the organisational complexity of digital industries and all those that take advantage of the Internet networks. It is also an alternative to the traditional analysis of the industry, branch, sector, and value chain (Miguel de Bustos 2014). Nevertheless, these analyses of the industrial economy are too linear and vertical for comprising the horizontality, mainstreaming, and the topology of the value networks that articulate the digital and cognitive economy.

That is why we resort to the metaphors of ecosystems and epigenetics –analysis of casual interactions amongst genes (Waddington 1941), in order to study the organisation, innovation and business models of large communication companies (Gómez-Uranga et al. 2014). These authors also use the same model for analysing the dynamics of nanotechnology clusters and comparing the German and the Basque and Catalonian cases.

Miguel de Bustos and Casado (2016) note that Arlandis and Ciriani (2010), Fransman (2007), Lombard (2007) and Mars et al. (2012) use ecosystems when describing aspects of organisations and layer models, as well as stages of communication technologies: hardware and software, networks, platforms and contents. Miguel de Bustos and Casado (2016) use the ecosystem model for categorising the large global infomediaries, articulated around news search platforms—Google— (Perrot 2011; Simon 2011), which are accessed from various devices (Apple), social networks (Facebook), and content intermediaries (Amazon).

The GAFA (Google, Apple, Facebook and Amazon) generate two-dimensional markets, financialization, traditional and innovative business models, mixed and hybrid economies, ecosystem competence, and bargained power in contents' life cycles. Platforms' owners, app developers, contents distributors and users are the pivotal aspects on which the ecosystem's architecture turns (Tiwana, 2013). Miguel de Bustos and Casado Nieto (2016) also agree in this respect when they show the GAFA's morphology, competing and collaborating with each other and other traditional communication groups.

A relevant aspect of the digital ecosystem architecture is the environment of infrastructure and networked relationships, which leads to cooperation, sharing, collaboration, conversation, and opened and adapted technologies of evolutionist business models (Santamaría 2010). The social media ecosystem connects with companies' value networks (Hanna et al. 2011) through multiple layers of relationships between brands and users (Larson and Watson 2011).

1.1 Digital Value Networks

The Social Network Sites, which arise from the technological tools of the web 2.0, have become communication and social relation platforms upon which media ecosystems are organised (Ellison and Boyd 2008; Beer 2008; Stenger 2009; Campos-Freire 2015). The networks are metamedia ecosystems composed by individual, public, and semi-public digital profiles, which establish relationships and exchanges of own and external contents flows, conversations, prescriptions, transactions, experiences and use of artificial intelligence devices with particular, social, commercial and institutional purposes. Metamedia are the tools, platforms are the networks and ecosystems are their architecture.

This new techno-cultural infrastructure multiplies the dimension and the value of people and companies' relationships, through the interconnections and interactions of their social capital. Thus, the strategic axis of the competitive advantage (Porter 1985) is moved from the value chain to the value network, since the latter is wider and more comprehensive (Kothandaraman and Wilson 2001; Peppard and Rylander 2006) for business models (Shafer et al. 2005). The business model is the strategy on how an organisation achieves its goals and income from the creation and acquisition of value.

Both the structure and all related to digital networks has some value, but value networks have a particular purpose. These networks may be groups, communities, and specialised platforms (education, business, services). That way they become digital value networks with multiple nodes and entry and exit points linking up with a complex diversity of players involved (Feng and Jason 2002).

A value network is a purposeful group of people and organisations that create economic and social assets through dynamic, tangible, and intangible complex exchanges (Allee 2009). The purpose of a value network is to generate economic and other benefits for their participants. Theory and analysis on value networks is a reliable and organic form to describe, analyse, assess, and integrate the organisational performance into complex environments. This switchover of value occurs through the social exchange of relations established by nodes, links, degrees of separation and interactions.

The network society is an open manner of social organisation that increases autonomy, self-representation, and self-organisation of people and companies. That organisational individualisation is structured around personal and corporate networks operated by digital networks, which have become reference platforms for all kind of purposes (Castells 2014: 25).

According to Castells (2014), the autonomy fostered by networks strengthens individualism and, at the same time, increases sociability. Also, these new social spaces foster private exploitation of data and dominance of control algorithms from search engines, platforms, and other intermediaries through information filter bubbles (Bozdag and Van den Hoven 2016). The competition focuses now on the capture, control, application and monetisation of value scattered by the networks, once diluted the competitive advantage of traditional organisations.

1.2 New Forms of Organisation

Evans (2015) claims that the re-acceleration of technological changes on information management (Big Data) and the networks is transforming organisations. There are two remarkable and opposed trends: (1) the deconstruction of value chains from traditional industries, and the polarisation of economies of scale around fragmentation; and (2) the mega-concentration. In this regard, Kotter (2015) notes that the new organisation for a constantly changing world should be a dual structure for transition and a networked constellation opened to the future, with strategies and polycentric, flexible, autonomous, virtual, and results-based ways of working (Thomson 2015).

The crisis of the value chain, the linear model of industrial economy based on concentration and integration, causes the vertical breakdown of some processes in small and medium-sized enterprises, which choose to outsource and establish partnerships, since some of their stages are not competitive nor valued. That provokes two types of polarization (Evans 2015): negative, which causes fragmentation, since it has lost value in economies of scale and experience; and positive, when megagroups and ecosystems converge on value networks.

The consequences of those changes lead to the shift from vertical to horizontal organisation; from the hierarchical chart to collaborative networks; from departments to project teams; and from integrated companies to clusters. Developments in large communication groups are oriented to these trends. Nevertheless, the more traditional and integrated they are, the more difficult to adapt themselves. In 2015, Google has already been transformed into Alphabet to gain more value and innovation. The four GADA are constellations of small companies operating as ecosystems. But the heavy organisational structures of traditional communication companies are more difficult to change.

Traditional media (press, radio and television) find it most difficult to adapt to changing conditions. Proof of this is the weak position in which their value chains and business models are in relation to the new providers of streaming platforms. The structural crisis and the circumstantial nervousness about the perception of changes, makes companies prototypes of Amara's law (1983), what is to say, overestimating the immediate effect of a technology and underestimating it in long term, reacting without an appropriate strategy which allows them to combine transition and innovation.

Traditional business models suffer the greater impact, since their value chains have been broken and not adapted to the new value networks. The decline in sales of printed press has weighed heavily on circulation and advertising revenues. The emigration from the linear TV-consumption—organized through programming and advertising-, to other on demand options, also breaks the traditional value chain of commercialisation, favouring the appearance of new streaming platforms (YouTube, Netflix, Hulu, Amazon). The economy of attention, which supports the traditional advertising model, is being replaced by the big data economy, which gives way to the programmatic advertising and the blockers' struggle.

Technological changes have also fostered new economic models based on social commons, though altruistic collaborations, open access and right of use with no exclusions on many communication and knowledge resources (Benkler 2014). That wave of innovation and changes fosters the collaborative economy, the prosumerism, the transparency and the governance as a more opened multilevel system.

The society is an ecosystem of tensions and complex interactions in which digital networks boost some aspects and make other complicated. Besides the exponential growth of communication and sociability flows—noted by Castells (2009, 2014), some experts draw attention to the false horizontality and superficial democratisation (Pitman 2007); to the disappointments in false revolutions (Morozov 2011); to the surveillance and prosumerism (Fuchs 2011); to the financialization, commodification, gif economy and free labour (Terranova 2000). We belong to a society under construction, organized and changed through networks and ecosystems. It is important to be aware of the value and diversity of that architecture, since it is waterway of our social capital.

Acknowledgments The results of this work belong to the research projects promoted by the network XESCOM (R2014/026 XESCOM), supported by the Regional Ministry of Culture, Education and Organisation from the "Xunta de Galicia"; by the R + D project—CSO2015-66543-P—belonging to the state Programme for Supporting Excellent Scientific and Technique Research, state subprogram of knowledge creation from the Spanish Ministry of Economy and Competitiveness, entitled "Indicators related to broadcasters governance, funding, accountability, innovation, quality and public service applicable to Spain in the digital context"; and by the Prometeo Program which belongs to the National Secretary of High Education, Science, Technology, and Innovation (SENESCYT) of the Ecuadorian government, and is developed at the Technical University of Loja and the Pontifical Catholic University of Ecuador—Ibarra headquarters.

References

Allee, V. (2009). Value creating networks: Organizational issues and challenges. *The Learning Organization Special Issue on Social Networks and Social Networking, 6*(6), 427–442. Retrieved from: http://citeseerx.ist.psu.edu/viewdoc/download

Amara, R., & Lipinski, A. J. (1983). *Business planning for an uncertain future: scenarios & strategies.* New York: Pergamon Press.

Arlandis, A., & Ciriani, S. (2010). How firms interact and perform in the ICT ecosystem? *Communication and Strategies, 79,* 121–141.

Beer, D. (2008). Social network(ing) sites. Revisiting the story so far: A response to Danah Boyd & Nicole Ellison. *Journal of Computer-Mediated Communication, 13,* 516–529, International Communication Association.

Benkler, Y. (2006). *The wealth of networks.* New Haven: Yale University Press. Retrieved from: http://www.benkler.org/wealth_of_networks

Benkler, Y. (2014). *Innovación distribuida y creatividad, trabajo colaborativo y el procomún en una economía en red.* BBVA, Openmind. Retrieved from: https://www.bbvaopenmind.com/articulo/innovacion-distribuida-y-creatividad-trabajo-colaborativo-y-el-procomun-en-una-economia-en-red/

Boley, H., & Chang, E. (2007). Digital ecosystems: Principles and semantics. In *Inaugural IEEE International Conference on Digital Ecosystems and Technologies, Australia*. Canada: National Research Council. Retrieved from: http://citeseerx.ist.psu.edu/viewdoc/download? doi=10.1.1.90.4199&rep=rep1&type=pdf

Bolter, J. D., & Grusin, R. (1999). *Remediation. Understanding new media*. Cambridge: MIT Press.

Bozdag, E., & Van den Hoven, J. (2016). Breaking the filter bubble: Democracy and design. *Ethics and Information Technology, 17*(4), 249–265.

Campos-Freire, F. (2015). Adaptación de los medios tradicionales a la innovación de los metamedios. *El Profesional de la Información, 24*(4), 441–450.

Castells, M. (2009). *Comunicación y poder*. Madrid: Alianza Editorial.

Castells, M. (2014). *El impacto de internet en la sociedad: una perspectiva global*. Retrieved from: https://www.bbvaopenmind.com/articulo/el-impacto-de-internet-en-la-sociedad-una-perspectiva-global/

Ellison, N., & Boyd, D. (2008). Social network sites: Definition, history, and scholarship. *Journal of Computer-Mediated Communication, 13,* 210–230. International Communication Association.

Evans, P. (2015). *De la deconstrucción a los big data: como la tecnología está transformando las empresas*. BBVA: In Reiventar la empresa en la era digital.

Feng, L., & Jason, W. (2002). Deconstruction of the telecommunications industry: from value chains to value networks. *Telecommunications Policy, 26*(9–10), 451–472.

Fransman, M. (2007). *The new ICT ecosystem*. Korokoro, Edinburgh: Implications for Europe.

Fuchs, C. (2011). Web 2.0, prosumption, and surveillance. *Surveillance & Society 8*(3): 288–309.

García Canclini, N. (1989). *Culturas híbridas: Estrategias para entrar y salir de la modernidad*. Mexico: Grijalbo.

García Canclini, N. (1999). *La globalización imaginada*. Barcelona: Paidós.

García Canclini, N. (2007). *Lectores, espectadores e internautas*. Barcelona: Gedisa.

Gómez-Uranga, M., Miguel de Bustos, J. C., & Zabala-Iturriagagoitia, J. M. (2014). Epigenetic economic dynamics: The evolution of big internet business ecosystems, evidence for patents. *Technovation, 34*(3), 177–189.

Hanna, R., Rohm, A., & Crittenden, V. L. (2011). We're all connected: The power of the social media ecosystem. *Business Horizons, 54*(3), 265–273. Retrieved from: https://www.bbvaopenmind.com/articulo/de-la-deconstrucion-a-los-big-data-la-tecnologia-esta-transformando-las-empresas/ and https://www.bbvaopenmind.com/articulo/la-organizacion-del-futuro-un-nuevo-modelo-para-un-mundo-de-cambio-acelerado/

Innis, H. (1952). *Changing concepts of time*. Toronto: University of Toronto Press.

Jenkins, H., & Deuze, M. (2008). Convergence culture. *Convergence: The international journal of research into new media technologies, 14*(1), 5–12. London: Sage Publications.

Kay, A., & Goldberg, A. (1977). Personal Dynamic Media. *Journal Computer, 10*(3)

Kothandaraman, P., & Wilson, D. T. (2001). The future of competition: Value-creating networks. *Industrial Marketing Managament, 30*(4), 379–389.

Kotter, J. P. (2015). *La organización del futuro: un nuevo modelo para un mundo de cambio acelerado*. BBVA: In Reiventar la empresa en la era digital.

Larson, K., & Watson, R. (2011). The value of social media: Toward measuring Social Media Strategies. In *ICIS 2011 Proceedings. Paper 10*. Retrieved from: http://aisel.aisnet.org/icis2011/proceedings/onlinecommunity/10

Latour, B. (2001). *La esperanza de Pandora, ensayos sobre la realidad de los estudios de la ciencia*. Barcelona: Gedisa.

Latour, B. (2012). *Enquête sur les modes d'existence: une anthropologie des Modernes*. París: La Découverte.

Lombard, D. (2007). *Le Village numérique mundial*. Odile Jacob: París.

Manovich, L. (2005). *El lenguaje de los nuevos medios de comunicación: la imagen en la era digital*. Barcelona: Paidós.

Manovich, L. (2008). *Software takes command.* New York: Georgetown University. Retrieved from: http://faculty.georgetown.edu/irvinem/theory/Manovich-Software-Takes-Command-ebook-2008-excerpt.pdf

Mars, M. M., Bronstein, J. L., & Lusch, R. F. (2012). The value of a metaphor: Organizations and ecosystems. *Organizational Dynamics, 41,* 271–280.

McLuhan, M. (1964). *Understanding Media: The Extensions of Man.* New York: New American Library.

Miguel de Bustos, J. C. (2014). *Ecosistemas y 'los GAFA'.* Retrieved from: http://nanoregior.com/?p=748

Miguel de Bustos, J. C, & Casado, M. A. (2016). GAFAnomy (Google, Amazon, Facebook and Apple): The big four and the b-ecosystem. In M. Gómez-Uranga, J. M. Zabala-Iturriagagoitia & J. Barrutia (Eds.), *Dynamics of big internet industry groups and future trends: A view from epigenetic economics,* Berlin: Springer. (En prensa).

Morozov, E. (2011). *The net delusion: How not to liberate the world.* London: Allen Lane, An Imprint of Penguin Books.

Norman, D. A., & Draper, S. W. (1986). *User centred system design: New perspectives on human-computer interaction.* New Jersey: Lawrence Erlbaum Ass.

Peppard, J., & Rylander, A. (2006). From value chain to value network: Insights for mobile operators. *European Management Journal, 24*(2–3), 128–141.

Perrot, A. (2011). Le numérique: Enjeux des questions de concurrence. Culture-médias & numerique: nouvelles questions de concurrence (s). *Concurrences, 3.*

Pitman, T. (2007). Latin American cyberprotest: Before and after the Zapatistas. In C. Taylor & T. Pitman (Eds.), *Latin American cyberculture and cyberliterature.* Liverpool: University Press.

Porter, M. E. (1985). *Competitive advantage.* New York: Free Press.

Postman, N. (2000). The humanism of media ecology. In: *Proceedings of the Media Ecology Association, vol 1* (10–27). Retrieved from: http://www.mediaecology.org/publications/MEA_proceedings/v1/humanism_of_media_ecology.html

Postman, N. (2004). The information age: A blessing or a curse?. *Quaderns del CAC, 34, XIII*(1), 17–25.

Santamaría González, F. (2010). *Una introducción a los ecosistemas digitales.* Retrieved from: http://fernandosantamaria.com/blog/2010/07/una-introduccion-a-los-ecosistemas-digitales/

Scolari, C. (2010). Ecología de los medios. Mapa de un nicho teórico. *Quaderns del CAC, 34,* 17–25.

Shafer, S. M., Smith, H. J., & Linder, J. C. (2005). The power of business models. *Business Horizons, 48*(3), 199–207.

Simon, P. (2011). *The age of the platform: How Amazon, Apple, Facebook, and Google Have Redefined Business.* Motion Publishing.

Stenger, T. (2009). Social network sites (SNS): do they match? Definitions and methods for social sciences and marketing research. In *XXIX INSNA conference in San Diego (EE.UU.).* Retrieved from: http://www.academia.edu/2521387/Social_Network_Sites_SNS_do_they_match_Definitions_and_methods_for_social_sciences_and_marketing_research

Terranova, T. (2000). Free labour: producing culture for the digital economy. *Social Text, 18*(2), 33–58.

Thomson, P. (2015). *Nuevas formas de trabajar en la empresa del futuro.* Retrieved from: https://www.bbvaopenmind.com/articulo/nuevas-formas-de-trabajar-en-la-empresa-del-futuro/

Tiwana, A. (2013). *Platform ecosystems: Aligning architecture, governance, and strategy.* The Netherland: Elsevier Science.

Waddington, C. (1941). Evolution of developmental systems. *Nature, 147,* 108–110.

Chapter 2
Internet, Mind and Communication: New Perspectives and Challenges

Xosé Rúas Araújo

2.1 Feelings and Computation

Knowing what people is thinking about, as well as recording their internal and external perception, are some of the main challenges in communication, marketing and advertising. In order to arrive to that conscious and unconscious thought, effort is being made to seek out new ways of exploring and observing senses through physiological recordings and psychological-emotional responses to the exposure to stimuli and analysis of texts, images, and audiovisual works.

The media evoke emotions through a simple description of reality or an instant —as the recent picture of a 3 years-old Syrian child lying face-down on a Turkish beach, but it may also arouse aesthetic emotions (Mar and Oatley 2008), as literature, music, visual and performing arts, through fictional constructions in cinema, television, and digital media.

Research on emotions has continuously increased since 1980, when psychological studies and theoretical analysis on human emotions appeared (Frijda 1986), expanding gradually to related areas, such as sociology, politic sciences, anthropology, communication, and cultural studies, among others (Lewis et al. 2008).

Within the emerging field of psychology of the media, the study of the role of emotions and their effects on the media and the entertainment industry increased gradually (Doveling et al. 2011). Different methods and measuring tools are being applied, with a view to find out what is going on inside people's "black box" (Lang 2011).

X. Rúas Araújo (✉)
Neurocommunication, Advertising and Politic Research Group (NECOM),
University of Vigo (UVigo), Vigo, Spain
e-mail: joseruas@uvigo.es

© Springer International Publishing Switzerland 2017
F.C. Freire et al. (eds.), *Media and Metamedia Management*,
Advances in Intelligent Systems and Computing 503,
DOI 10.1007/978-3-319-46068-0_2

Information (as a bit, as a gene, as a computable basic unit) is the beginning of everything (Gleick 2012). Life is basically information, and the evolution has not been stopped at the physiological process, but it has continued with the development of systems and processes, which are necessary in the development of the intelligence and complex lives.

In that process is included the "memes" system, term coined by Richard Dawkins (2000: 21) in order to designate the ideas circulating on society. He pointed out that, if genes are the functional units of heredity, then "memes" are the functional units of cultural heredity and collective knowledge. Thus, as genes spread from person to person through sperm, "memes" move from brain to brain through imitation.

"Memes"—pictures, ideas, quotes and slogans—are imposed on individuals creating similar (opinion) trends to that produced by the space colonization of insect pests, epidemics, or the bodies invasion by cells, following the clustering patterns described by the expert in biomedical and social models, Joshua M. Epstein (2006).

The analysis of digital social networks, in which large part of articles of this book are focused, is inspired on the media ecology, a term coined by Neil Postman to define the study of the media as environments. It aims at showing, through the combination of both biological and technological perspective, a step forward in the evolution of species (media, traditional and digital), which are related in the same space and ambiance (Campos and Rúas 2015) in their struggle for survival.

Consequently, neuroscientists, involved with physical brains, started to realize that they could better understand how the brain functions if they had in mind environments and processes driven by various social structures. For their part, researchers in social sciences also realized that traditional techniques and tools in neurosciences might contribute to the assessment and objective recording of attention and perception of users, consumers and voters.

The brain is not an isolated entity from the outside world, and our behaviour and thoughts are not only the result of a single and conscious unit-mind-, but it is also a social brain, as Gazzaniga (1993) named his book. The philosophy based on placing the man as the measure of all things, that social being by nature, takes on new meanings. The study of emotions constitutes an epistemological turning point in the field of humanities and social sciences, as stressed in one of the chapter of the book.

Hence the appropriateness of the use of qualitative and quantitative methods and the contrast between various techniques and tools, in the search for triangulation, and the defence of interdisciplinarity as the reaffirmation and epistemological constant of the regrouping and sharing of knowledge, as a principle of meeting and communication between disciplines, where each one brings its problems, concepts, and research methods. It is also a valid perspective for the understanding of metamedia, languages, and metalanguages of the so-called network society and hypertextuality, in its evolution towards the digital brain, which means, in the words of Small and Vorgan (2009), a brain's development according to the ICT.

At present, computer scientists and neuroscientists are trying to disassemble and re-assemble the brain, neuron to neuron, through the so-called "reverse engineering the brain" (Kaku 2014). So far, the most they have managed to rebuild—partially— was the brain of a worm and a fly, and IBM scientists have simulated the 4.5 per cent of neurons and synapses in the human brain through the macro-computer Blue Gene. Nonetheless, the most difficult part is still to come—connecting the brain to the world of sensations, memories and emotions, languages, and culture.

The challenge is even greater if we consider the existence of many different situations that can trigger the same emotion, increasing exponentially the possible combinations and responses. And, although a typical robot can reach a level 1 of consciousness, which describes insects and reptiles (robots are not totally aware that they are robots), if androids want to access to level 2, they have to be able to create a model and world simulation in relation to others, multiplying the number of group members by the emotions and gestures they use to communicate. In short, it is a question of being able to recognize feelings in our facial expressions, gestures, and tones of voice; in other words, in our verbal and non-verbal communication.

In this regard, in the 70s efforts were made to develop a protocol and coding system, cataloguing, and classification of the facial action (Ekman and Friesen 1976; Ekman 2003) to determine which muscular actions were related to the different kinds of emotions, starting from the seven basic emotions (anger, fear, disgust, contempt, joy, sadness, and surprise), with a strong empirical support from that time on (Matsumoto and Ekman 2010).

Also, clinical research on neurosciences provided new avenues for experimental research on marketing and communication, with the implementation of X-ray in computerized tomography (CT), which allowed moving from a flat image to a three-dimensional model (González-Álvarez 2012: 254); the functional magnetic resonance (MRI), which observes the oxygenated blood in different parts of the brain with neuroimaging tools (PET, positron emission tomography); the electroencephalogram (EEG), placing electrodes on the head; biofeedback techniques, which show the physiological reactions in the body, measuring heart rates (Rúas et al. 2015); and the dermal conductance of the skin, in addition to other tools such as the "eye-tracking", whose details are described in one of the chapters of this book.

Finally, there are tools for analysing and processing texts, language, and discourses, among which is the lexicometrics, using new IT developments (Rúas and Pérez 2015). Indeed, one of the current discussions are focused on the different perception and understanding between printed and online reading. The scanning of people's brains while they were making searches and reading online and printed texts show that book readers have much more activity in brain regions associated with language, memory and visual processing, but not so much in prefrontal regions associated with decision-making and troubleshooting (Carr 2014, 2012).

The conclusion of this study is that the constant concentration is more difficult when working online, specially given the distraction caused by the existence of multiple links, which require constant coordination and decision-making capacity and, in consequence, provoke a distraction and a mental workload. McLuhan (2009) had already identified that new media were changing our thoughts and senses and, ultimately, our view of the world and us—"the medium is the message", with all the power and threat this statement suggested.

Knowledge involves more than just seeking information and requires encoding and decoding data and experiences stored in personal memories. Our working and processing memory, such as a computer's hardware and software, help us to think, fast and slow (Kahneman 2015).

Another study by Evans (2008), in which he analyses academic articles published in scientific journals over 60 years, concluded that the transition from traditional research to online searching led to a decrease of science and erudition. The tyranny of search engines and databases, which reinforce the prevailing opinion—the most cited, makes people follow scripts written by others and event reproduce contents of books and articles extracted in a partial manner.

It seems to be no room for interpretation, criticism and discussion, as well as subjective and aesthetic judgements, contemplation, encouragement of creativity and imagination, within the industrial and capitalist essence of the Internet, the speed, efficiency, and consumption.

2.2 Technology, Culture and Democracy

Often, the quality of communication through the Internet and social networks is only examined from the calculation of text-messages that are sent and registered and, sometimes, answered—mostly as propaganda, without analysing the content, tone, context and environment in which they are produced. The Internet technology is unable by itself to create a fairer and free world, and to cause a transformation of rights and people´s welfare, if this is not accompanied by social, cultural, and political changes. The problem of trying to explain the media and the Internet only from the Internet side, leads us to the Internet-centrism (Morozov 2012), and to a technological determinism, a reduced form to describe the present and predict the future without the necessary perspective that should feed scientific knowledge.

People able to access to new technologies says a lot about the difference between a first and a third world, in digital terms, and a Wide Web of various speeds and different levels of access to information. President Barack Obama had already seen it on his first visit to Google headquarters, noting the large swaths of darkness on the map of the finder they were representing, he said, the disconnected corners of an interconnected world (Beas 2011: 25). The main idea of using technology is to enhance the quality of democracy and to have all government information in an accessible online format, providing all administrative and governmental production (e-Government, Open Government).

However, three years before the investiture of US President Obama, Wikileaks came on the scene, disseminating "secret" documents (the quotation marks respond to the pretext, sometimes unjustified, of internal security's safeguarding). The initiative quickly became a threat to the credibility of North-American foreign policy, the government, and the net governance—formerly know as "net neutrality". This last concept is related to the discourse, sometimes naïve, sometimes cynical, of the supposed neutrality of the network, used as an argument to prevent regulation and maintain the arbitrariness of power. The revelations of Julian Assange and the persecution to Edward Snowden—the young man that revealed the US espionage, considered a betrayer and a hero, are good examples of this idea, as well as the "blackouts" decreed by the owners of the networks in some dictatorships, sacrificing the exercise of freedom of expression in favour of commercial interests. Behind the Arab Spring there was also the harsh winter of Western companies selling surveillance and censorship technology to dictatorial regimes.

The recent Apple's fight with the FBI, which requested to unlock an iPhone used by a gunfire's author for alleged association with Islamic terrorism, has reopened the debate on the boundary between security and freedom, since this might set a precedent and open the door to future requirements by the US government.

Similarly, that own vision of Internet-centrism leads many researchers to mix up cyberactivism with cyber-utopianism and any virtual intention with real actions. Many real achievements arose from many virtual purposes, but few online political campaigns may achieve success in isolation and not accompanied by a series of offline actions. The so recurrent example of Obama's election campaign, considered revolutionary from the digital point of view, was aware that any online action would have no sense if it were not accompanied by the consequent offline mobilization. Obama's team took full advantage of Big Data potential and database access through the use of new and old technologies, such as the sending of customized emails and the phone calls ("N2N" campaign, neighbour to neighbour); the official Obama's social network (my.barackobama.com); the organic coordination through the media department; and a digital military headquarters to mobilize an army of fans through the global network but with the slogan "keep it local, keep it real" (Harfoush 2009, 94). All the same time, the old tradition and the American pyramidal structure (one person, five calls or contacts) were not forgotten. It was put into effect long before the discovering of the six degrees of separation, door-to-door visits, and the talking-snacks with contacts and links provided virtually and then in person.

Even when Obama pledged to participate in virtual forums where users and readers voted and chose—becoming content editors, he had to deal with uncomfortable situations that led him to show caution when dealing with this form of digital populism. Experience is showing us the mistake of separating society, culture and politics from technology; analysing the relevance of Facebook and Twitter without looking beyond Facebook and Twitter; replacing moral values with technical values; as well as understanding technique without ethics.

References

Beas, D. (2011). *La reinvención de la política. Obama, internet y la nueva esfera pública.* Barcelona: Península.

Carr, N. (2014). *Atrapados. Cómo las máquinas se apoderan de nuestras vidas.* Madrid: Taurus.

Carr, N. (2012). *Qué está haciendo internet con nuestras mentes?.* Taurus: Superficiales. Madrid.

Gleick, J. (2012). *La información: Historia y realidad.* Barcelona: Crítica.

Dawkins, R. (2000). *El gen egoísta: las bases biológicas de nuestra conducta.* Barcelona: Salvat.

Doveling, K., Scheve, C., & Konijn, E. (Eds.). (2011). *The Routledge handbook of emotions and mass media.* London: Routledge.

Ekman, P. (2003). *El rostro de las emociones.* Barcelona: RBA.

Ekman, P., & Friesen, W. V. (1976). Measuring facial movement. *Journal of Environmental Psychology, 1,* 56–75.

Epstein, J. M. (2006). *Generative social science: Studies in agent-based computational modeling.* New Jersey: Princeton University Press.

Evans, J. A. (2008). Electronic publication and the narrowing of science and scholarship. *Science, 321,* 395–399.

Frijda, N. H. (1986). *The emotions.* Cambridge: Cambridge University Press.

Gazzaniga, M. (1993). *El cerebro social.* Madrid: Alianza.

González Álvarez, J. (2012). *Breve historia del cerebro.* Barcelona: Crítica.

Harfoush, R. (2009). *Yes we did. Cómo construímos la marca Obama a través de las redes sociales.* Barcelona: Gestión 2000.

Kahneman, D. (2015). *Pensar rápido, pensar despacio.* Barcelona: DeBolsillo.

Kaku, M. (2014). *El futuro de nuestra mente.* Barcelona: Debate.

Lang, A. (Ed.). (2011). *Measuring psychological responses to media.* London: Routledge.

Lewis, M., Haviland-Jones, J. M., & Barrett, L. F. (Eds.). (2008). *Handbook of emotions.* New York: Guilford.

McLuhan, M. (2009). *Comprender los medios de comunicación: las extensiones del ser humano.* Barcelona: Paidós.

Mar, R. A., & Oatley, L. (2008). The function of fiction is the abstraction and simulation of social experience. *Perspectives on Psychological Science, 3,* 173–192.

Matsumoto, D., & Ekman, P. (2010). Subjective Experience and the Expression of Emotion in Man. In G. F. Koob, M. Le Moal, & R. F. Thompson (Eds.), *Encyclopedia of behavioural neuroscience.* Oxford: Academic Press.

Morozov, E. (2012). *El desengaño de internet.* Barcelona: Destino.

Rúas, J., Punín, M. I., Gómez, H., Cuesta, P., & Ratté, S. (2015). Neurociencias aplicadas al análisis de la percepción: Corazón y emoción ante el Himno de Ecuador. *Revista Latina de Comunicación Social, 70,* 401–422.

Rúas, J., & Pérez, J. (2015). Neuroscience applied to the text and images analysis. *10th Iberian Conference on Information Systems and Technologies (CISTI)* (pp. 1–5), Portugal.

Small, G., & Vorgan, G. (2009). *El cerebro digital. Cómo las nuevas tecnologías están cambiando nuestra mente.* Barcelona: Urano.

Chapter 3
Immediacy and Metamedia. Time Dimension on Networks

Valentín Alejandro Martínez Fernández

Abstract Immediacy has become the dominant future of the networked society. In this new digital universe, everything is "installed" in present time; past and future are blurred to build a constant present. Everything takes place in a constant transition around a Moebius strip, the temporal space. A space where the timelessness of a "suspended life" is also built, and where the "right to oneself" no longer exists. In the digital world, people do not belong to themselves, they simply "are". "To be", though, does not mean a control over individuality, as in every part the whole is reflected, and this whole is made up of those parts included in the idea of instant.

Keywords Time · Technologies · Networked society · Immediacy

Our present society is calling for immediacy—not for urgency-; for the moment as the essential living space; for the instant as the reference framework; for the nano-reduction of message's feedback and the immediacy in its viral process. Everything is at present. The temporal distance is delimited to now.

We are subjected to the power of now, as Bertman notes when summarising the idea of that present unconnected to any time dimension, which replaces long-term with short-term, length with immediacy; permanence with transience, memory with sense, vision with momentum.

The "now" is imposed due to the new dimension of present, which acts, according to Innerarity (2009: 22), as a colonizer of a future to which parasitizes and then absorbs. The "now" blurs the past and leads it to a reconstruction that converts that past into a present moment, a reminiscent instant.

Past and future became blurred and disjointed islands. They are pure contingency as regards the essentiality of the active present, which gives reason to the digital swarm conceptualised by Han (2014: 16), which represents the "social mass", where any "unequivocal hierarchy separates the sender from the receiver". Also, "everyone is sender and receiver, consumer and producer"; it its that *produser*

V.A. Martínez Fernández (✉)
iMARKA Research Group, University of A Coruña (UDC), Corunna, Spain
e-mail: valentin.martinez@udc.es

© Springer International Publishing Switzerland 2017   19
F.C. Freire et al. (eds.), *Media and Metamedia Management*,
Advances in Intelligent Systems and Computing 503,
DOI 10.1007/978-3-319-46068-0_3

to which Dahlgren refers (2012: 49) when discussing the new profile of content consumer-producer, usually channelled by metamedia. The networks have given us the control over information, as well as its access, creation and dissemination.

3.1 Presence and Immediate Present

The digital medium is, above all, a medium of presence. It does not have more temporality than an immediate present and, within it, communication flows are clearly recognised, since information is produced, sent and received without any intermediary (Han 2014: 33), giving rise to the sunset of media mediation.

Networks and digital media do not have more filter than the expressed willingness of senders and receivers. The role of mediation is diluted in the network and becomes an essential element for hyperconnectivity.

The "general demediatization" (Han 2014: 34–35), concludes an era of "representation", making way for "presence" and "co-presentation."

And everything is depicted in that physical space, in which the digital world is reflected, that instant screen to which Márquez refers (2015: 10) as "the most important prostheses of our lives", since within it a present truth is flowing through all the layers of the current daily nature. From a "world-screen" described by Serrano (2013: 48), since the screen is the whole and the part; the developed hologram of a reality (Ballard 2008: 62), in a constant rebuilding, perceived as unique by the individual, who is living in a ubiquitous digital community in which shows a dependent behaviour.

Reality is mainly "something that is being seen" and perceived by sight, since the vision allow us to see, to check reality, while sometimes it is something elusive, blurred, and unclear on all our thinking (Mora 2012: 17).

To that extent we are faced with a mobile and instantaneous screen. Instantaneous is its essence, its "being in itself", since it not only allow us to communicate immediately with others; capture pieces of "our" reality; and access at any time to the information we want to include so far in our vital construct, apart from the place where it takes place and through a new dimension of global cyber-time. This is about what William Gibson called cyberspace, that truly transnational construction, where boundaries have little relevance, and by which information flows are global and have led to a world where knowledge is dispersed, where information is available, places are accessible, and communications are instantaneous, interdependent, and linked (Innerarity 2013: 33–35).

According to Giddens (2002: 64), the interconnection does not mean order, as an interconnected world is also a runaway world by the effect of immediacy. In this regard, Virilio (1997: 52) notes that the globalization of the present—known as real time—should be recognised: the ubiquity, immediacy, and instantaneous. These are but "the implementation of the divine" from a *hologramatic* approach, if virtuality and total vision and power are added. It is what Mora (2012: 132) summarizes in a Deus ex machina. On mobile screens, the future seems to be singular illusions subtly integrated into a single space-time.

3.2 An Hyperconnected Context

The mobile screen is the channel whereby the final function of hyperconnectivity becomes a fact. And that hyperconnectivity is the result of the ICT, which have created a global and networked platform in which citizens around the world have access to immediate, continuous, and non-stop information and knowledge.

As Martínez-Barea remarks (2015: 13–14), the hyperconnected world has been created for whom "the planet is a big social network" in which citizens are constantly online. The hyperconnectivity has confirmed the assertion by Baudrillard (1992: 11), who noted that every event is worldwide spread, to which it could be added the term of immediacy. When something happens, it is reported at the same time.

The network is therefore a three-dimensional space with reciprocal links, whose value is, according to Jarvis (2010: 45) "multiplied with the use", while it could be also added that the tag of "real time" gives the final value. The emphasis in the network is on real time, as Carr (2009: 211) notes, on simultaneity, eventuality, and subjectivity.

Information and Communication Technologies enable exchanges and generation of communication flows in real time, creating a feeling of immediacy (Lipovetsky 2014: 66). It seems that everything is based on the carpe-diem philosophy. It is what the sociologist Elzbieta Tarkowska calls "synchronic humans", mentioned by Bauman (2007: 145) when he talks about "presentist culture", conformed by individuals only living in the present and not paying attention to past experiences and future consequences of their actions.

Han (2015: 61–62) understand the ICT as the annulment of the time interval, as it removes any barrier to immediate access. Intervals are "abolished in favour of a total proximity and simultaneity". It therefore disappears any time and spatial distance. Immediacy becomes passion, since everything that is not present does not exist. Thus, "spatial and time intervals opposed to present are deleted". And he concludes, "there are only two states: nothing and present".

There is no pleasure without the experience of immediacy, as Lipovetsky (2007: 104) notes, and without it we only find unease, anxiety, and frustration, as we feel displaced from the free ICT environment to the present, where hypervelocity, direct access and immediacy are fiercely imposed. Something that, in Ottone's opinion (2011: 36) is emphasized in the case of young people, the digital natives, for those there is only immediacy.

Immediacy is one of the essential features of hypermodernity, which works well under the logic of continuous recycling of the past (Charles 2014: 35), and whose symbiotic precedent is the post-modernity. Thus it started an unprecedented social temporality, characterised by the primacy of the here and now (Lipovetsky 2014: 54). There the antinomy of the present (Sartre 2013: 186) is reflected, as it gives the present the condition of what it is, as opposed to the future, which is not yet present, and the past, which is no longer present.

The past in hypermodernity is constantly reconstructed in the present and integrated in it, and future is based on recollections perceived and understood at the moment of their present evocation.

The constant present is the time dimension of the digital universe. A wraparound space where social media become parallel and never-ending worlds, as in the graphic adventure *Moebius*, and interconnected by neurons, giving rise to a collective intelligence. This overcomes the postmodern concept of public opinion, in which that Library of Babel described by Jorge Luis Borges finds place, and in which there is that "catalogue of catalogues". That is the place where nothing is external and everything is a key piece for universal knowledge, which exists inasmuch it is accessible, a "model which makes history obsolete as the eternity of that library may be the consequence of the dissolution of past and future, to melt into a permanent present" (Palma 2010: 57).

The digital universe has created a new social life where individuals are constantly opened to other people's gaze. There is no chance to scape from other people's scrutinizing eyes. There is not even the "forget of the self".

We are living in public, permanently connected to a "public square" in which it is difficult to maintain privacy (Reig 2012: 168). A virtual space in which we are immediately seen and perceived, and we are subjected to that permanent present, even far beyond our physical life.

We find ourselves before a "digital panopticon", allowing people to observe without being seen and have the feeling of being observed without being watched, as the social theorist Jeremy Bentham made with his panopticon in order to influence social behaviour (Turkle 1997: 311).

For the *homo digitalicus* there is not anymore "safety distance" referred to by the social philosopher Hannah Arendt when she talked about privacy. That is "being oneself" against the permanent exhibition of people (Cohen 2013: 137).

3.3   Instantaneity and Convergence

We have not been and we won´t be on the network. We are there and, as a consequence, we are, we are "being there", as Heidegger (2008: 200) notes when talking about "being in the world".

On the network, all remains unchanged in constant entropy, in which the temporal disorder is actually order (Morin 2007: 91). Nothing is destroyed; everything is transformed in a constant present. And that changeability is a constant of the imperishable. The change is a feedback of the irremovable, where the beginning and the end become blurred, as an emulation of the myth of the eternal return stated by Nietzche (2014: 531) in his work *The Gay Science*, in the sense that nothing is forgotten, everything is in a "permanent remain".

We live in an instant convergence and we are immersed in that "convergence culture" pointed out by Jenkins (2008: 185). That is the result of a "complex process" that implies changes in producing and consuming social media and alters the relation between existing technologies, industries, markets, genres, and public; and whose main actors are participative audiences (López and Ciuffoli 2012: 49), whom Rheingold (2004: 58) called *smart mobs*, composed by people with common interests and objectives and, thanks to the hyperconnectivity, are able to contact, join groups, and interact.

While convergence not only responds to technological elements, it is basically a mental set-up. It is an idea, projected through an attitude, which arises from the generation of collective intelligence, made within an unchangeable present that, in turn, makes it transparent, since transparency is the main consequence of the control over presence and the present.

References

Ballard, J. F. (2008). *Fiebre de guerra*. Córdoba: Berenice. ISBN 978-84-996-7565-2.

Baudrillard, J. (1992). *La ilusión del fin. La huelga de los acontecimientos*. Barcelona: Anagrama.

Bauman, Z. (2007). *Vida de consumo*. Madrid: Fondo de Cultura Económica.

Carr, N. (2009). *El gran interruptor. El mundo en la red, de Edison a Google*. Barcelona: Deusto.

Charles, S. (2014). El individualismo paradójico. Introducción al pensamiento de Gilles Lipovetsky. In G. Lipovetsky & S. Charles (Eds.), *Los tiempos hipermodernos* (pp. 13–49). Barcelona: Anagrama.

Cohen, D. (2013). *Homo economicus, el profeta (extraviado) de los nuevos tiempos*. Barcelona: Ariel.

Dahlgren, P. (2012). Mejorar la participación: La democracia y el cambiante entorno de la web. In S. Champeau & D. Innerarity (Eds.), *Internet y el futuro de la democracia* (pp. 45–68). Barcelona: Paidós.

Giddens, A. (2002). *Un mundo desbocado. Cómo está modificando la globalización nuestras vidas*. Madrid: Taurus.

Han, B-Ch. (2014). *En el enjambre*. Barcelona: Herder.

Han, B-Ch. (2015). *El aroma del tiempo*. Barcelona: Herder.

Heidegger, M. (2008). *El ser y el tiempo*. México: Fondo de Cultura Económica.

Innerarity, D. (2009). *El futuro y sus enemigos*. Barcelona: Paidós.

Innerarity, D. (2013). *Un mundo de todos y de nadie*. Barcelona: Paidós.

Jarvis, J. (2010). *Y Google ¿cómo lo haría?*. Barcelona: Planeta.

Jenkins, H. (2008). *Convergence culture. La cultura de la convergencia de los medios de comunicación*. Barcelona: Paidós.

Lipovetsky, G. (2007). *La felicidad paradójica. Ensayo sobre la sociedad del hiperconsumo*. Barcelona: Anagrama.

Lipovetsky, G. (2014). Tiempo contra tiempo o la sociedad moderna. In G. Lipovetsky & S. Charles (Eds.), *Los tiempos hipermodernos* (pp. 53–109). Barcelona: Anagrama.

López, G., & Ciuffoli, C. (2012). *Facebook es el mensaje*. Buenos Aires: La Crujía.

Márquez, I. (2015). *Una genealogía de la pantalla*. Barcelona: Anagrama.

Martínez-Barea, J. (2015). *El mundo que viene* (3ª ed.). Barcelona: Planeta.

Mora, V. L. (2012). *El lectoespectador*. Barcelona: Seix Barral.

Morin, E. (2007). *Introducción al pensamiento complejo*. Barcelona: Gedisa.

Nietzsche, F. (2014). *La ciencia jovial*. Madrid: Gredos.

Ottone, E. (2011). *Gobernar la globalización*. Chile: Universidad Diego Portales.

Palma, D. A. (2010). *Borges.com*. Buenos Aires: Biblos.

Reig Hernández, D. (2012). *Socionomía*. Barcelona: Deusto.

Rheingold, H. (2004). *Multitudes inteligentes: La próxima revolución social*. Barcelona: Gedisa.

Sartre, J.-P. (2013). *El ser y la Nada*. Buenos Aires: Losada.

Serrano, P. (2013). *La comunicación jibarizada. Cómo la tecnología ha cambiado nuestras mentes*. Barcelona: Península.

Turkle, S. (1997). *La vida en la pantalla. La construcción de la identidad en la era de Internet*. Barcelona: Paidós.

Virilio, P. (1997). *El cibermundo: La política de lo peor*. Madrid: Cátedra.

Chapter 4
Journalism for Metamedia: Tools and Metrics for Quality and Ethics

Xosé López García

Abstract Journalism is facing great challenges due to a major restructuring caused, to a large extent, by the emergence of communication and information technologies. While the number of players has been increased, and hence the competitiveness within an increasingly complex scenario, journalism preserves its essence while trying to meet the demands of the digital arena.

Keywords Journalism · Immersive journalism · ICT · Hi-tech journalism · Elements of journalism

4.1 Time of Change

Every day, journalism fosters its own change. The evolution of the networked society is defining news practices and trends in new media, that is to say, metamedia. Two trends come along with these changes: the essential remains, and the innovation emerges. While the elements of journalism remain, metamedia are opening up new dimensions for journalistic practice, which searching a more communicative efficiency and communication models integrated in the networked society. Multimedia narratives, experimentation on new formats, and immersive initiatives illustrates the recent trends of journalism. These emerging initiatives include data journalism, immersive journalism, slow journalism, real-time

X. López García (✉)
Novos Medios Research Group, University of Santiago de Compostela (USC), Santiago de Compostela, Spain
e-mail: xose.lopez.garcia@usc.es

© Springer International Publishing Switzerland 2017
F.C. Freire et al. (eds.), *Media and Metamedia Management*,
Advances in Intelligent Systems and Computing 503,
DOI 10.1007/978-3-319-46068-0_4

journalism, glocal journalism, drone journalism, and hi-tech journalism. We are living in a moment for experimentation and innovative practices.[1]

There is no doubt that journalism is undergoing a profound restructuring (Boczkowski 2004; Casero Ripollés 2012), which has accelerated convergence processes and production and multi-platform challenges (García Avilés and Carvajal 2008). Also, journalism's professional practice is marked by the daily search for a sustainable future. Renewed practices, encouraged by labs of some of the main media and by results from the scientific community, are benchmarks that allow us to understand the evolution of this communication technique, which has proved its worth in democratic societies. Texts included in this chapter show the existence of renewed practices, and analyse some of their effects.

These practices come from the hand of current technologies within a communication context where news uses and consumption are changing. The result of changing processes and the development of information technologies lead to the configuration of a complex and highly dynamic communication arena. As the complexity of processes is increased, the noise and the overabundance of messages are multiplied.

Many scholars have focused their work on analysing the present changing digital scenario (Jenkins 2006; Scolari 2012), in order to provide knowledge on trends and challenges. That context of technological changes in the third millennium has influenced on the new journalistic practices, with challenges related to multimedia narratives (Deuze 2004), and the integration of users' participation (Carpentier 2011).

As changes have happened faster, dangers have been related to the disregard of journalism's ethical principles and quality. As a consequence, many works have focused their interest on how to address ethical (Maciá Barber and Herrera 2010) and quality challenges (Mompart et al. 2015). Questions related to the implementation of journalism rules and quality return to be in public discussions and in challenges of journalism for metamedia.

New processes increase the complexity of the journalistic activity, determined to take advantage of the possibilities to tell better non-fiction stories. But the path requires a lot of experimentation, innovation and monitoring of the characteristics of the work, in order to draw attention to the mistakes and possible measures to ensure a good journalism in the digital era.

---

[1]The research is part of the results obtained in a research project, promoted by the Ministry of Economy and Competitiveness. The title is "Innovation and development of Spanish cybermedia. Architecture of news interactivity on multiple devices: news formats, conversation and services". Reference: CSO2012-38467-C03-03. It is also carried out under the international network XESCOM, and the support for research of the Xunta de Galicia.

4.2 Interest in Journalism

Metamedia have increased the interest in journalism, journalistic practices, and pieces of news. The ubiquity of products, which are disseminated on multiple platforms, makes the work of journalists more visible. While the media have been multiplied in an increasingly dynamic ecosystem, a large part of them have a strong presence in online forums and many citizens make comments and give their opinion.

After more than a decade of the emergence of social networks, discussion on their use in journalism has virtually disappeared. The debate now centres on how to use them, since all cybermedia produce pieces for these networks. And also, there is discussion on verification systems, strategies for increasing impact and, in short, getting the best from them.

It seems that platforms may play a relevant role in journalism. As new tools come in the digital scene, the journalistic practice introduces them in production routines. Results of daily activity's tests will be key in determining their level of visibility. But professionals should assume that they have to be constantly updated in skills and abilities, in order to tackle technological and communication challenges.

Digital technologies, which make working easier, are a powerful tool for improving the exercise of journalism (García Avilés 2007). Also, they require journalists to be constantly updated. While the essence of the journalistic activity is the messages production, there is no doubt that it is now more complex and requires a mastery of new technologies.

Metamedia-understood as new media with renewed properties (Manovich 2013)-, involve a challenge in the search for new paths for a combination of journalism as a social technique with the use of the new online dimensions. Journalism is now concerned about gaining the future every day, with initiatives able to contribute with added value within a context where communication and information are ubiquitous.

4.3 Conclusions

Journalism is more alive than ever. Connectivity and mobility are feeding current initiatives, which are marked by high technologies, automation of processes, and settings offered by virtual reality.

But the essence—the elements of journalism—remains, since they daily show their utility for the construction of better-informed societies. And the emergent initiatives provide with vitamins for experimentation and with added value for news processes.

Journalism for metamedia has moved beyond concerning about tools, which are constantly coming up; and metrics, which set rates of psychological pressure for professionals; to the old discussions on the quality of messages and the fulfilment of journalism's ethical principles. Old and new elements are walking together within the background of journalism for metamedia.

References

Boczkowski, P. J. (2004). The processes of adopting multimedia and interactivity in three online newsrooms. *Journal of Communication, 54*, 197–213.

Carpentier, N. (2011). *Media and participation*. Bristol: Intellect.

Casero Ripollés, A. (2012). Contenidos periodísticos y nuevos modelos de negocio: Evaluación de servicios digitales. *El Profesional de la Información, 21*(4), 341–346.

Deuze, M. (2004). What is multimedia journalism?. *Journalism Studies, 5*(2), 139–152. Retrieved from: http://jclass.umd.edu/classes/jour698m/deuzemultimediajs.pdf. Accessed at February 6, 2016.

García Avilés, J. A. (2007). Nuevas tecnologías en el periodismo audiovisual. *Revista de Ciencias Sociales y Jurídicas de la Universidad de Elche. I*(2), 59–75.

García Avilés, J. A., & Carvajal, M. (2008). Integrated and cross-media newsroom convergente: two models of multimedia news production: The cases of Novotécnica and La verdad multimedia in Spain. *Convergence, 14*(2), 223–241.

Jenkins, H. (2006). *Convergence culture: Where old and new media collide*. New York: New York University Press.

Maciá Barber, C., & Herrera, S. (2010). La excelencia informativa: Dilemas éticos y retos profesionales del periodista. *Cuadernos de Periodistas, marzo de 2010*. Retrieved from: http://www.apmadrid.es/images/stories/doc/vapm20100422175214.pdf. Accessed at February 5, 2016.

Manovich, L. (2013). *Sofware takes comand*. New York: Bloomsbury Academic.

Mompart, J. L., et al. (2015). Los periodistas españoles y la pérdida de calidad informativa: El juicio profesional. *Comunicar, XXIII*(45), 143–165. Retrieved from http://www.revistacomunicar.com/index.php?contenido=detalles&numero=45&articulo=45-2015-15. Accessed at February 6, 2016.

Scolari, C. A. (2012). Comunicación digital. Recuerdos del futuro. *El Profesional de la Información, 21*(4). Retrieved from: http://repositori.upf.edu/handle/10230/25653. Accessed at February 6, 2016.

Part II
New Media in the Digital Era

Chapter 5
Public Information Services in the Digital Era

Miquel de Moragas Spà

Abstract The text proposes to review the changes that are taking place in the paradigm of communication (broadcast, production, distribution, storage) and its possible impact on the construction of new models of public service media in the digital era.

Keywords Media · Communication policies · Public service · Digital communication

Pontevedra, September 1999. Fifteen years ago, a conference was held on public television models in the 21st century. In that occasion, I had the opportunity to ask the General Director of RTVE, Pío Cabanillas, about how they were planning to respond to the Internet. The question impressed him. He recognised they had no plan in this regard.

But even at that time, many scholars pointed out the need to address the new strategic challenges of TV and radio public services; among them, their responsibility to be the engine of the information system in the digital era.

Since then, changes have occurred really fast. For instance, the Olympic Games of Barcelona in 1992 were carried out without the Internet; it was the time of videotext and teletext. Today it seems impossible to organize a large event without technological tools.

The main challenges of the present research on communication—hence the relevance of the work of the International Research Network on Communication Management, XESCOM-, are to make an accurate diagnosis of the changes resulting from the new ecology of communication, with a view to develop communication policies adapted to the new arena.

M. de Moragas Spà (✉)
Spanish Association of Researchers on Communication (AE-IC), Barcelona, Spain
e-mail: miquel.demoragas@uab.cat

© Springer International Publishing Switzerland 2017
F.C. Freire et al. (eds.), *Media and Metamedia Management*,
Advances in Intelligent Systems and Computing 503,
DOI 10.1007/978-3-319-46068-0_5

5.1 Reinterpret the Tipology of the Media

This new media ecology requires a continuing theoretical exercise for defining the typology of the media and for interpreting changes produced in the paradigm of communication, as made by McLuhan when he wrote Gutenberg Galaxy and Manuel Castells when he published Communication and Power.

As an introduction, six aspects should be highlighted:

5.1.1 Technological Convergence in the Media

There is a need to analyse the convergence between different media (press, radio, television, cinema), and other communication tools (mobile phones, cameras, tablets, big screens, automation). This requires a new media economy and democratic communication policies.

5.1.2 New Thresholds on Communication Spaces

New technologies are resizing the former boundaries of communication: live remote broadcasting (globalization); direct access to multiple information sources (simultaneity, instantaneity); networks and nodes (creation of communities). This determines a new global-local dialectical, which does not exclude any of the two areas and offers new opportunities for minority cultures.

5.1.3 Ease of Images and Multimedia Production

The digitization has multiplied the possibilities to capture and process images and to develop software for producing multimedia contents, which has a major impact not only on audiovisual production, but also on commonly used languages in society. Consider, for instance, the youngest skills (digital natives) for multimedia production, using new narratives based on images and hyperlinks.

5.1.4 New Players and New Media

Other relevant consequence of these changes is the transformation of communication actors (institutional issuers). In the age of the Internet, political, economic, and social institutions (governments, telecom companies, financial institutions, companies); but also civic movements (NGO, cultural organizations, sport clubs, churches, universities) are becoming the new "media".

In the earliest stages of this process, between 1996 and 2006, the media occupied intermediate and low positions in the list of the most visited websites. But more recently, a recovering trend of the media is evident, since they understand the benefits associated to their know-how in producing and disseminating information online. As Castells pointed out, the most relevant companies think of remarketing the independent mass self-communication based on the Internet. Thus, large groups control the fundamental nodes between the media and the online sphere.

5.1.5   New Audiences and Forms of Association

Changes in communication are also evident when analysing receivers' nature. The audience fragmentation has been initiated with the multiplication of TV channels and the new cable and satellite platforms, but the Internet represents a qualitative change in that direction. The "mass communication" has moved to "mass customization", what is to say, the possibility of providing custom information (related to contents, time and place of consumption) to a massive number of people.

The Internet—"the medium is the message"—should be also analysed in terms of its effect on social organizations, on communities. It is still too early to foresee the future of the relation between social movements and the Internet, but the experiences show the potential influence of the new networks. The Internet enables the creation of new spaces and forms of communities in the cyberspace.

5.1.6   New Forms of Storing and Retrieving Information

Digital communications not only change dissemination methods and create new local and global spaces, but also lead to new forms of storing and retrieving contents. Under these conditions, media and other professionals (doctors, lawyers, economists, professors) started to focus their work on information production to make it rapidly available for users.

This is a key aspect of the paradigm shift. In the digital era, where channels and content offers are multiplied, the centre of gravity of communication power tends to be pushed towards content dissemination and production. This situation had already begun to manifest in the case of the audiovisual industry, but it becomes more evident with the emergence of the Internet.

5.2   New Communication Policies in the Digital Era

Special attention must be paid to the social and political consequences of all these changes. In this regards, five points with a direct impact on communication policies should be highlighted:

1. The way in which the former centrality (authority and power) of media dissemination is shifting towards content production. This may result in a shift of priorities of communication policies moving from the regulation in dissemination to the product management in four main areas: information, culture, formation and entertainment.
2. One of the priorities of cultural and communication policies must be to provide original and quality contents to the great digital content memory. Contents should be tailored to citizens needs (culture, education, information, entertainment).
3. New forms of managing cultural identities should be considered within a context of multiple links between the local and the global, giving value to each identity in the global arena. This may be applicable to various strategic sectors, from cultural industries to other phenomena such as tourism and events organization.
4. The aforementioned circumstances will make cultural, education, and communication policies to be blurred. At the end of the 20th century, with the emergence of digitization and globalisation processes, it takes place an integration or synergy between these policies. Cultural policies should consider their communications and meet the needs of the new cultural industries. Communication policies should address policies but also all development factors, social welfare and public participation, essential elements for cultural policies.
5. And eventually, communication policies could not longer refer exclusively to the mass media, but also to the convergence between welfare, education, health, culture, and telecommunication policies—formerly autonomous-. All institutions, not only the media, become producers and communication mediators, which will also redefine public information services in the digital era.

5.3   Content Production: New Focus in Public
Communication Policies

The shift from the media to metamedia is produced in a context of general context in communication paradigms, a change that has a fundamental axis: the new centrality of content production in the whole communication system.

This requires a new conception of public communication/information services in the 21st century, which will be no longer focused on the audiovisual sector but on a more complex structure.

The power of communication lies no longer with the process of disseminating (rights, licenses, and permissions for broadcasting), but with the production of contents and the storing of information, prepared for a direct access by users.

An "information for all" society could only be constructed with a new production policy, including contents "for all social needs". Public information systems should represent an alternative to the commercial logic based on the pay per view, as well as to the society of the "toll communication". If the only logic of information society ends up being commercial, then we'd only have access to cost-effective information.

The new ecology of communication—from the media to metamedia—calls for a...

5.4 Redefinition of the Role of Public Communication Services

The new ecology of communication requires, first of all, a review of the more conventional functions attributed to public broadcasters (guarantees related to democracy, aesthetics, identity, and development). In addition to these traditional roles, the new public media have to address the challenges defined by the rapid transformation of the communication system in the digital era:

- Guaranteeing a universal access to contents
- Guiding and mediating with the large supply of information
- Collaborating with the most reduced areas (local, communal, proximity)
- Being the engine of convergence processes between communication and other social sectors (culture, education, health, social welfare)
- Being the engine of the public information system in the digital arena, opening up a new panorama for democratic communication policies.

From now on, content production accessible to everyone will be under the responsibility of all public and private institutions that, thanks to the new technologies, have the possibility to create their own information systems. All institutions and, especially, public bodies, have now the possibility—and the responsibility—to exploit the resources resulting from the digitization, which allow institutions to produce, store, and distribute information.

Those institutions that, by their nature, produce contents (universities, scientific institutes, registers, museums, theatres and cultural institutions, even football clubs), are becoming media and metamedia thanks to digital technologies.

5.4.1 Who Can Take the Lead in These Processes?

In communities with reduced populations and languages, such as Galicia and Catalonia, the leadership—not the exclusivity—of these processes correspond to the communication institutions, which are the best prepared for this function of producing, storing and distributing information and, more specifically, to public broadcasting services, now turned into "metamedia".

Chapter 6
Content Curation in Digital Media: Between Retrospective and Real-Time Information

Javier Guallar

Abstract The concept of content curation and its application in journalism are introduced. Then, the use of content curation in digital media is analysed using 15 examples of articles that were produced with this technique. The examples are classified according to whether the information in them is retrospective, recent, current or real-time.

Keywords Digital media · Online newspapers · Content curation · Journalists · Content curators

6.1 Theoretical Framework

6.1.1 What Is Content Curation?

The origin of the concept and the speciality of content curation is generally attributed to a paper by Rohit Bhargava, a marketing lecturer and professional, published at the end of 2009 and entitled: Manifesto For The Content Curator: The Next Big Social Media Job Of The Future? In this paper, the author stated:

> In the near future, experts predict that content on the web will double every 72 hours. The detached analysis of an algorithm will no longer be enough to find what we are looking for. To satisfy the people's hunger for great content on any topic imaginable, there will need to be a new category of individual working online. Someone whose job it is not to create more content, but to make sense of all the content that others are creating. To find the best and most relevant content and bring it forward. The people who choose to take on this role will be known as Content Curators (Bhargava 2009).

Bhargava highlighted a problem that, in his opinion, is caused by the large amount of digital information available on the internet. This information is characterized by its continuous, unstoppable growth and rapid obsolescence, and people

J. Guallar (✉)
University of Barcelona (UB), Barcelona, Spain
e-mail: jguallar@gmail.com

© Springer International Publishing Switzerland 2017   37
F.C. Freire et al. (eds.), *Media and Metamedia Management*,
Advances in Intelligent Systems and Computing 503,
DOI 10.1007/978-3-319-46068-0_6

can easily become overloaded with it, a situation known as "infoxication". In this context, Bhargava argues, the systems of algorithms that are generally used to locate information (that is, Google) do not alone have enough capacity to meet people's information needs. Consequently, content curation is a new approach; a new solution. The emphasis of this new approach is the content curators' ability to "make sense" of the content that others have created.

The speciality of content curation has spread rapidly, in a matter of years, from this initial proposal associated with the field of digital marketing, to other disciplines and specializations such as library and information science, journalism and education.

In addition to Bhargava, the creator of the concept, key authors on content curation include Steven Rosenbaum, a leading disseminator and author of the bestseller Curation Nation (Rosenbaum 2011); Robin Good, who is perhaps the most popular content curator, known particularly for his work on Scoop.it and a very interesting practical bibliography (Good 2010); and Pawan Deshpande, who, from a marketing perspective, has contributed various methods and content curation techniques (Deshpande 2013).

But are we just talking about a filtering or content-selection system? The definition of content curation given by the author, Guallar, and Leiva-Aguilera in the book "El Content Curator" aims to clarify this issue by bringing together the range of approaches in a comprehensive, integrated view of the speciality:

> Content curation is a system used by a specialist, a content curator, on behalf of an organisation or individual. It is based on continuously searching for, selecting, making sense of and sharing the most relevant content from several online information sources about a specific topic or set of topics and a specific area or set of areas. The content is chosen for a specific audience either online or in other contexts, such as an organisation, has added value and therefore engages its audience/users (Guallar and Leiva-Aguilera 2013).

The same book also proposes a content curation method called the four S's system, which is based on four stages: search, select, sense-making and share. It stresses that the third stage of sense-making is the key, as it is what distinguishes content curation from a simple social recommendation or from pure dissemination of contents. In this process, curators must add value to the information that they have previously selected, by offering their personal touch, their style or their own "voice".

6.1.2   Content Curation in Journalism

As stated above, content curation has been applied to a range of disciplines and professions. In the field of journalism, a leading article by McAdams (2008) was published just a few months before Bhargava's Manifesto, and discussed curation as a future activity in journalism. Since then, various specialists have referred to the use of content curation in this field, including Robin Good (mentioned above), Sternberg (2011), Bradshaw (2013), Guerrini (2013), Bakker (2014) and Días Arias (2015). In a previous paper, we differentiated between a general and specific association between content curation and journalism. In terms of the general association, we stated:

> The journalist looks for current information for a specific audience. To that end, sources are researched and content is obtained and checked. Therefore, journalists are professionals with content creation and information management abilities: they search for sources and verify news. What we say is equally applicable to other professionals who work in the newsroom of a media outlet, and particularly to press documentalists, whose speciality essentially involves managing journalistic information (searching for and verifying news, archiving, documentary analysis or tagging) and also includes content creation. None of the previously described tasks is alien to the basic concept of curation, which means that there is already a general association of journalists and press documentalists with content curators (Guallar 2014).

Beyond this general association, we are interested in what content curation contributes to journalism specifically. In other papers (Guallar 2014, 2015, 2016), we have investigated types of curation products in current journalism, to identify and demonstrate the practice of journalistic curation. In this paper, we present a set of examples of journalistic curation, taking into account the time range of the curated information and other variables.

6.2   Methods

We analysed content selected randomly from the Spanish digital press in 2015, and occasionally drew from other media, such as French cybermedia during the night of the Paris terrorist attacks in November 2015. Representative examples of current journalistic curation were presented, taking into account three variables:

The time frame of the curated information. We distinguished between retrospective information from previous months or years; recent information from the last few days; current information from the last few hours; and real-time information. The time frame was the main variable used to classify the examples.

The type of product used. We indicate whether the article was a news item, a topical issue, a timeless topic, a research article, a documentary product (such as a biographical profile), or a "live" event, among others.

The type of source used. We identify different sources, including archives, websites, cybermedia and social media such as Twitter and YouTube.

6.3   Results: Examples of Journalistic Curation Products

Below, we present and discuss several examples of articles based on content curation that were published in 2015 in a range of digital media. They are grouped according to the time frame of the curated information.

6.3.1   Curation of Retrospective Information

Example 1 is a classic example of the use of sources of journalistic information in an article for publication in digital press. Articles of this type have commonly been written since the start of digital press, and the main sources are official websites and other cybermedia (Table 6.1).

Examples 2 and 3, which are written in different styles, represent a further step in the variety of sources curated in an article of the same type as in Example 1. The authors used their own and other journalistic sources, and a wide range of internet sources, including specialized blogs and Twitter accounts. In Example 3, online sources are the focus of the work of curation/documentation, and the rest of the sources serve as a support.

Example 4 is an excellent piece of investigative journalism based on curation of information from social networks such as Instagram, Twitter and Pinterest, complemented with other internet sources. The variety of sources used is notable, as well as the number of links to them throughout the article.

Example 5 is a classic product of journalistic documentation: a biographical profile. If, as in this case, it is written after the person's death it is an obituary. The given example differs from classic biographical profiles in journalism because it is based on the curation of videos that are available on YouTube. This format is increasingly common today.

Example 6 is representative of content curation articles on topics like leisure and travel that are not necessarily topical. The sources used in these pieces are timeless, or not associated with a news item. These may be classic sources of digital information, in this case, the websites of the jazz clubs that are described, or social media, including videos of concerts from YouTube.

6.3.2   Curation of Recent or Current Information

Examples 7 and 8 refer to curation of recent information (from the last few days or weeks). Example 7 is based exclusively on Twitter, and specifically on a current internet phenomenon: "memes" or typically humorous material published by internet users on variations of a topic; in this case, an election campaign. Example 8 is a research/ informative article on a current political issue. It is based on the

Table 6.1 Examples of curation of retrospective information

Product	Type	Sources
1. Historia y libros para vacunarse contra la islamofobia, eldiario.es, 21/11/2015 http://www.eldiario.es/cultura/historia/Historia-libros-vacunarse-islamofobia_0_453905100.html	Informative article on a topical issue	– Official websites – Cybermedia
2. El 'ciberyihadismo' en ocho preguntas clave, El español, 17/11/2015 http://www.elespanol.com/ciencia/20151117/79992036_0.html	Informative article on a topical issue	– Official websites – Cybermedia – Own archive – Specialized blogs – Twitter
3. Cómo colarle un Nobel falso a los medios, El Mundo, Enredados, 08/10/2015 http://www.elmundo.es/enredados/2015/10/08/5616853d22601d614e8b45e8.html	Informative article about a topical issue	– Cybermedia – Own archive – Websites – Specialized blogs – Twitter
4. Radiografía de la #thinspiration, una peligrosa apología de la delgadez extrema, eldiario.es, Hoja de router, 19/06/2015 http://www.eldiario.es/hojaderouter/internet/thinspiration-thinspo-anorexia-bulimia-internet-Twitter_0_399310527.html	Research article	– Cybermedia – Websites – Blogs – Instagram – Twitter – Pinterest
5. Los 7 vídeos de Les Luthiers que nos hicieron reír hablando de ciencia y filosofía, El País, Verne, 21/08/2015 http://verne.elpais.com/verne/2015/08/21/articulo/1440171749_316809.html	Biographical profile (obituary) in videos	– YouTube
6. Los mejores clubes para escuchar jazz en Berlín, El español, 20/11/2015 http://www.elespanol.com/ocio/viajes/20151120/80741937_0.html	Article on a timeless topic (leisure)	– Websites – YouTube

Source Prepared by the author

curation of numerous sources of information: mainly pieces published in the media by the people who are analysed in the article, and tweets by the same people. In addition to the digital press and Twitter, other types of websites, blogs and Facebook are curated (Table 6.2).

The next two examples are related to the curation of current information, which is defined as information produced in the last few hours. This time frame is very appropriate for the written press, accustomed as it is to producing daily editions. Examples 9 and 10 have similar characteristics to 7: they are based on what is being said about a specific news item on social networks, including Twitter in this case. These examples are highly representative of a kind of content curation that is proliferating and can lead to a wide variety of articles: from short pieces (Example 9) to longer, in-depth ones (Example 10) that may be based on tweets found by hashtag searches.

6.3.3   Curation of Information in Real-Time

As the consumption of information on digital networks accelerates, the generation of journalistic reports must also be speeded up. The wide variety of information that is generated continuously on the internet leads media outlets to step up their live news production. Content curation has space for expansion in this area. Some platforms that are specialized in live content curation, such as Storify, are very useful in journalism. They can be used to create a journalistic account of what is happening that includes content from social networks. Digital media sometimes publish news in real-time in open profiles on platforms such as Storify (examples 11 and 12), but increasingly (examples 13, 14 and 15), they integrate the "Storify philosophy" directly into products published on their own websites (Table 6.3).

In Example 11, the curation is based exclusively on content from Twitter about specific elections—a political news event. Example 12 has similar characteristics to 11, but describes film awards—a cultural event. In this case, much more varied sources were used, including real-time information from Twitter, retrospective information obtained mainly from the archives of the media outlet, and YouTube.

The final examples (13, 14 and 15) correspond to live narration of a major event on a media outlet's website. In this case, the event was the tragic Paris terrorist attack on 13 November. Many of the main Spanish and French media channels consulted by the author whilst events were unfolding that night used similar live accounts, whilst some also used content curation of information revealed and discussed on social networks. In these latter cases, Twitter was once again a major source of information on the "real-time web". It is the most effective channel for journalist-curators to obtain information about an event that is underway and offer it immediately. We have selected three examples of the media outlets that, in our opinion, carried out the most comprehensive, commendable work that night. Although the documents faithfully reflect the final result of how these events were

Table 6.2 Examples of curation of recent or current information

Product	Type	Sources
7. Los mejores memes de las elecciones municipales. El Periódico, 25/05/2015 http://www.elperiodico.com/es/noticias/redes/mejores-memes-las-elecciones-municipales-4217911	Article on a topical issue on the internet	– Twitter
8. Antologia del #PressingCUP: els millors tuits, articles i vídeos crítics amb l'esquerra independentista. Crític, 22/11/2015 http://www.elcritic.cat/actualitat/antologia-del-pressingcup-els-millors-tuits-articles-i-videos-critics-amb-esquerra-independentista-6811	Research or informative article on a topical issue	– Cybermedia – Websites – Blogs – Twitter – Facebook
9. 7 tuits que resumen el Madrid-Barcelona. El Huffington Post, 21/11/2015 http://www.huffingtonpost.es/2015/11/21/tuits-bernabeu_n_8617862.html?utm_hp_ref=spain	Article on a topical issue on the internet	– Twitter
10. Batalla de hashtags en el Hemiciclo: #HayFuturo contra #LaEspañaReal. El Mundo, 24/02/2015 http://www.huffingtonpost.es/2015/11/21/tuits-bernabeu_n_8617862.html?utm_hp_ref=spain	Article on a topical issue on the internet	– Twitter

Source Prepared by the author

Table 6.3 Examples of curation of information in real-time

Product	Type	Sources
11. Elecciones catalanas 2015. Última hora sobre unos comicios que marcarán el futuro de Cataluña. El País, Storify, 27/09/2015 https://storify.com/el_pais/elecciones-catalanas-2015	Live account of an event, with content curation in Storify	– Twitter
12. Así hemos contado la Gala de los Premios Goya 2015. El Confidencial, 08/02/2015 https://storify.com/elconfidencial/sigue	Live account of an event, with content curation in Storify	– Own archive – Twitter – YouTube
13. Série d'attaques terroristes à Paris, au moins 120 morts, état d'urgence décrété. Libération, 13/11/2015 http://www.liberation.fr/france/2015/11/13/fusillade-dans-le-10e-arrondissement-de-paris_1413313	Live account of a major event	– Twitter
14. Attentats du 13 novembre. Les dernieres informations en direct. Le Monde, 13/11/2015 http://www.lemonde.fr/societe/live/2015/11/18/attentats-du-13-novembre-les-dernieres-informations-en-direct_4812231_3224.html	Live account of a major event	– Twitter
15. Directo. La última hora de los atentados de París. El País, 13/11/2015 http://internacional.elpais.com/internacional/2015/11/13/actualidad/1447451281_752084.html	Live account of a major event	– Twitter

Source Prepared by the author

related, obviously a product of these characteristics is most clearly perceived during the live transmission itself, whilst the event is unfolding.

6.4 Conclusions: Eight Points on Content Curation in Digital Media

We can draw the following conclusions from the above examples.

1. An analysis of the content of digital media today reveals that content curation does indeed play a role in digital journalism. It can be found in different forms and at different levels of intensity: from a complementary or supporting element, to a core element that is prominent in, or even the main feature of, the journalistic product.
2. Journalistic curation brings sources that are external to the media outlet into the foreground: these external sources or content are the pieces used to construct the curation product. Without them there is no curation.
3. Media outlets should adopt the approach of assessing and valuing contents that they themselves have not created. This is one of the keys to current internet culture, based on the idea of seeing, recommending and sharing, and is something that has not always been grasped by traditional media.
4. External contents are not enough on their own. They must be given meaning (sense-making) and incorporated into the media outlet's account. This is the main task of journalist-curators, who validate, prioritize and give value to content when they "explain it". Without this role, there is no curation.
5. Journalists must develop skills related to selecting content from sources outside the media outlet, and contextualization skills (sense-making) for these contents.
6. In all the time frames that journalists tackle when they create a piece (information from the past, the last few days, the last few hours, in real-time), there has been an explosion of content. This is particularly true of real-time or current information. Therefore, journalistic curation should provide a response that is suited to the readership in all of these time ranges.
7. The traditional specialization of journalistic documentation, whether undertaken by journalists themselves, specialized professionals or documentalists, dealt with past or retrospective information. Now this task should be extended to current information and even to real-time information, which is of increasing interest to the public. Journalistic documentation, which we could now define as journalistic curation, involves working in all time contexts and with all kinds of sources.
8. To conclude, the objective of Bhargava, "to make sense of all the content that others are creating" is perfectly valid and stimulating in journalism at the present time. We could even ask whether journalism has any meaning today if it does not fulfil this role, in a society in which there are numerous conversations on digital networks. Content curation presents a series of opportunities and challenges for the media and for journalists.

Acknowledgments This study was made possible with the support given by the Spanish Agency for the Management of University and Research Grants (AGAUR) of the Government of Catalonia to the consolidated research group "Cultura i Continguts Digitals" (2014 SGR 760).

References

Bakker, P. (2014). Mr. Gates returns: Curation, community management and other new roles for journalists. *Journalism Studies, 15*(5), 596–606.

Bhargava, R. (2009). Manifesto for the content curator: The next big social media job of the future? *Rohitbhargava.com*. Retrieved September 30 from http://www.rohitbhargava.com/2009/09/manifesto-for-the-content-curator-the-next-big-social-media-job-of-the-future-.html

Bradshaw, P. (2013). Journalism is curation: Tips on curation tools and techniques. *Online Journalism Blog*. Retrieved September 30 from http://onlinejournalismblog.com/2013/09/30/curation-tools-tips-advice-journalism

Bradshaw, P. (2016). Curation is the new obituary: 8 ways media outlets marked Bowie's life and death. *Online Journalism Blog*. Retrieved January 11 from http://onlinejournalismblog.com/2016/01/11/curation-is-the-new-obituary-8-ways-media-outlets-marked-bowies-life-and-death/

Deshpande, P. (2013). Six content curation templates for content annotation. *The Curata blog*. Retrieved August 13 from http://www.curata.com/blog/6-content-curation-templates-for-content-annotation/

Díaz Arias, R. (2015). Curaduría periodística, una forma de reconstruir el espacio público. *Estudios del mensaje periodístico, 21*. Retrieved from http://revistas.ucm.es/index.php/ESMP/article/view/51129

Good, R. (2010). Real-time news curation—The complete guide part 4: process, key tasks, workflow. *Master New Media*. Retrieved September 29 from http://www.masternewmedia.org/real-time-news-curation-the-complete-guide-part-4-process-key-tasks-workflow

Guallar, J. (2014). Content curation in journalism (and journalistic documentation). *Hipertext, 12*. Retrieved from http://raco.cat/index.php/Hipertext/article/view/275781/364537

Guallar, J. (2015). Curación de contenidos en los medios digitales. *I Simposio Internacional XESCOM Gestión de la comunicación*, Facultad de Ciencias Sociales y de la Comunicación, University of Vigo, Campus Pontevedra. Retrieved November 28 from http://eprints.rclis.org/28614

Guallar, J. (2016). Curación de contenidos en el periodismo digital. *Seminario Digidoc*. Barcelona: Pompeu Fabra University. Retrieved January 28 from http://eprints.rclis.org/28866/

Guallar, J. & Leiva-Aguilera, J. (2013). *El content curator. Guía básica para el nuevo profesional de internet*. Barcelona: Editorial UOC, in the collection "El profesional de la información", n. 24, 162 pp. ISBN 978-84-9064-018-0.

Guerrini, F. (2013). *Newsroom curators & independent storytellers: content curation as a new form of journalism*. Reuters Institute for the Study of Journalism, University of Oxford. Retrieved from https://reutersinstitute.politics.ox.ac.uk/publication/newsroom-curators-and-independent-storytellers

McAdams, M. (2008). Curation and journalist as curators. *Teaching Online Journalism*. Retrieved December 3 from http://mindymcadams.com/tojou/2008/curation-and-journalists-as-curators

Rosenbaum, S. (2011). *Curation nation: How to win in a world where consumers are creators* (284 pp.). McGraw-Hill. ISBN 978-0-07-176039-3.

Sternberg, J. (2011). Why curation is important to the future of journalism. *Mashable*. Retrieved March 10 from http://mashable.com/2011/03/10/curation-journalism

Chapter 7
The Relationship Between Mainstream Media and Political Activism in the Digital Environment: New Forms for Managing Political Communication

Andreu Casero-Ripollés

Abstract Social media are introducing changes into the relationship between the mainstream media and online political activism. In a new context characterized by citizen empowerment, social movements are promoting new models for interacting with journalists. This chapter aims to analyze this phenomenon. To that end, a qualitative methodology has been used, based on in-depth interviews with online political activists and journalists affiliated with the 15M Movement in Spain. The results illustrate the emergence of a new strategy grounded in the overturn of mediatization.

Keywords Political communication · Social media · Political activism · Mediatization · Journalism

7.1 Introduction

Social media are leading to numerous transformations in the field of political activism and in social movements. This idea encompasses all digital technologies that allow active citizen participation in communicative processes and that include elements such as social network sites (Facebook), micro-blogging services (Twitter and Weibo), video-sharing platforms (YouTube), photography platforms (Instagram and Flickr), and blogs (Fuchs 2014). The existence of these platforms has a three-fold impact on political activism. It affects their organizational dynamics, increasing effectiveness and making interconnections possible between geographically dispersed people and groups. It also influences mobilization, reducing its cost and allowing for the emergence of new mechanisms for political participation, including civic monitoring (Feenstra and Casero-Ripollés 2014). Additionally, it has a direct

A. Casero-Ripollés (✉)
Jaume I University of Castelló (UJI), Castelló, Spain
e-mail: casero@uji.es

© Springer International Publishing Switzerland 2017
F.C. Freire et al. (eds.), *Media and Metamedia Management*,
Advances in Intelligent Systems and Computing 503,
DOI 10.1007/978-3-319-46068-0_7

impact on activist communication strategies, involving new strategies for production, dissemination, and information management.

This chapter aims to analyze the changes in communication in online political activism that social media are producing. Despite the breadth of this phenomenon, we focus only on the communications dimension, setting aside issues related to organizing and mobilization. We specifically analyze the transformations that are introduced in the relationship between political subjects, journalists, and the mainstream media. To that end, we use a qualitative methodology based on in-depth interviews with political activists and journalists associated with the 15M Movement in Spain.

7.2   Literature Review: The Relationship Between Journalism and Activism

The literature has identified traditional strategies used by activists and social movements when interacting with journalists and the mainstream media. One of the primary approaches is formulated by Rucht (2004), who asserts that interactions are based on the quadruple "A": attack, abstention, alternative, and adaptation. The first two are aspects of the same phenomenon: activists' and journalists' mutual suspicion. This mistrust is based on three primary motivations. On the one hand, the social movements and players associated with social change believe that the mainstream media contribute to the maintenance and perpetuation of capitalism and its structures. This belief stems from a post-Althusserian Legacy (Cammaerts 2012) that perceives the media as ideological apparatuses of the State. As such, activists think of themselves and the media as opposing camps pursuing different, even antagonistic, goals.

On the other hand, activists argue that media coverage of protests and social change is strongly biased (Gitlin 1980) because media companies defend their political and economic interests, leading to the prevalence of the politicization and commodification of news about social movements. According to activists, this prevalence results in the lack of credibility and independence of the mainstream media. Consequently, activists do not consider the mainstream media to be valid venues for disseminating their claims and demands. Hence, 74 % of U.S. activists and 49 % of Latin American activists believe that the mainstream media threaten democracy and social justice (Harlow and Harp 2013).

Finally, social movements mistrust journalists because they do not grant them visibility in the news. The media agenda is hardly oriented towards activists' issues, which are frequently ignored and silenced. When such issues do make the news, the "protest paradigm" tends to dominate media coverage. Thus, journalists present protests as a spectacle, highlighting sensational information related to violence, visible drama, and deviant or uncharacteristic behavior (Lee 2014). In doing so, the media promote the de-legitimization and even the demonization of activists and social change (McLeod 2007).

As a result of this suspicion, social movements move between attack and abstention. The first strategy involves actively criticizing the role played by the media, denouncing bias and negativity in the news. This strategy leads to a relationship built on confrontation and conflict, within which activists view journalists as enemies. Abstention involves adopting the opposite position. Faced with limited awareness of social change within the mainstream media, social movements refuse to engage with journalists. In so doing, they give up on influencing the media and use their own channels and supporters to disseminate their demands among a broader audience. The result is a lack of contact and engagement between journalists and activists.

The third "A" is alternative. It comes about with the emergence of radical or alternative media that occupy alternative spaces for communication (Downing 2001). Their news agendas are nourished by the issues and complaints of social movements, offering stories that are generally excluded or silenced. They express the interests of those who are dominated, who question the structures of capitalist exploitation and who offer alternative formulas for development and possibilities for living that are suppressed by the discourse and agenda of the mainstream media (Fuchs 2010). The Internet has given way to the revitalization of alternative media, paralleling the rise of political activism (Fernández-Planells et al. 2014). In this vein, the self-made media created during protests to disseminate demands coexist with alternative and independent media that sympathize with causes of protesters while remaining separate from them. Among those, the monthly publication madrid 15M (http://madrid15m.org), created by citizen assemblies associated with the 15M Movement, stands out. It is followed in importance, in Spain, by Diagonal (http://diagonalperiodico.net), Directa (https://directa.cat), and La Marea (http://www.lamarea.com).

Recent studies make clear that despite their vitality and growth, these two forms of alternative media are insufficient for giving social visibility to political activism and that the mainstream media continue to play a key role in creating such visibility (Micó and Casero-Ripollés 2014). In the case of 15M, the activists themselves have made this point clear (Casero-Ripollés 2015).

Finally, the fourth strategy is adaptation, in which activists adhere to the mediatization of politics (Mazzoleni 2008). This idea stems from the fact that the mainstream media, especially television, occupy a key position in the field of political communication, granting politics visibility and social representation. They are citizens' primary sources of information (Strömbäck 2008), which makes them one of contemporary society's central institutions for political socialization. Additionally, the mainstream media use their own language, narratives, and rules, based on features such as sensationalization, simplification, and personalization. Similar to activists, social actors and politicians adapt to the media's requirements.

7.3 Method

The methodology is based on a qualitative approach that uses semi-structured interviews. This method allows us to understand the how and the why of the communication strategies that online political activists use in relating to the mainstream media. The full sample includes 20 activists and 7 journalists. The selection criteria for the first group included three characteristics: (a) subjects with connections to online political activism (b) who were directly responsible for conceiving of or managing protest communications and (c) who were directly associated with the 15M Movement, which developed in Spain in 2011. The interviewees had participated in 15M in several Spanish cities, including Madrid, Barcelona, Seville, Valencia, and Castellón. Journalists were selected between who had covered the 15M. Face-to-face interviews were conducted over two periods: (1) between July and October 2011 and (2) between November 2014 and April 2015. The interviews lasted between 60 and 90 mins and were manually coded, without the use of computer software.

7.4 Results

7.4.1 Towards Change in the Framework of Relationships

The 15M Movement represents a change in online political activists' understanding and perceptions of the mainstream media. According to the interviews, the activists stopped regarding journalists as "enemies" and began considering them "sources to be influenced" (Activist 1). This metamorphosis stems from being convinced that the media are the most effective tool to (a) amplify the visibility of activism, (b) increase the social dissemination of the activists' demands, and (c) grant them political legitimacy.

This change in mindset has led to a significant debate within 15M. The activists interviewed are conscious that such relationships involve risks, but they also believe that the mainstream media are presently one of the primary spaces for political socialization of citizens. This perspective has generated tension within this new social movement. Debates have formed around two positions. On one side are those who argue that journalists cannot be trusted due to suspicions between the two groups. On the other are those who defend the need to interact with journalists due to their important position in the fields of political socialization and communication. Activists are faced with a complex dilemma: to adapt to the logic of the media and accept the rules of mediatization or to squander the potential amplifier that is the mainstream media and their potential to disseminate activists' demands among broad swaths of citizens. What is to be done?

The response of 15M online activists to this question was to establish a new framework for relating to the media, seeking to influence them but without following their criteria. Doing so involved the launch of a new relationship strategy: the overturn of mediatization, which involves breaking from the principles of news management and not adopting the rules of the mediatization of politics without rejecting a relationship with the media.

7.4.2   The Three Dimensions of the Overturn of Mediatization

The overturn of mediatization has three dimensions. The first is depersonalization, which involves two tactics: on the one hand, the absence of stable spokespeople responsible for interacting with the media and, on the other hand, to avoid personalization, the absence of externally identifiable movement leaders. Accordingly, 15M's online political activists eliminated spokespeople and media leaders.

This elimination resulted in confusion in the media, whose expectations were not met. Journalists found themselves without valid interlocutors, making it more difficult for them to access information. Additionally, activists failed to issue unified or centralized messages. Indeed, their messages had a high degree of disparity, and no unifying version existed. This elevated discursive and thematic diversity made news production enormously complicated. The interviews illustrate the great confusion and bewilderment that the strategies of online political activists generated in journalists:

> It was disconcerting. You would get the phone number of a supposed spokesperson, would speak with him, and two days later, he would not take your calls. (Journalist 4).

> It was disorienting for us to speak with five different sources and hear five different explanations. (Journalist 1).

The factor that prevented journalists from abandoning the topic and excluding it from the news agenda was the huge importance of information about the protests and the high degree of sensationalism involved in the mobilization, which involved a street occupation by thousands of people in several Spanish cities. As political activists attest in the interviews, the tent cities were "a media product of the first order" (Activist 4) that perfectly fit with the media's standards of newsworthiness. "We had what they wanted," asserted one of 15M's communications leaders during our interview (Activist 4).

The second dimension of the overturn of mediatization practiced by 15M was the rejection of formal mechanisms for managing reporter relations. Online political activists did not use the usual conventional formulas to interact with and contact the media. They organized very few press conferences, maintained reduced personal contact with journalists, and offered almost no informational materials such as press releases, photographs, or videos. This strategy did not mean that they did not value communications or the mainstream media. Neither did it mean that they were not organized with regard to communication. Indeed, they created communication committees composed of people with journalism and communications degrees, launched their own alternative media, created content for dissemination on social media, and organized digital campaigns to generate trending topics.

Formal mechanisms for interacting with journalists were rejected because online political activists were convinced that the models of traditional media relations did not work and were not appropriate for achieving their goals. This led those in charge of 15M's communications to the conviction that it was necessary to establish a new framework for relationships. To that end, it was essential to employ the third component of the overturn of mediatization.

The third component consisted of relocation, searching for a change in the relationships and exchanges between online political activists and journalists. Activists sought to move journalists onto social media. The digital environment is the natural habitat for new social movements that are nourished by a "cultural logic of networking" (Juris 2008) and that are based on the principles of connective action (Bennett and Segerberg 2012). By removing journalists from their cultural logic and routines and drawing them into their own territory, online political activists sought to disorient journalists. They were confident that by doing so, they would gain the advantage in their relationships with journalists. Additionally, they considered that their greater technological competence in the use of digital platforms also gave them an advantage. One interviewee expressed this idea clearly:

> Although journalists can enter social networks and use them well, they rarely reach the level of sophistication and have cyber domains because we're very connected with hacker culture. (Activist 5).

Placing relationships with journalists in the social media space was a key element that granted online political activists "comfort" and "ease" and made the overturn of mediatization possible. Web 2.0 permitted members of 15M to use strategies to attract media coverage to their demands. In particular, this was achieved by using social media campaigns to draw attention, working to generate trending topics (TT) about the protests and their agendas and demands. By these means, activists sought to put in motion the reversed agenda-setting, mediated by the Internet (Sung-Tae and Young-Hwan 2007). This idea can be observed clearly in the in-depth interviews:

> The idea was to get strong TT, which would oblige the media to look at what we were talking about—what the TT were, so that we could get our content into the media. Because we had come to understand that they would not speak about us. (Activist 2).

Together with these advantages, shifting from interactions with journalists to social media platforms also involved risks for the management of online political activists' communications. The porousness of social networks made it difficult to control the messages of activists during the 15M Movement, and the great number of profiles, blogs, and websites launched by the movement on digital platforms led to dispersion (Micó and Casero-Ripollés 2014). This gave rise to public contradictions and allowed undesirable information to reach journalists. Online political activists used the metaphor of Gruyere cheese to explain this problem:

For example, any controversy or controversial thing said in a list is forwarded to various groups. Can you be sure of knowing who is there? So, it is like Gruyere cheese, as you are fully exposed by the wide circulation of your message. It is easy to give information to a group that gives it to another group so that eventually it gets to the press. (Activist 1).

7.5 Conclusion

Using social media, online political activists are introducing numerous innovations into the management of political communication. The primary innovation is the overturn of mediatization. This strategy consists of breaking with news management principles and refusing to adopt the rules of the mediatization of politics, all without abandoning media relationships. The 15M Movement in Spain represents a highly significant example of this trend.

Using this communications strategy allows new digital social movements to gain and generate influence. First, this strategy involves having a greater capacity to interact with journalists and obtain more favorable media coverage, which allows activists to position their issues and frames in the media agenda. In this manner, they can gain influence. Additionally, the high degree of innovation introduced by this activist communications model can quickly influence other, more institutional actors in political communication, such as political parties and governments. In this manner, "organizational hybridization" may advance (Chadwick 2007), especially because it is likely that conventional political organizations will add new logics, rooted in the field of social movements and social media, to their news management systems. This dynamic can create greater influence for activism in the rest of the field. Understanding and mastering these new, activism-based news management strategies will become key to innovations in political communication, broadly defined.

Acknowledgments This chapter is part of the I+D CSO2014-52283-C2-1-P project, funded by the Spanish Government's Ministry of Economics and Competitiveness (Ministerio de Economía y Competitividad—MINECO). It is also part of a research project titled "Social media and political activism on the Internet: towards redefining the links between communication and democracy in the digital era," a beneficiary of the First BBVA Foundation Call for Proposals for Researchers, Innovators, and Cultural Creators, granted in 2014. The BBVA Foundation takes no responsibility for the opinions, remarks, or content included in this chapter or for the findings derived from the study, which are the complete and total responsibility of the author.

References

Bennett, L. W., & Segerberg, A. (2012). The logic of connective action: Digital media and the personalization of contentious politics. *Information, Communication & Society, 15*(5), 739–768.
Cammaerts, B. (2012). Protest logics and the mediation opportunity structure. *European Journal of Communication, 27*(2), 117–134.

Casero-Ripollés, A. (2015). Estrategias y prácticas comunicativas del activismo político en las redes sociales en España [Communication strategies and practices of political activism on social media in Spain]. *Historia y Comunicación Social, 20*(2), 245–260.

Chadwick, A. (2007). Digital network repertoires and organizational hybridity. *Political Communication, 24*(3), 283–301.

Downing, J. (2001). *Radical media*. Thousand Oaks, CA: Sage.

Feenstra, R., & Casero-Ripollés, A. (2014). Democracy in the digital communication environment: A typology proposal of political monitoring processes. *International Journal of Communication, 8*, 2448–2468.

Fernandez-Planells, A., Figueras-Maz, M., & Freixa, C. (2014). Communication among young people in the #spanishrevolution: Uses of online-offline tools to obtain information about the #acampadabcn. *New Media & Society, 16*(8), 1287–1308.

Fuchs, C. (2010). Alternative media as critical media. *European Journal of Social Theory, 13*(2), 173–192.

Fuchs, C. (2014). *Social media: A critical introduction*. London: Sage.

Gitlin, T. (1980). *The whole world is watching: Mass media in the making and unmaking of the new left*. Berkeley, CA: University of California Press.

Harlow, S., & Harp, D. (2013). Alternative media in a digital era: Comparing news and information use among activists in the United States and Latin America. *Communication and Society, 26*(4), 25–51.

Juris, J. S. (2008). *Networking futures: The movements against corporate globalization*. Durham, NC: Duke University Press.

Lee, F. L. (2014). Triggering the protest paradigm: Examining factors affecting news coverage of protests. *International Journal of Communication, 8*, 2318–2339.

Mazzoleni, G. (2008). Mediatization of politics. In W. Donsbach (Ed.), *The international encyclopedia of communication*. Malden, MA: Blackwell.

McLeod, D. M. (2007). News coverage and social protest: How the media's protect paradigm exacerbates social conflict. *Journal of Disputes Resolution, 1*, 185–194.

Micó, J. L., & Casero-Ripollés, A. (2014). Political activism online: Organization and media relations in the case of 15M in Spain. *Information, Communication & Society, 17*(7), 858–871.

Rucht, D. (2004). The quadruple "A": Media strategies of protest movements since the 1960s. In W. Van de Donk, B. Loader, P. Nixon, & D. Rucht (Eds.), *Cyberprotest: New media, citizens, and social movements* (pp. 29–54). London: Routledge.

Strömbäck, J. (2008). Four phases of mediatization: An analysis of the mediatization of politics. *The International Journal of Press/Politics, 13*(3), 228–246.

Sung-Tae, K., & Young-Hwan, L. (2007). New functions of Internet mediated agenda-setting: Agenda-rippling and reversed agenda-setting. *Korea Journalism Review, 1*(2), 3–29.

Chapter 8
Jihad Online: How Do Terrorists Use the Internet?

Raphael Cohen-Almagor

Abstract Terrorism is designed to attract attention to the terrorist's cause and to spread fear and anxiety among wide circles of the targeted population. This paper provides information about the ways terrorists are using the Internet. The threat of terrorism is real and significant. As the Internet became a major arena for modern terrorists, we need to understand how modern terrorism operates and devise appropriate methods to forestall such activities.

Keywords Al-Qaeda · Terror · ISIS · Jihad · Encryption · Social responsibility

8.1 Introduction

The aim of this paper is to update and supplement my 2012 paper (Cohen-Almagor 2012) about the ways radical, terrorist Islamists exploit the Internet. Modern terrorism is diffused into cells in different parts of the world. The Internet plays a crucial part in maintaining connections between those cells. Indeed the Internet is a multipurpose tool and weapon. It brings together like-minded people and creates a forum for them to discuss and exchange ideas. The Internet has been exploited by terrorists to deliver instructions and plans, to prepare dangerous operations and to launch violent attacks on designated targets.

8.2 How Do Terrorists Use the Internet?

Information Terrorist organizations share knowledge globally via the Internet. Information on sensitive targets and potential state weaknesses can be easily attained via the communication systems. Google Earth covers major parts of the

R. Cohen-Almagor (✉)
University of Hull, Hull, UK
e-mail: R.Cohen-Almagor@hull.ac.uk

© Springer International Publishing Switzerland 2017
F.C. Freire et al. (eds.), *Media and Metamedia Management*,
Advances in Intelligent Systems and Computing 503,
DOI 10.1007/978-3-319-46068-0_8

globe. Hamas was said to use Google Earth to plan its operations (Tamimi 2007). Al-Qaeda maintains a database that contains information about potential American targets (Thornton 2010; Coolsaet 2011; Neumann 2012). Public transport routes and timetables, maps of building sites, their opening times and their layout are readily available. The Internet can also be used to disclose code names and radio frequencies used by security services.

Propaganda and Indoctrination The Al-Qaeda and ISIS networks have been successful in their use of audio-visual propaganda, producing pre-recorded video-tapes, computer games and music in order to spread radical ideology and to reach sympathizers across the globe (Gruen 2006; UNODC 2012). ISIS strictly monitors the Internet and media in places under its control to ensure its dominance. An internal US State Department assessment from 2015 held that the Islamic State's violent narrative—promulgated through thousands of messages each day—has been winning the social media propaganda warfare (Mazzetti and Gordon 2015).

On jihadi websites, the attacks on the World Trade Center in New York were displayed as part of an assault on the U.S. economy. There are video clips on YouTube that encourage the launch of attacks and suicide bombing. Some include English translation (Thomas 2003). Other images show how Muslims are being victimized by American, British, Israeli and Western soldiers while yet others show images of jihadi bravery, how jihadists kill and maim their enemies.

Already in 1996, Babar Ahmad—a twenty-two-year-old computer whiz and mechanical engineering student at Imperial College in London—launched a website called Azzam.com, in honor of Abdallah Azzam, the founder of Maktab al-Khidamat and the mentor to Osama bin Laden and Ayman al-Zawahiri. The site was dedicated to promoting Islamist fighters in Bosnia, Chechnya, and Afghanistan. It quickly became a prominent and influential platform for Islamist militants because it posted firsthand news reports from amateur correspondents around the world. International news organizations, including the BBC, often cited dispatches from Azzam.com and its sister websites when reporting on events in Chechnya and Afghanistan. According to terrorism expert Evan Kohlmann, this was the very first al-Qaeda website. Even in its nascency, it was professional. While it was not technically sophisticated, still it was professional looking, attracting the attention of jihadists (Forest 2004a; Schmidt 2004).

The Internet is used to target young people. Terrorist organizations use cartoons, popular music videos and computer games to get their attention (UNODC 2012). In 2003, Hezbollah began online promotion of a computer game simulating terrorist attacks on Israeli targets. The computer game, called *Special Force*, was developed by the Hezbollah Central Internet Bureau, and its development took two years. The game, played in Arabic, English, French and Farsi, was based on actual Hezbollah battles with Israeli forces. It placed the player in different Hezbollah operations against Israelis and players could gained points by killing Israeli leaders (Harnden 2004). The game ended with an exhibit of Hezbollah "martyrs"—fighters killed by Israel. The message to users was: "Fight, resist and destroy your enemy in the game of force and victory" (WorldNetDaily 2003). Mahmoud Rayya, a member of

Hezbollah, noted in an interview for the *Daily Star* that the decision to produce the game was made by leaders of Hezbollah, and that "In a way, *Special Force* offers a mental and personal training for those who play it, allowing them to feel that they are in the shoes of the resistance fighters" (*Ibid.*).

Today the prospect of establishing a Caliphate is appealing to thousands of jihadists. The slick *Inspire* magazine, published online in English, has become popular in jihadi circles (Lemieux et al. 2014). In May 2015, ISIS released the ninth issue of *Dabiq*, its online English- and multi-language magazine.[1] The magazine is opened with the following statement:

> As the crusaders continue to reveal their intense hatred and animosity towards Islam through their relentless bombing and drone campaigns on the Islamic State, a new breed of crusader continues shedding light on the extent of their hatred towards the religion of truth. This breed of crusader aims to do nothing more than to anger the Muslims by mocking and ridiculing the best of creation, the Prophet Muhammad Ibn 'Abdillāh (sallallāhu 'alayhi wa sallam), under the pretext of defending the idol of "freedom of speech".[2]

Thus, it is the duty of each and every good Muslim to support the cause of Allah and punish those who dare to insult the Prophet. *Dabiq* calls upon Muslims to carry their duty and perform the virtues of jihad.

Networking Jihadi websites enable like-minded people to engage in a power struggle against the western enemy (Atwan 2006; AIVD 2012). Jihadi forums provide them with friends and support. In the forums they share their fantasies and aspirations with their online friends. The networking and exchange provide them with a sense of belonging to a greater community with the common cause and benchmark of Islam. The forums prove the existence of the ummah, the Muslim nation (Sageman 2008). Among the popular jihadi forums were/are: al-Qimmah, Atahadi, al-Jihad al-Alami, al-Fajr, al-Furqan, al-Hanein, Al-Luyuth al-Islamiyyah, al-Maark, al-Malahem, al-Medad, al-Shamukh, at-Tahaddi, as-Ansar, Hanein, and The Mujahideen Electronic Network. Some of them are comprised of tens of thousands of people. www.shawati.com had more than 31,000 registered members. www.kuwaitchat.net had more than 11,000 members (ICT 2012; Qin et al. 2008).

A 2015 George Washington University report documents 71 people in the United States charged with crimes related to the Islamic State since March 2014: 40 % were converts to Islam, defying any ethnic profile. They were young, with an average age of 26; overwhelmingly American citizens or legal residents. 14 % were women. Nearly all of them had spent hours on the Internet boasting their feelings about the Islamic State and engaging with English speakers from other countries. Nearly all of them were arrested after their online posts drew the attention of the F. B.I. (Shane et al. 2015). Consider, for instance, Ali Amin, a 17-year-old Virginia resident who was gradually drawn into ISIS virtual world, associated with other

---

[1] http://jihadology.net/category/dabiq-magazine/.

[2] http://jihadology.net/category/dabiq-magazine/.

supporters and subsequently provided logistical support to ISIS, instructing them how to transfer funds secretly and drove an ISIS recruit to the airport (Posner 2015).

Psychological warfare The Internet is a major tool for terrorists to bypass mainstream media sources when attempting to use psychological warfare. Al-Qaeda and ISIS regularly publish videos that are designed to evoke fear. Al-Qaeda's media department as-Sahab ("the Clouds")[3] produced Osama bin-Laden audio and video tapes (Stratfor 2006), aiding Al-Qaeda in its international propaganda campaign. Another jihadi media organization, Al-Fajr Media Center, turned insurgency into a courageous journey, mayhem and violence into inspiring music videos.[4] There are thousands of jihadi-terrorist publications and videos that show gore and violence, hostage taking, suicide bombings, bomb explosions, operation tactics and religious radical oratory inciting bloody jihad (Salem et al. 2008; Lappin 2011; Lemieux et al. 2014).[5]

In May 2004, the shocking video of al-Zarqawi decapitating Nicholas Berg was downloaded millions of times on the Internet and stimulated copycat beheadings by other groups (Nye 2005). At some point, leaders of al-Qaeda felt that the beheadings videos had exhausted their impact and stopped producing them. Part of the terrorist tactics is to surprise by changing the mode of operations. Much like changes in fashion, terrorists decide on the most appropriate method for a given time and once they feel that the method is exhausted, they will opt for another fashionable tactics. Thus, at one point they may opt for hijacking or bombing airplanes. Then they may resort to suicide bombings, to hostage taking, to planting bombs in strategic places, to launching attacks on civilians, to rocket fire, or to cyberterror. They may mix their methods, and at the same time may highlight their "signature" method to serve as a tactical contagion point for their followers. In 2014, ISIS has turned to make itself radically distinct from other jihadi organizations by their shocking and revolting group beheadings.

The Palestinian issue is a global concern, facilitated by the Internet far more than it does by conventional media. Jihad Online was a high profile, Arabic language, Palestinian jihad-oriented site, synthesized Islamic Jihad and pro-Palestinian ideologies. It posted photos of shahid suicide murderers. The site was created in 2002 and was registered in Beirut (Bunt 2003). Laskar Jihad, based in Jakarta, was allegedly connected with al-Qaeda, and proactively applied the Internet as a means of disseminating its ideology in Indonesian and English. The site was regularly

---

[3]http://www.globaljihad.net/; http://www.liveleak.com/view?i=dea_1175200207.

[4]http://www.liveleak.com/view?i=2a9_1207598912; http://www.liveleak.com/view?i=775_12075 99624.

[5]West Point (2008) published an unsigned document dated on December 10, 2008. The document is labeled as top secret, and is basically an organizational chart of the media committee of an unidentified jihadi organization in Iraq. This media committee contains media section, filming and documenting section, preaching section and the section of techniques and internet. Each section is divided to division and subdivisions with details that show the duties of each individual within the hierarchy.

updated with news, articles and photographs. The site contained details of the organization's paramilitary activities in the Moluccan Islands against Christian population (Bunt 2003).

Socialization and motivation Those who become violent jihadis can be divided into four main groups: Truth seekers, identity seekers, revenge seekers, and thrill seekers. Those jihadists are emotionally-driven by images from Iraq, Afghanistan and Palestine, where Muslims suffer humiliation inflicted on them by western powers, and by images of terror attacks carried out in Europe and in the United States (Rosen 2006; Kern 2015).

Security expert Thomas Hegghammer distinguished between the jihadi "mother sites" and the jihadi Internet message boards. The "mother sites" are run by people who get their material directly from the ideologues or operatives. On the jihadi Internet message boards one could find the political and religious discussions among the sympathizers and potential recruits. Among the most important message boards for al-Qaeda sympathizers were al Qal'ah (the Fortress), al Sahat (the Fields), and al Islah (Reform). Those message boards provided links to the "information hubs," where new radical-Islamists texts, declarations, and recordings were posted. Those hubs existed in the "communities" sections of Yahoo!, Lycos, and other popular Internet gateways (Wright 2004).

Fund raising Funding is essential for terrorist operations. Terrorist organizations raise funds via the Internet by making email appeals or through their websites; by selling goods through their websites; through associated side businesses; through fraud, gambling, or online brokering (Gruen 2004; Guiora 2011); and through online organizations that resemble humanitarian charity groups. ISIS revenues come primarily from oil, from major donors, from their extensive criminal enterprises, and from small donations (Levitt 2014).

Hizb ut-Tahir has established websites from Europe to Africa that ask supporters to assist the cause of jihad. Like those sponsored by fighters in Chechnya, the websites displayed numbered bank accounts to which supporters could contribute. Another example of terrorist fund-raising on the Internet comes from Lashkar e-Tayba (LeT), one of the most violent terrorist groups in Kashmir, serving as the terrorist wing of the Markaz Dawa-Wal-Irshad, an Islamic fundamentalist organization of the Wahhabi sect in Pakistan[6] (Cline 2008; Subhani 1996). It is a well-organized and well-funded terrorist group that has trained thousands of mujahedeen, sending them to Afghanistan, Chechnya, Kashmir, Bosnia, Kosovo, Iraq, and the Philippines (Weimann 2006a: 137).

---

[6]Wahhabism is an Islamic movement dating back to the 18th Century and named after its founder, Muhammad ibn Abd al-Wahhab. It is against innovation; true Muslims must adhere solely and strictly to the original beliefs set forth by Muhammad. Wahhabism is anti-secularism and anti-modernism. Its goal, according to Bin Laden's interpretation, is to achieve Muslim world domination. Wahhabism is the dominant Islamic tradition on the Arabian peninsula. *GlobalSecurity.org*, http://www.globalsecurity.org/military/world/gulf/wahhabi.htm.

Spreading tactics—Instructions and online manuals In 2004, al-Qaeda issued a chilling manual directed at new volunteers who were "below the radar" of counter-terrorist authorities and who could not undergo formal training in terrorist techniques. The manual encouraged the use of weapons of mass destruction (Burke 2004). *Al-Battar* ('the sharp-edged sword') was an online al-Qaeda journal that appeared regularly since 2003. It served as a virtual training camp, providing readers with instruction in weapons handling, explosives, kidnapping, poisoning, cell organization, guerrilla warfare, secure communications, physical fitness, the making a suicide-bomb belt (even down to the correct thickness of the cloth), how to plant land mines, how to detonate a bomb remotely with a mobile phone, topography, orientation (including crossing a desert at night), map reading and survival skills (Atwan 2006; Kennedy 2006; Spencer 2004). The courses were accompanied by statements and speeches of al-Qaeda leaders. Would be jihadis were urged to follow the virtual courses at home or in groups, practice the instructions, obtain firearms and maintain a high level of fitness in preparation for taking steps to join the mujahedin.

Some of the issues focused on single, important issues. For instance, the September 2004 issue of *Al-Battar*[7] focused on how to properly conduct kidnapping operations. It provided detailed instructions as to requirements for conducting kidnapping, stages of public kidnapping, how to deal with hostages, and what security measures should the kidnappers take (Al-Muqrin 2004). Al-Qaeda Targeting Guidance specifies types of targets inside cities, economic targets, the purpose of human targets, the advantages and disadvantages of operation against cities (Intelcenter 2004).

Mustafa Setmariam Nasar, known as Abu Mus'ab al-Suri, wrote *The Global Islamic Resistance Call*. This sort of strategist/ideological 1600-page magnum opus, was written over a two-year research. This meticulous study draws on the experience of prior conflicts to explain how global terror network should organize, finance, recruit, train and operate. *The Call* gives practical instructions for mass murder and urges wannabe jihadists to act independently. It is aimed to transform al-Qaeda from a terrorist network into a truly global movement of individuals organized in small cells ("detachments"), thus making the movement impermeable to U.S. and allied counterterrorism efforts. According to Nasar, the key to jihadi success is to make the global terror network looser, meaner and more resilient. In order to defeat the United States, jihad should spread to all parts of the world and recruit far more warriors. The fighting should be conducted in three stages, each stage with the appropriate weapons and guerrilla warfare methods. The first stage is of exhaustion which includes assassinations, raiding, ambushes and explosion operations. This is followed by the balancing stage, where the jihadis move to great strategic attacks with great battles. Finally, the stage of decisiveness and liberation

---

[7]Al Battar is an alias of Sheikh Yousef Al-Ayyiri, an al-Qaeda leader in Saudi Arabia and bin Laden's personal bodyguard who was killed in a clash with Saudi security forces.

after some sectors of the army join the guerrilla warriors and bring Muslim victory (Al-Suri 2007).

The manuals were put into use. Mohammad Sidique Khan, Shehzad Tanweer, Hasib Hussain and Jermaine Lindsay who were responsible for the death of 56 people and the injury of some 700 others on July 7, 2005 downloaded the instructions on how to build their bombs from the Internet (Hewitt 2008; Cohen-Almagor 2012).

Many password-protected forums refer to extensive literature on explosives. Tutorials provide insights on viruses, hacking stratagems, encryption methods, anonymity guidelines, and use of secret codes. Bomb-making knowledge is available on jihadi websites in the form of detailed, step-by-step video instructions showing how to build improvised explosive devices. Online instructions played a critical role in the March 2004 Madrid bombings, the April 2005 Khan al-Khalili bombings in Cairo, the July 2006 failed attempt to bomb trains in Germany, the June 2007 plot to bomb London's West End and Glasgow, and the April 2013 Boston Marathon bombing, in which Dzhokhar and Tamerlan Tsarnaev allegedly placed pressure cookers close to the finish line (Sageman 2008; Dastagir 2013; Weimann 2014a, b).

In addition to manuals and diagrams, training videos are common among terrorist websites. For example, in early June 2005, a contributor to the militant Arabic language web forum *Tajdid Al-Islami* posted a series of training videos for beginner mujahideen that included discussions on basic fitness, ninja arts, proper uniform, and communication techniques. As a follow-up, in August 2005, the contributor posted a seven-part lesson on how to use a handheld, portable global positioning system (GPS) receiver (Forest 2004b).

Arms Trade Terrorist organizations are using social networking sites to buy weapons. Facebook has been hosting online arms bazaars, offering weapons ranging from handguns and grenades to heavy machine guns and guided missiles. The weapons include many distributed by the United States to security forces and their proxies in the Middle East. These online bazaars have been appearing in regions where the Islamic State has its strongest presence (Chivers 2016). A recent report documented 97 attempts at unregulated transfers of missiles, heavy machine guns, grenade launchers, rockets and rifles, used to disable military equipment, through several Libyan Facebook groups since September 2014 (Jenzen and Rice 2016).

Recruitment Online recruiting has increased exponentially due to the increased popularity of social networking sites. They attract interested jihadists who play a critical role in identifying potential radicals and alert about suspected others (Kunt 2006; Witte et al. 2009). The Islamic State has published videos to help in its recruitment of foreign fighters, some of these videos in the form of hyped music clips: come to ISIS-controlled territory; fight for the cause; kill the infidels and feel great about death. Allah promises virgins and violence (The Clarion Project 2015).

Ziyad Khalil's path to terror is illustrative. Khalil enrolled in a computer science course in Columbia College in 1995. He became a Muslim activist in college and created a number of radical websites. At some point, his activities were noticed by al-Qaeda. Later Khalil became the organization's procurement officer in the USA (Verton 2003). He bought a satellite phone used by Osama bin Laden to plot the 1998 bombings of U.S. embassies in Tanzania and Kenya.[8]

In recent years we witness thousands of jihadists who volunteer to fight for ISIS. Information gathered about those people show that there is no one unifying explanation for their endeavour. Many are genuinely committed to the Caliphate project. Some of them are pious while others are not. Many have troubled histories while others would have had good prospects had they stayed in their countries. Some were driven by the humanitarian suffering of their brethren (say the Syrian people who rebel against the Assad dictatorship) while others were seeking thrill and adventure (Neumann 2015).

Planning of activities and coordination Al-Qaeda and ISIS have been using the Internet in planning and coordinating terror attacks. The prime example is the September 11, 2001 attack that prompted the "war on terror". Thousands of code-word and encrypted messages that had been posted in a password-protected area of a website were found by federal officials on the computer of arrested al-Qaeda terrorist Abu Zubaydah (Weimann 2006b; Cohen-Almagor 2010). More recently, the *NY Post* (2013) revealed that American intelligence intercepted discussion on a jihadi website about launching attacks against American targets. Consequently, the USA closed 19 diplomatic posts across Africa and the Middle East for more than a week.

8.3   Conclusions

Diffusion in terrorist locations is made possible by Internet communications. Jihadi texts and videos are available for people who seek such guidance. Extreme religious ideologies are spread through websites and videotapes accessible throughout the world (Bobbitt 2008; Glasser and Coll 2005). Great reverence is paid to the views of the militant leadership (Windrem 2015). The digital legacy of the jihadist ideologue cleric Anwar al-Awlaki continues to inspire countless plots and attacks (Shane 2015; Goodwin 2015). Fatwas of religious sages legitimize and endorse violence. Already in 2002 Middle East expert Paul Eedle had warned:

> The Website is central to al-Qaeda's strategy to ensure that its war with the U.S. will continue even if many of its cells across the world are broken up and its current leaders are killed or captured. The site's function is to deepen and broaden worldwide Muslim support, allowing al-Qaeda or successor organizations to fish for recruits, money and political backing. The whole thrust of the site, from videos glorifying September 11 to Islamic legal arguments justifying the killing of civilians, and even poetry, is to convince radical Muslims that, for decades, the U.S. has been waging a war to destroy Islam, and that they must fight back (Edle 2002).

---

[8]https://cryptome.org/usa-v-ubl-02.htm.

Some Western Internet Service Providers and web-hosting companies have been hosting terrorist sites and helped the cause of jihad. Some do it knowingly while others did it inadvertently. A British company, ServerSpace Ltd and American company Vault Network Ltd provide Internet services to Hezbollah's www.almanar. com.lb. The French Desibox Sas provides Internet access to Hezbollah's www. jihadbinaa.org. The French Online SAS and the American Servepath provide Internet services to Hezbollah's www.almahdiscouts.net. The American HostGator LLC provides access to www.inbaa.com which is affiliated to the Hezbollah. American XLHost.com hosts Hezbollah's www.alnour.com.lb. American Global Net Access Llc provides Internet services to Hezbollah's www.alahednews.com.lb. American eNom Inc services Hezbollah's www.alshahid.org. American ISP Gogrid Llc services Hezbollah's www.aljarha.net (The Meir Amit Intelligence and Terrorism Information Center 2013). InfoCom Corporation in Texas hosted terrorist organizations and also served to launder money. Large amounts of money came from Saudi Arabia and the Gulf states to sponsor Hamas activities (The Investigative Project on Terrorism, no date; Katz 2003; Whitetaker 2001).

According to the security senior officials and experts, blocking and shutting websites is not a viable option. What is required is more responsibility, more monitoring and more structure.[9] YouTube, Facebook, Twitter and others have the technological abilities to scrutinize their servers to ensure that content that violate their terms of services (ToS) will not be posted. Terrorist content surely violates their ToS thus it should not be allowed publicity. What needed is awareness and willingness to put the capabilities into effect (Cohen-Almagor 2015). Social responsibility should influence ISPs and web-hosting companies to verify that they do not become hubs for terror. Socially responsible Internet intermediaries are opposed to terrorism. They understand that associating themselves with terror is bad for business. Unfortunately others either support terror or do not seem to care. The struggle is hard and long. The end of the battle of ideas between the ideology of destruction and the liberal ideologies of live-and-let-live is not yet in sight.

Acknowledgments I thank Sabela Direito Rebollal, Diana Lago Vázquez and José Rúas Araújo for their kind assistance in bringing this paper to print.

References

AIVD. (2012). *Jihadism on the Web. A breeding ground for Jihad in the modern age.* Retrieved from https://english.aivd.nl/publications/publications/2012/02/14/jihadism-on-the-web
Al-Muqrin, A. (2004). Al-Battar, the al-Qaeda Manual on kidnapping. *Netwar04.org.* Retrieved from http://netwar04.blogspot.com/2004/09/al-battar-al-qaeda-manual-on.html
Al-Suri, A. M. (2007). The global Islamic resistance call. In B. Lia (Ed.), *Architect of global Jihad: The life of Al-Qaida strategist Abu Mus'ab al-Suri.* London: Hurst.
Atwan, A. B. (2006). *The secret history of al-Qaeda.* Berkeley: University of California Press.

---

[9]Interviews with security senior officials and experts in the USA, Israel and the United Kingdom.

Bobbitt, P. (2008). *Terror and consent*. New York: Knopf.

Bunt, G. R. (2003). *Islam in the digital age*. London: Pluto Press.

Burke, J. (2004). Al-Qaeda launches online terrorist manual. *The Guardian*. Retrieved from http://www.guardian.co.uk/technology/2004/jan/18/alqaida.internationalnews

Chivers, C. J. (2016). Facebook groups act as weapons bazaars for militias. *NY Times*. Retrieved from http://www.nytimes.com/2016/04/07/world/middleeast/facebook-weapons-syria-libya-iraq.html?emc=edit_th_20160407&nl=todaysheadlines&nlid=33802468&_r=1

Cline, A. (2008). Wahhabism and Wahhabi Islam: How Wahhabi Islam differs from Sunni, Shia Islam. *Islamdaily.org*. Retrieved from http://www.islamdaily.org/en/wahabism/6259.wahhabism-and-wahhabi-islam-how-wahhabi-islam-diff.htm

Cohen-Almagor, R. (2010). In internet's way. In M. Fackler & R. S. Fortner (Eds.), *Ethics and evil in the public sphere: Media, universal values and global development* (pp. 93–115). Cresskill, NJ: Hampton Press.

Cohen-Almagor, R. (2012). In internet's way: Radical, terrorist Islamists on the free highway. *International Journal of Cyber Warfare and Terrorism, 2*(3), 39–58.

Cohen-Almagor, R. (2015). *Confronting the internet's dark side: Moral and social responsibility on the free highway*. NY, Washington, DC: Cambridge University Press, Woodrow Wilson Center Press.

Coolsaet, R. (2011). *Jihadi terrorism and the radicalization challenge*. Surrey: Ashgate.

Dastagir, A. E. (2013). Internet has extended battlefield in war on terror. *USA Today*. Retrieved from http://www.usatoday.com/story/news/nation/2013/05/05/boston-bombing-self-radicalization/2137191

Edle, P. (2002). Terrorism.com. *The Guardian*. Retrieved from https://www.theguardian.com/technology/2002/jul/17/alqaida.g

Forest, J. J. F. (2004a). Training camps and other centers of learning. In J. J. F. Forest (Ed.), *Teaching terror: Strategic and tactical learning in the terrorist world*. Lanham, MD: Rowman & Littlefield Publishers.

Forest, J. J. F. (2004b). Introduction. In J. J. F. Forest (Ed.), *Teaching terror: Strategic and tactical learning in the terrorist world*. Lanham, MD: Rowman & Littlefield Publishers.

Glasser, S. B. & Coll, S. (2005). The web as weapon. *The Washington Post*. Retrieved from http://www.washingtonpost.com/wp-dyn/content/article/2005/08/08/AR2005080801018.html

Goodwin, L. (2015). San Bernardino attacks latest example of Anwar al-Awlaki's deadly legacy. *Yahoo* (December 23). Retrieved from https://www.yahoo.com/politics/san-bernardino-attacks-latest-example-1327003987255350.html

Gruen, M. (2004). White ethnonationalist and political Islamist methods of fund-raising and propaganda on the internet. In R. Gunaratna (Ed.), *The changing face of terrorism*. Singapore: Marshall Cavendish.

Gruen, M. (2006). Innovative recruitment and indoctrination tactics by extremists: Video games, hip-hop, and the world wide web. In J. J. F. Forest (Ed.), *The making of a terrorist: Recruitment, training, and root causes*. Westport, CT: Praeger.

Guiora, A. N. (2011). *Homeland Security*. Boca Raton, FL: CRC Press.

Harnden, T. (2004). Video games attract young to Hizbollah. *The Telegraph*. Retrieved from http://www.telegraph.co.uk/news/worldnews/middleeast/lebanon/1455011/Video-games-attract-young-to-Hizbollah.html

Hewitt, S. (2008). *The British war on terror*. London: Continuum.

ICT Monitoring Group. (2012). *In the depths of Jihadist web forums: Understanding a key component of the propaganda of Jihad*. Herzliya: International Institute for Counter Terrorism (ICT). Retrieved from https://www.ict.org.il/Article.aspx?ID=211

Intelcenter. (2004). Al-Qaeda targeting guidance. *Intelcenter*. Retrieved from http://www.intelcenter.com/Qaeda-Targeting-Guidance-v1-0.pdf

Jenzen, N. R. & Rice, G. (2016). The online trade of light weapons in Libya. *Security Assessment in North Africa, 6*. Retrieved from http://www.smallarmssurvey.org/fileadmin/docs/R-SANA/SANA-Dispatch6-Online-trade.pdf

Katz, R. (2003). *Terrorist hunter: The extraordinary story of a woman who went undercover to infiltrate the radical Islamic groups operating in America*. New York: Harper Collins.

Kennedy, M. (2006). How terrorist learn. In J. J. F. Forest (Ed.), *Teaching terror: Strategic and tactical learning in the terrorist world*. Lanham, MD: Rowman & Littlefield Publishers.

Kern, S. (2015). Inside the mind of a jihadist. *Gatestone Institute*. Retrieved from http://www.gatestoneinstitute.org/5587/jihadist-mind

Kunt, K. (2006). Osama bin Laden fan clubs, jihad recruiters build online communities. *USA Today*. Retrieved from http://usatoday30.usatoday.com/tech/news/2006-03-08-Orkut-al-qaeda_x.htm

Lappin, Y. (2011). *Virtual Caliphate*. Washington, DC: Potomac Books.

Lemieux, A. F., Brachman, J. M., Levitt, J., & Wood, J. (2014). Inspire magazine: A critical analysis of its significance and potential impact through the lens of the information, motivation, and behavioral skills model. *Terrorism and Political Violence, 26*, 354–371.

Levitt, M. (2014). *Terrorist financing and the Islamic state*. Retrieved from http://www.washingtoninstitute.org/policy-analysis/view/terrorist-financing-and-the-islamic-state

Mazzetti, M. & Gordon, M.R. (2015). ISIS is winning the social media war, U.S. concludes. *NY Times*. Retrieved from http://www.nytimes.com/2015/06/13/world/middleeast/isis-is-winning-message-war-us-concludes.html?emc=edit_th_20150613&nl=todaysheadlines&nlid=33802468&_r=0

Neumann, P. (2012). *Countering online radicalization in America*. Washington, DC: Bipartisan Policy Center.

Neumann, P. (2015, February 19). *Remarks on terror on the internet*. Washington, DC: The White House Summit to Counter Violent Extremism.

Nye, J. (2005). How to counter terrorism's online generation. *The Financial Times*. Retrieved from http://belfercenter.hks.harvard.edu/publication/1470/how_to_counter_terrorisms_online_generation.html

Posner, E. (2015, December 15). ISIS gives us no choice but to consider limits on speech. *Slate*. Retrieved from http://www.slate.com/articles/news_and_politics/view_from_chicago/2015/12/isis_s_online_radicalization_efforts_present_an_unprecedented_danger.2.html

Post Staff Report. (2013, August 15). Al Qaeda fighters planning and coordinating attacks on secret chat rooms and internet message boards. *NY Post*. Retrieved from http://nypost.com/2013/08/15/al-qaeda-fighters-planning-and-coordinating-attacks-

Qin, J., Zhou, Y., Reid, E., & Chen, H. (2008). Studying global extremist organizations' internet presence using the dark web attribute system. In H. Chen, E. Reid, J. Sinai, A. Silke, & B. Ganor (Eds.), *Terrorism informatics*. New York: Springer.

Rosen, N. (2006, February 19). Iraq's Jordanian Jihadis. *NY Times*. Retrieved from http://www.nytimes.com/2006/02/19/magazine/iraqs-jordanian-jihadis.html?_r=0

Sageman, M. (2008). *Leaderless Jihad: Terror networks in the twenty-first century*. Philadelphia: University of Pennsylvania Press.

Salem, A., Reid, E., & Chen, H. (2008). Content analysis of Jihadi extremist groups' videos. In H. Chen, E. Reid, J. Sinai, A. Silke, & B. Ganor (Eds.), *Terrorism informatics*. New York: Springer.

Schmidt, S. (2004). British citizen indicted on U.S. terrorism charges. *The Washington Post*. Retrieved from http://www.washingtonpost.com/wp-dyn/articles/A13269-2004Oct6.html

Shane, S. (2015, December 18). Internet firms urged to limit work of Anwar al-Awlaki. *NY Times*. Retrieved from http://www.nytimes.com/2015/12/19/us/politics/internet-firms-urged-to-limit-work-of-anwar-al-awlaki.html?emc=edit_th_20151219&nl=todaysheadlines&nlid=33802468&_r=0

Shane, S., Apuzzo, M., & Schmitt, E. (2015). Americans attracted to ISIS find an 'echo chamber' on social media. *NY Times*. Retrieved from http://www.nytimes.com/2015/12/09/us/americans-attracted-to-isis-find-an-echo-chamber-on-social-media.html?emc=edit_th_20151209&nl=todaysheadlines&nlid=33802468&_r=1

Spencer, R. (2004). Al-Qaeda's online training camp. *Jihad Watch*. Retrieved from http://www.jihadwatch.org/2004/01/al-qaedas-online-training-camp.html

Stratfor. (2006). As-Sahab: Al Qaeda's nebulous media branch. *Stratfor Publications*. Retrieved from https://www.stratfor.com/analysis/sahab-al-qaedas-nebulous-media-branch

Subhani, A. J. (1996). Wahhabism. *Naba Organization*. Retrieved from http://www.al-islam.org/wahhabism/

Tamimi, A. (2007). *Hamas: A history*. Oslo: Olive Branch Press.

The Clarion Project. (2015). ISIS launches 'viral' english recruitment pop video. *The Clarion Project*. Retrieved from http://www.clarionproject.org/news/exclusive-isis-launches-viral-english-pop-video

The Investigative Project on Terrorism. (no date). *Mousa Abu Marzook profile*. Retrieved from http://www.investigativeproject.org/profile/106/mousa-abu-marzook

The Meir Amit Intelligence and Terrorism Information Center. (2013). *Terror and the internet*. Retrieved from http://www.terrorism-info.org.il/he/article/20488

Thomas, T. L. (2003). Al Qaeda and the internet: The danger of "cyberplanning". *Parameters*, *33* (1). Retrieved from http://www.iwar.org.uk/cyberterror/resources/cyberplanning/al-qaeda.htm

Thornton, H. L. (2010). *Countering radicalism with a "virtual library of freedom"*. Virginia: The Project on International Peace and Security.

UNODC, UN Office on Drugs and Crime. (2012). *The use of the internet for terrorist purposes*. New York: United Nations.

Verton, D. (2003). *Black ice: The invisible threat of cyber terrorism*. California: McGraw-Hill.

Weimann, G. (2006a). *Terror on the Internet: The new arena, the new challenges*. Washington, DC: Institute of Peace Press.

Weimann, G. (2006b). Virtual training camps: Terrorists' use of the internet. In J. J. F. Forest (Ed.), *Teaching terror: Strategic and tactical learning in the terrorist world*. Lanham, MD: Rowman and Littlefield Publishers.

Weimann, G. (2014a). *New terrorism and new media*. Washington, DC: Woodrow Wilson International Center for Scholars.

Weimann, G. (2014b). Virtual packs of lone wolves. *The Wilson Center*. Retrieved from https://medium.com/its-a-medium-world/virtual-packs-of-lone-wolves-17b12f8c455a

West Point. (2008). Organizational chart of the media committee of an unidentified Jihadi Group in Iraq. Retrieved from https://www.ctc.usma.edu/posts/organizational-chart-of-the-media-committee-of-an-unidentified-jihadi-group-in-iraq-original-language

Whitetaker, B. (2001). Pulled plug on 500 Arab/Muslim websites day before Jetliner attacks. *Rense.com*. Retrieved from http://www.rense.com/general13/jet.htm

Windrem, R. (2015, July 25). Dead Cleric Anwar al-Awlaki still sways terror wannabes. *NBC News*. Retrieved from http://www.nbcnews.com/news/us-news/dead-cleric-anwar-al-awlaki-still-sways-terror-wannabes-n397506

Witte, G., Markon, J., & Hussain, S. (2009). Pakistani authorities hunt for alleged mastermind in plot to send N. Virginia men to Afghanistan to fight U.S. troops. *The Washington Post*. Retrieved from http://www.washingtonpost.com/wp-dyn/content/article/2009/12/12/AR2009121201598.html?sid=ST2009121002234

WorldNetDaily. (2003). Hezbollah's new computer game. *WorldNetDaily*. Retrieved from http://www.worldnetdaily.com/news/article.asp?ARTICLE_ID=31323

Wright, L. (2004). The terror web. *New Yorker*. Retrieved from http://www.newyorker.com/magazine/2004/08/02/the-terror-web

Chapter 9
Spanish General Elections, Microdiscourses Around #20D and Social Mobilisation on Twitter: Reality or Appearance?

Estrella Gualda

Abstract The nature of collective mobilisation has profoundly changed with the rise of Web 2.0 and collaborative platforms such as Twitter, Facebook, YouTube, and Instagram. Specifically, Twitter—a public microblogging service—is clearly helping to inform, discuss, announce, and disseminate many types of political, social, or cultural protest or collective mobilisation. The recent decline of bipartisanship in Spain has captured significant attention through the recent sociopolitical debates and messages circulating on Twitter. This chapter is based on an extraction of tweets before, during and after the recent electoral process in Spain. We mined tweets published from 23 November 2015 to 8 February 2016 (the elections took place on 20 December 2015). For the mining process, we used the string '20D' as a search criterion through which tweets about the electoral process in Spain were identified. The dataset for this chapter consists of a sample of 28,261 tweets containing '20D'. They were extracted with the help of NodeXL. Our objective through this work is to identify the microdiscourses that appear with '20D', analyse the co-hashtags relationships, and attempt to identify patterns about the types of messages that are mainly disseminated through Twitter. We also detected different communities of hashtags and visualised them through Gephi, which helps us understand how political and social discourses are articulated through visual and textual narratives.

Keywords Twitter · Social networks · Spanish general elections · Co-hashtags · Microdiscourses

E. Gualda (✉)
"Estudios Sociales e Intervención Social" Research Group,
University of Huelva (UHU), Huelva, Spain
e-mail: estrella@uhu.es
URL: http://www.eseis.es; http://www.cieo.pt

E. Gualda
Research Centre for Spatial and Organizational Dynamics,
University of Algarve, Faro, Potugal

© Springer International Publishing Switzerland 2017   67
F.C. Freire et al. (eds.), *Media and Metamedia Management*,
Advances in Intelligent Systems and Computing 503,
DOI 10.1007/978-3-319-46068-0_9

9.1 Introduction

The current relevance of social media in the processes of building sociopolitical realities has not gone unnoticed. Twitter, as a microblogging platform, is well known today as an important source of information for journalists, politicians, sociologists, activists, and a wide variety of other social agents, experts and citizens. Part of our reality is constantly being built through the permanent flow of information and conversations that circulate through social media, mass media and other varied informal and formal social spaces and organisations. The old 'chicken or the egg causality dilemma' is especially difficult to solve at the present time given all the data circulating in real and virtual spaces. It is extremely complicated to understand exactly how the processes are constituted. On the other hand, the nature of collective mobilisation has profoundly changed with the rise of Web 2.0 and collaborative platforms such as Twitter, Facebook, and Instagram. Twitter, as a public microblogging service, is clearly helping to inform, discuss, announce, and disseminate many types of political, social, or cultural protest or collective mobilisation. The recent decline of bipartisanship in Spain—the result of factors such as the financial crisis and the negative perception of politicians and institutions combined with political corruption (Massey and Rustin 2015; Torcal 2014)—has captured significant attention in the debates and messages circulating through Twitter. In this chapter, our aim is to contribute to the general understanding of how Twitter works with these types of sociopolitical processes.

9.2 Building Sociopolitical Realities Through Microdiscourses on Twitter

Near a century ago, in their classical book The Child in America: Behavior Problems and Programs, Thomas and Thomas related an interesting story:

> ...the warden of Dannemora prison recently refused to honor the order of the court to send an inmate outside the prison walls for some specific purpose. He excused himself on the ground that the man was too dangerous. He had killed several persons who had the unfortunate habit of talking to themselves on the street. From the movement of their lips he imagined that they were calling him vile names, and he behaved as if this were true. If men define situations as real, they are real in their consequences (Thomas and Thomas 1928: 572).

Here, we see the beginning of the famous Thomas's Theorem in sociology, which is attributed to William Isaac Thomas (Merton 1995). This chapter tries to connect with and revisit this important theorem through our analysis and reflections based on microdiscourses that are emerging on Twitter. Do interactions, information and conversations on social media, and particularly on Twitter, create realities?

How real are they? What types of discourses were found in the last political campaign? Did they define sociopolitical realities? Did they produce narratives through time with an effect on real reality? As Beltrán (1982) stated, inspired in Thomas's Theorem, real reality and the appearances of reality are real in their effects. Thus, following Beltran's approach, aspects of daily life are as they are, but they can appear to be different thing, as if dressed up. This may imply real consequences, which drives us to consider the real effects of diverse actors' conversations on Twitter.

Thomas and Thomas (1928) also established that contexts are fundamental in the definition of situations. Refreshing this very important idea, in terms of situations, we have to consider the interactions that occur in social media between online and offline processes (again, the chicken or the egg dilemma)—and specifically on Twitter, the interactions or public conversations on this platform (chains of tweets, mentions, RT, etc.). At a more micro level, in the context of a particular tweet, the study of hashtags that share a common space introduces new questions that are especially focused on this chapter to explore patterns of sociopolitical microdiscourses on Twitter.

Thomas and Thomas's situational approach (1928) is applicable to the analysis of Twitter. It is important to note the way in which individuals interpret their situations and how this impacts behaviour. Sociopolitical realities that are discussed on Twitter can be interpreted as objective or subjective realities. Strong components of irony, humour—among others—are usually present. However, individuals tend to filter these realities through their personal situations and experiences. Thus, behaviour is better studied in connection with their context. On Twitter, several contexts exist that also change significantly through time.

9.3   Objectives

Our objective in this study is to identify the microdiscourses that appear around #20D (20 December general elections in Spain), analyse the co-hashtag relationships (co-occurrence of hashtags in the same tweets), attempt to discover patterns in the discourses and delimiting the types of messages that are mainly disseminated through Twitter on this topic. Does the co-hashtag network offer information to help us interpret the political participation of activists, politicians, citizens and media on Twitter? Other aims of this study are to learn about how different hashtags are combined in the context of '20D' conversations and the type of discourses to which they refer. Do they allude to slogans, places, dates, people, or something else? Additionally, through this research, we wish to obtain a better understanding of activity on Twitter that is linked to sociopolitical mobilisations in Spain regarding the past general elections.

9.4   Methods

9.4.1   Data Collection, Filtering and Data Processing

Data were extracted from Twitter, from which we collected 28,261 tweets with '20D' and used this search string to mine data with the help of NodeXL Professional.[1] In this chapter, we focus on an extraction of tweets published before, during and after the recent electoral process in Spain. The tweets ranged from 23 November 2015 to 8 February 2016 (the general elections in Spain took place on 20 December 2015). The samples were taken at seven different times. For the mining process, we used '20D' as the search criterion, through which tweets on the electoral process in Spain were identified. The dataset for this chapter consists of a sample of 28,261 tweets. The tweets were processed with the help of a combination of Excel macros and the SPSS programme to separately extract the hashtags included in each tweet and present them in a readable format for a co-hashtag network analysis with Gephi.[2]

9.4.2   Final Dataset Basic Description

Our dataset, which was prepared for the co-hashtag network, was ultimately composed of 1283 vertex or nodes and 9245 edges or ties. In total, 20.5 % of tweets were original, 71.9 % were mentions (RT), and 7.6 % were replies to other tweets beginning with @. Of all the tweets collected, 62.8 % included a URL in the tweet, suggesting the importance of in-out Twitter connections.

9.4.3   Co-hashtag Networks or Visualisation of Microdiscourses with Gephi

For a better understanding of microdiscourses that emerge from the co-occurrence of hashtags in the context of particular tweets, we decided to visualise them through a co-hashtags or co-words network (Rieder 2012; Gualda et al. 2015). For this purpose, we used the open-source visualisation and exploration software Gephi (Bastian and Heymann 2009). One of the strategies, apart from measuring the centrality of each node (degree, weighted degree, betweenness, etc.), was to visualise the way hashtags can be grouped into clusters or communities by their proximity, which helped us think sociologically about the idea of 'microdiscourses'. Chart 9.1 was

---

[1]NodeXL Pro Website: https://nodexl.codeplex.com/.

[2]https://gephi.org/.

designed with Gephi using the following steps: the weighted degree for setting the node size was calculated, and the Force Atlas 2 Layout was used to achieve good visualisation of the nodes (hashtags). Previously, other layouts were tested. The communities of nodes were drawn in greyscale as a requirement for this book. The communities of co-hashtags were created after calculating the modularity of the network. Nodes belonging to the same group were painted in the same colour. Force Atlas 2 is a continuous graph layout algorithm for Network Visualisation, which is useful for 'an all-around solution to Gephi users' typical networks (scale-free, 10–10,000 nodes)', which is 'a force-directed layout that simulates a physical system in order to spatialise a network. Nodes repulse each other like charged particles, while edges attract their nodes, like springs. These forces create a movement that converges to a balanced state. This final configuration is expected to help the interpretation of the data' (Jacomy et al. 2014: 1–2). The basic idea is that pairs of nodes are more likely to be connected if they are both members of the same communities and are less likely to be linked if they do not share a community.

9.5 Microdiscourses About '20D' in Spain: Semantic or Politics?[3]

Approaching discourses on Twitter about the general elections in Spain (20D) allow us to learn slightly more about activity in cyberspace, which is currently a key issue for Web Social Science in terms that are discussed by Ackland (2013) and allows us to identify what can help support and improve a political campaign or propel a candidate to success. At the same time, this method allows us to actively listen to discourses, debates, conversations, etc. that are taking place on Twitter. A visual approach to microdiscourses about the pre- and post-electoral process that shows co-hashtag relationships is a semantic method of dealing with the topic that is also complex and addresses meanings, social representations and imaginaries. Chart 9.1 includes a co-hashtag network with hashtags that were included in the same tweet in which the string '20D' was found. The visual representation briefly shows that there are several communities of hashtags. To make sense of this data, we have to connect them with the most recent political campaign debates and events. We found interesting patterns in the methods used by political parties to build their discourses during and after the campaign, but we also found other social mobilisation expressions on Twitter in the same discursive context.

Microdiscourses of hashtags about the 20 December elections in 2015 represented different political parties and different methods of sociopolitical mobilisation on Twitter. In the lower left corner is a group of hashtags that was strongly associated with the new leftist party Popular Unity, in which IU [United Left] is also

---

[3]Original hashtags are in Spanish. In brackets there is a translation into English.

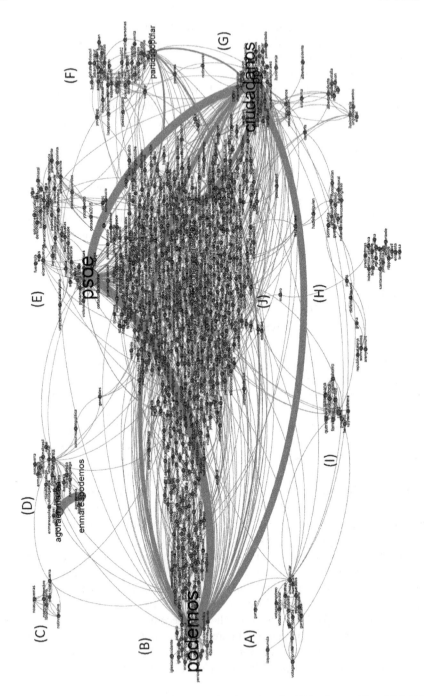

Chart 9.1 20D co-hashtag networks. *Source* Author based on the '20D' dataset (Nov 2015–Feb 2016)

integrated (A). Alberzo Garzón was the presidential candidate, so it is not unusual to find messages that use his name or surname (#AhoraConAlberto [#NowWithAlbert]) that support him at times and at other times evoke a sense of belonging (#garzoners). Some hashtags include funny or impressive expressions of political support (#votagarzonprimo [#votegarzoncousin]). The main demand or slogan hashtag is included in the dataset (#porUnNuevoPais [#forANewCountry], or similarly, #QueremosUnNuevoPais [#WeWantANewCountry]). Of course, hashtags also declare the adhesion to or support for his party (#votaunidadpopular [#votepopularunity], #yomesumoup [#iadheretopu], #unidadpopular [#popularunity]). #Unidadpopular[#PopularUnity] ranked 18th in betweenness centrality out of 1 283 nodes. Campaign on Twitter is impregnated with a critical, combative, tenacious tone; they are also imbued with a humorous and positive spirit (#garzonarrasa [#garzontriumphs], #LaCuevaDeGarzon [#GarzonsCave]). Some of these hashtags were trending topics in Spain.

In the middle of the graph to the left, we found an important community of hashtags connected to Podemos party (B), which has the highest value of betweenness centrality in our dataset and the second highest value for weighted degree. Pablo Iglesias is the primary political leader, and different hashtags can be seen with his name (#iglesiaspresidente [#iglesiaspresident], #yovoyaverapablo [#IgoToSeePablo], #yocontratoapablo [#IhirePablo]), his party (#Podemos [#WeCan]), or his partners (#EnComuPodem [#InCommonWeCan]). Other hashtags incorporating 'Podemos' can also be seen in other communities in our network (D, #Enmareapodemos [#InTideWeCan]).

Assertive hashtags are also continuously present (#Podemos [#WeCan], #sisepuede [#YesItIsPossible]), and other ideas show trust and optimism in change as well as defiance. Assertiveness narratives are optimistic regarding the possibility of winning the elections (#remontada [#comeback], #remontadapodemos [#podemoscomeback], #podemosremontarcontigo [#wecanovercomewithyou], #vamosaganar [#wewillwin], #ilusion [#illusion], #sisepuede [yesitispossible]). The expected result of mobilisation is #ganarelfuturo [#gainthefuture], #ganarelpais [#gainthecountry], #porelcambio [#forchange], #repartiendofuturo [#distributingfuture]. The image of a collective political project including others to build an 'us' is also present (#unpaiscontigo [#acountrywithyou], #porvosotros [#foryou], #contigopodemos [#withyouwecan]). There is a critical approach to topics that are seen across campaigns (#quenoteconfunda [#donotbeconfused], with regard to corruption cases in Spain).

On the far left corner of Chart 9.1, we found several hashtags that created a microdiscourse supporting an international campaign that appeared after 13 November 2015 #ParisAttacks (#NoASusGuerras [#NotToYourWars], #NoEn NuestroNombre [#NotInOurName], #NoALaGuerra [#NotTotheWar], #NotInMy Name). These hashtags show a pacifist stance that were incorporated in the campaign debates (C). They also manifest a strong idea of belonging using pronouns ('YourWars', 'MyName', 'OurName'). This discourse clearly marks the difference between 'they', 'the others', and 'me' or 'we'. Previous hashtags are strongly linked among them, sharing the space of a 140 characters tweet, but these tweets are linked to other more politically oriented tweets that appeared with the Spanish campaign. They were supported mainly by the leftist parties Unidad Popular-IU, Podemos, EnMarea [InTide], UPyD [Union, Progress and Democracy], and others that proposed not being involved in a possible forthcoming war following the #ParisAttacks.

The most important hashtags in this dataset were #EnMarea [InTide], #AgoraEnMarea [#NowInTide], or #EnMarea (D), which were connected to Podemos and Galicia. They form a dense microdiscourse that produces a clear narrative around collective mobilisation. Metaphorically, the idea of a 'Marea' [Tide] expresses this idea of a 'collective' working toward the same aims as compared to individual approaches. #EnMarea ranked 11th in betweenness centrality and 20th in weighted degree, out of 1283 nodes. In the case of other hashtags, betweenness is not as important as weighted degree, as seen in the case of #Agoraenmarea, #Enmareapodemos [#InTideWeCan] (in 5th and 6th).

Microdiscourses about the Socialist party PSOE (E) are found on the upper right side. They were organised around the idea that Spain has been divided, which the PSOE will reunite (#elcambioqueune [#thechangethatjoins]). The idea of change is also incorporated into other hashtags (#elcambiovotapsoe [#thechangevotespsoe]). Change also implies a break from the right (#cortaconladerecha [#cutoffwiththeright]). Hashtags that disseminate the campaign slogans are also extremely relevant (#unfuturoparalamayoria [#afutureforthemajority], #unpresidenteparalamayoria [#apresidentforthemajority]). Pedro Sánchez (PS) was the presidential candidate for PSOE, and his campaign is full of support and auto-identification messages (#Yoconpedro [#IwithPedro], #pedropresidente [#pedropresidents], #pspresidente [#pspresident]) or encouragement to vote (#votapsoe [#votepsoe]), in addition to party pride (#orgullosocialista [#socialistpride]). Other discourses appeared after the elections regarding a #pactodeizquierdas [#left-wingpartiespact] or even those criticizing the actions of other parties (#podemosnoquiere [#podemosdonotwant]).

People's Party (PP) in Spain is placed in the middle right section of Chart 9.1 (F). The campaign tried to reinforce the value of expertise in governing and also seemed to symbolically delimit 'us'and 'others'. With #españaenserio [#Spainseriously], the idea of a reliable and responsible 'we' is subtly presented, confronting 'others' that are less serious or who are highly experienced in politics, as is recurrently discussed in comparison to Ciudadanos [Citizens] or Podemos. #LaEspañaQueQueremos [#TheSpainThatWeWant] or #ParticiparEnSerio [#SeriouslyParticipate] are hashtags that try to suggest stability and seriousness. #PP, #PartidoPopular [#PeoplesParty], and #Rajoy are respectively ranked at 5th, 14th, 19th and 20th in betweenness centrality and attained high-ranked positions in weighted degree. Slogans such as #votapp [#votepp], support messages such as #vamospp [#comeonpp], or most auto-identificative messages such as #yovotopp [#Ivotepp] or #yoestoyconrajoy [#IamwithRajoy] also belong to this community.

The microdiscourses on the Ciudadanos Party [Citizens-Party of the Citizenry] (G) are placed on the right side of the graph. The presidential candidate, Albert Rivera, is protagonist of several hashtags (#yovotoalbert [#Ivotealbert],

#albertrivera, #albertpresidente [#albertpresident], #albertmipresidente [#albert-mypresident]). Other slogans claiming the vote (#votaciudadanos [#votecitizens]) and incorporating emotions in the discourse (#votaconilusion [#votewithillusion], #ilusionnaranja [#orangeillusion]) are well-represented in the campaign slogan (Vota con Ilusión). Interestingly, the betweenness centrality measures placed Ciudadanos in second, Ciudadanos ranked third for weighted degree position.

Other communities in Chart 9.1, seen at the bottom of the graph, represent other political options: Catalonian parties (H), and UPYD (I) also have representative hashtags for their campaigns. Finally, a large number of hashtags are placed in the middle of the graph (J) that do not belong to differentiated communities. This large community represents an important number of conversations, debates and information on Twitter around generic hashtags that were used during the campaign (#elecciones [#elections], #elecciones2015 [#2015elections], #eleccionesgen-erales2015 [#2015generalelections]) or particular debates during the campaign that were sometimes promoted by particular media (#tv3, #a3media, #elpaisdebate, #7deldebatedecisivo [#7dthedecisivedebate], etc.). In these generic hashtags, it was very common to introduce opinions and ideas of different candidates, slogans or ideas in the same tweet, which is the origin of the mixed community of co-hashtags.

9.6 Conclusions

Though campaign priority topics differed according to the party, some common patterns emerged after the analysis, especially concerning elements that are recurrent and connected with regard to the microdiscourses posed on Twitter about the recent Spanish general elections held on 20 December 2015. Communities of co-hashtags had previously been composed of short, condensed messages that were connected in tweets or through conversations on Twitter on a temporal basis. The name of the party, the main candidate, the campaign slogan and other optimistic, supportive, assertive and empowering messages are typical features of political microdiscourses found on Twitter in a context of political campaigns. Identification and belonging to the party are also promoted through Twitter, in which similar patterns and structures were found. Important strategies also included reducing long messages through metaphors, symbols or other abbreviated ways of disseminating complex ideas. All these components as well as the importance of symbolic and emotional discourses allow us to support to the idea of 'microdiscourse' on Twitter, which gives credence to this line of research for better understanding sociopolitical and mobilisation processes. Answering the question posed in the previous section, microdiscourses on Twitter about the 20D Elections go beyond semantics; a combination of factors contributes to the conversation.

Future studies will focus on other sides of the discourse, and those that are focused on collective and individual actors that produce the discourses for tweeting are especially important. Additionally, significant interest can be found in studying the temporal and interactional dynamic production of tweets that generates conversations in virtual space over time.

It will be more difficult to determine which part of our reality is being built and influenced by Twitter and other media, as the information provided in the social media impact journalists, policy makers, politicians, experts and other individual or collective actors. Although tweets are real, an appearance of reality can be created (for example, through bots or spammers). Closing with previous arguments (Thomas and Thomas 1928; Beltrán 1982), realities or appearances of realities can produce some type of effect in real reality by generating news and giving others more opportunities to understand the political messages promoted through this medium. Good examples of this are trending topics and other data generated with the help of experts on Web 2.0 (hackers, community managers, tweet stars, cyberactivists, resonators, or others type of actors involved in social media (see in Congosto 2015; Peña-López et al. 2014; Fernández Prados 2012, for recent examples in Spain). Some of these experts are today fully incorporated in the political campaigns.

Acknowledgments I am very grateful to the XESCOM Network for their invitation to the I XESCOM. International Symposium on Communication Management, From Media to Metamedia, November 27th and 28th, 2015, which was a great opportunity to discuss the first draft of this chapter in the excellent intellectual atmosphere fostered by the organisers.

References

Ackland, R. (2013). *Web Social Science: Concepts, Data and Tools for Social Scientists in the Digital Age.* London: Sage.
Bastian, M., Heymann, S., & Jacomy M. (2009). *Gephi: An open source software for exploring and manipulating networks.* Third International AAAI Conference on Web and Social Media. Retrieved December 15, 2015 from http://www.aaai.org/ocs/index.php/ICWSM/09/paper/view/154
Beltrán, M. (1982). La realidad social como realidad y apariencia. *Revista Española de Investigaciones Sociológicas, 19,* 27–54.
Congosto, M. L. (2015). Elecciones Europeas 2014: Viralidad de los mensajes en Twitter. *Redes. Revista Hispana de Redes Sociales, 26*(1), 23–52. Retrieved September 23, 2015 from http://revistes.uab.cat/redes/article/view/v26-n1-congosto/529-pdf-es
Fernández Prados, J. (2012). Ciberactivismo: conceptualización, hipótesis y medida. *Arbor, 188* (756), 631–639.
Gualda, E., Borrero, J. D., Carpio, J. (2015). La 'Spanish Revolution' en Twitter (2): Redes de hashtags y actores individuales y colectivos respecto a los desahucios en España. *Redes. Revista hispana para el análisis de redes sociales, 26*(1), June 1–22. Retrieved September 23, 2015 from http://revistes.uab.cat/redes/article/view/v26-n1-gualda-borrerodiaz-carpio
Jacomy, M., Venturini, T., Heymann, S., & Bastian, M. (2014). ForceAtlas2, a continuous graph layout algorithm for handy network visualization designed for the Gephi software. *PLoS ONE, 9*(6), e98679. doi:10.1371/journal.pone.0098679
Massey, D., & Rustin, M. (2015). Elections and political change. Soundings: *A Journal of Politics and Culture,* (59), 12. Retrieved January 15, 2016 from https://www.lwbooks.co.uk/soundings/59

Merton, R. K. (1995). The Thomas Theorem and The Matthew Effect. *Social Forces, 74*(2):379–424.

Peña-López, I., Congosto & Aragón (2014). Spanish Indignados and the Evolution of the 15 M Movement on Twitter: Towards Networked Para-Institutions. *Journal of Spanish Cultural Studies, 37–41.* Retrieved November 20, 2014 from http://dx.doi.org/10.1080/14636204.2014.931678

Rieder, R. (2012). The refraction chamber: Twitter as sphere and network. *First Monday, 17*(11). Retrieved January 20, 2016 from http://firstmonday.org/ojs/index.php/fm/article/view/4199/3359

Thomas, W. I., & Thomas, D. S., (1928). The methodology of behavior study. *The child in America: Behavior problems and programs* (pp 553–576). New York: Alfred A. Knopf.

Torcal, M. (2014). The decline of political trust in spain and portugal: Economic performance or political responsiveness? *American Behavioral Scientist, 58*(12), 1542–1567. Retrieved January 20, 2016 from http://dx.doi.org/10.1177/0002764214534662

Chapter 10
Video Game Screens: From Arcades to Nintendo DS

Israel Márquez

Abstract Video games have become one of the most important and profitable mediums of the modern era. They are even considered the most influential form of popular expression and entertainment in today's broader culture (Jones in The meaning of video games: gaming and textual strategies. Routledge, New York, 2008). Video games have been studied from different points of view (narratological, ludological, sociological, etc.) but not from a "screenological" one, that is, focused on the screen. Following the idea of "screenology" (de Kerckhove in The architecture of intelligence. Birkhäuser, Basel, 2001; de Kerckhove in Inaugural lecture of the UOC 2005–2006 academic year, 2005; Huhtamo in Elements of screenology: toward an archeology of the screen, 2004), this paper aims to provide a first look at how changes in size, design and location of video game screens have transformed the videoludic experience.

Keywords Video games · Screens · Screenology · Public space · Private space

10.1 Introduction

Video games have become one of the most important and profitable mediums of the modern era. They are "arguably the most influential form of popular expression and entertainment in today's broader culture" (Jones 2008: 2). The video game has an important place in media and cultural history because it was "the first medium to combine moving imagery, sound, and real-time user interaction in one machine, and so it made possible the first widespread appearance of interactive, on-screen worlds in which a game or story took place" (Wolf 2008: 21).

Although it may seem obvious, video games need a screen to exist. Without screen they would be other games but not video games. The screen allows us to see

I. Márquez (✉)
Juan de la Cierva Research Grant Program,
Open University of Catalonia (UOC), Catalonia, Spain
e-mail: isravmarquez@gmail.com

© Springer International Publishing Switzerland 2017
F.C. Freire et al. (eds.), *Media and Metamedia Management*,
Advances in Intelligent Systems and Computing 503,
DOI 10.1007/978-3-319-46068-0_10

the game and follow the action. It allows us to move our avatars through the space of the game and to do things inside it. The screen is "where the action is" (de Kerckhove 2001: 36) and video games are basically "actions":

> If photographs are images, and films are moving images, then *video games are actions*. Let this be word one for video game theory. Without action, games remain only in the pages of an abstract rule book. Without the active participation of players and machines, video games exist only as static computer code. Video games come into being when the machine is powered up and the software is executed; they exist when enacted (Galloway 2006: 2).

Video games have been studied from different points of view (narratological, ludological, sociological, etc.) but not from a "screenological" one, that is, focused on the screen. Following the idea of "screenology" (de Kerckhove 2001, 2005; Huhtamo 2004), this paper aims to provide a first look at how changes in size, design and location of video game screens have transformed the videoludic experience.

10.2 The Public Screen: Arcades

The first video game screens available for the general public were the screens of the arcade or coin-operated machines. The earliest forms of video games emerged out of laboratories and research centers in the 1960s and 1970s. These games, however, were not accessible to the public and only a few people really played them. These first forms of video game screens (or proto-video game screens) were located in private and institutional spaces and they were more "university games" than accessible entertainment games (Lowood 2009).

During the early 1970s, video arcade games were the first games that could be used by the general public. An arcade machine was basically a standing cabinet equipped with a screen and a set of controls, commonly a joystick for as many players as the game allows, plus action buttons. Arcades rapidly became "dedicated spaces for games, gamers and game play" (Tobin 2014: 127), especially for the kids: "Too poor to shop and too young to drink, kids could gather and socialize outside of school and away from their parents amid the glow of coin-op machines" (Gich 2015).

The arcade screen was a public and social screen. While someone was playing a video arcade game, groups of players and onlookers would gather around the screen watching and commenting the action. People not only were able to interact with the screen space but also with other people. Some arcade machines were even equipped with two or more joysticks (games like Atari's *Indy 800* allowed 8 people to play a color racing game), so players could play together on the same screen in a competitive or cooperative manner. Arcades and their screens "both actually facilitated video game play through their technical use and also allowed multiplayer video game play to occur, encouraging it by defining that space as one in which you can go up to a stranger and play with him or her" (Tobin 2014: 127).

Arcades were the first social spaces for video game play, game culture and gaming practices, and video arcade screens were the first video game screens to be really available and accessible to the public—hence their historical and "screeno-logical" importance.

10.3   The Private Screen: Video Game Consoles and Home Computers

Throughout the 1980s, the popularity of "going to the arcade" started to decline. Playing at home through gaming consoles or home computers became a new way to enjoy video games without ever leaving one's home. As Wolf points out, if video arcade games were the first computers that could be used by the general public, "home game systems became the first computers to enter people's homes", helping to build "a positive, fun, and user-friendly image of the home computer, introducing it as a recreational device instead of a merely utilitarian one" (Wolf 2008: 21).

The emergence of domestic game consoles opens a new chapter in the history of the video game screens. The classic video arcade screen is then replaced by a well-known screen: the TV screen. Game consoles transform the traditional TV screen in a new space for gameplay. Through game controllers attached to the game console, traditional TV "viewers" become "players". We don't "watch" the TV screen, we "play" with it: "Don't watch TV tonight. Play it!", as could be read in one Atari print ad (Fig. 10.1).

Along with home gaming consoles, personal computers became more accessible to the general public during the 1980s. With the introduction of computer games in people's homes, the computer screen became a new screen to play video games. As a result, people began to stay at home in front of their TV or computer screens, playing and enjoying video games in a more solitary, intimate and reflexive way. New video game genres such as adventure or strategy games emerged, embracing a slow and cerebral gameplay style which requires exploration, puzzle-solving, complex thinking, or strategy skills. Gaming, thus, moved away from the social space of the arcades into the living room and then into one's bedroom, reinforcing the idea (or myth) that playing video games is mainly an isolated and solitary activity.

Modern game consoles and home computers, however, have transformed the private screen into a "connected" and social screen which allows new forms of online interaction and socialization. Although the user still plays at home in front of the TV or the computer screen, he or she can be digitally connected to other people. TV or computer screens are no longer private screens but social screens. They encourage the cooperation, collaboration and/or competition between players in a digital way. The physical multiplayer video game play of the arcades becomes virtual and the screen transforms itself in a new space for interaction and social-ization through video games.

Fig. 10.1 "Don't watch TV tonight. Play it!" *Source* www.atarimania.com

10.4   The Mobile Screen: Handheld Game Consoles, Smartphones and Tablets

After arcades, gaming consoles and home computers, the latest and most recent manifestation of video games are mobile games. They form a new type of video game screen, a mobile/portable screen represented by handheld game consoles, smartphones and tablets. These devices allow players to play video games as they walk. We no longer have to go to a specific place to play a game (arcades), stay at home to plug in the game console to the TV screen (home consoles) or open our personal computer. We can carry the console and games with us and play with them anytime, anywhere. With handheld game consoles, smartphones and tablets we can play games both at the street and at home, in public and private spaces. The screen thus becomes a kind of hybrid screen, a public-private screen we can use in different places.

The handheld console is actually a kind of tiny arcade machine. The screen, buttons and controls are integrated into a single device, but unlike the arcade machine, the handheld game console integrates all these elements in a tiny version we can carry in our pocket and use when we want. As Huhtamo (2004) points out, "Handheld devices are personal, attached to the body of the user like clothing, jewellery or a wallet [...] While we leave our TV sets and PlayStations behind from time to time, the portable small screens have become permanent extensions of the user-owner's body."

The handheld game screen is, like the video arcade screen, a pure videoludic screen, that is, a type of screen specifically designed for playing video games. This is different from other video game screens such as TV screens, computer screens or smartphone screens. These screens allow the use of video games but they were not specifically created for it. Only arcades and handheld console screens were designed mainly for watching and playing video games.

A very interesting manifestation of the mobile game screen is the Nintendo DS. The DS (Dual Screen) format was already used in Nintendo's Game and Watch series. These devices, however, offered single games rather than multiple games, being more handheld electronic games than handheld game consoles. The Nintendo DS popularized the DS format introducing a new kind of screen, a many-sided screen "that can be held in one hand: a mobile screen, a double screen, a touchscreen, and a wired or connected screen" (Verhoeff 2012: 23). Smartphones and tablets are also many-sided or hybrid screens. As the DS, they offer an example of multiple screen models. They are mobile screens, touchscreens and wired or connected screens, but they don't have the double screen format. They offer all these screen models in a single, portable, and small (smartphone) or medium (tablet) screen.

With these new types of screens the videoludic experience transforms itself into something hybrid, a (re)mix of mobility, tactility and connectivity. We can play (alone or connected with others) while we move, touching the screen with our own fingers. It constitutes a very different experience from that of "going to the arcade" or to stay at home in front of the TV or the computer screen. We can play "in transit", while walking or waiting for a bus. These portable small screens become a permanent extension of the user's body, creating ludic spaces and moments in both public and private spaces.

10.5 Conclusions

The screen "is becoming our principal connection to information. It is becoming the main cognitive interface" (de Kerckhove 2005: 6). As the importance of screens in contemporary world increases, the task of understanding their social and cultural roles becomes urgent. Screenology, thus, can be understood as a new field of research and "a way of relating different types of screens to each other and assessing their significance within changing cultural, social and ideological frames of reference" (Huhtamo 2004).

Video games can be studied from a "screenological" and media archeological point of view. As this article has shown, changes in size, design and location of video game screens have affected the gameplay experience across time. "Going to the arcade" and playing games on a video arcade screen constitutes a very different experience than stay at home playing on a TV or a computer screen. The video arcade screen can be understood as a public screen because it stands at the social space of the arcades. Some video arcade screens were equipped with two or more joysticks, allowing players to play together in a cooperative or competitive manner. But even the single player games were social because they were played in a social space where people could interact with other people.

On the other hand, playing at home on TV or computer screens constitutes a very different videoludic experience. These screens can be understood as private screens because gameplay usually takes place in intimate spaces such as bedrooms. Modern game consoles and home computers, however, allow multiplayer gameplay through online connections. Internet have transformed the private screen of game consoles and home computers in a "wired" or "connected" screen through which people can interact with other people in a digital manner. The online multiplayer mode of modern game consoles and home computers recreates in some way the physical multiplayer gameplay of the arcades, encouraging it by defining the screen space as one in which you can go online, meet people (friends or strangers) and play with them in a competitive or cooperative manner.

Mobile screens such as handheld game consoles, smartphones and tablets are also wired or connected screens because they allow us to play with other people through digital networks. They are also portable, tactile and, in the case of the Nintendo DS, double screens. With these small and portable screens gameplay becomes a pervasive and ubiquitous experience not limited to a specific place, whether public (arcade) or private (home). Portable screens are permanent extensions of the user's body in today's gaming culture, allowing players to play anytime, anywhere. They are ubiquitous screens through which we are not only able to play video games but to do several other things (take photographs, listen to music, watch videos, surf the Internet, etc.). They are a kind of "meta-screen": the screen of the new meta-media era.

References

de Kerckhove, D. (2001). *The architecture of intelligence*. Basel: Birkhäuser.
de Kerckhove, D. (2005). The biases of electricity. In *Inaugural lecture of the UOC 2005–2006 academic year*. Retrieved January 18, 2016, from http://www.uoc.edu/inaugural05/eng/kerckhove.pdf
Galloway, A. R. (2006). *Gaming. Essays on algorithmic culture*. Minneapolis: University of Minnesota Press.
Gich, E. (2015). Playing loud in quiet spaces. *Kill Screen*, Retrieved November 20, 2015, from https://killscreen.com/articles/playing-loud-quiet-spaces/
Huhtamo, E. (2004). Elements of Screenology: Toward an Archeology of the Screen. *ICONICS: International Studies of the Modern Image* 7. Retrieved January 10, 2016, from http://gebseng.com/media_archeology/reading_materials/Erkki_Huhtamo-Elements_of_Screenology.pdf
Jones, S. E. (2008). *The meaning of video games: Gaming and textual strategies*. New York: Routledge.
Lowood, H. (2009). Videogames in computer space: The complex history of pong. *IEEE annals of the history of computing, 1058–6180, July–September, 2009*, 5–19.
Tobin, S. (2014). Arcade mode: Remembering, revisiting, and replaying the american video arcade. In L. A. Freeman, B. Nienass, & R. Daniell (Eds.), *Silence, screen, and spectacle: Rethinking social memory in the age of information*. New York: Berghahn.
Verhoeff, N. (2012). *Mobile screens. The visual regime of navigation*. Amsterdam: Amsterdam University Press.
Wolf, M. J. P. (2008). *The video game explosion: A history from PONG to playstation and beyond*. London: Greenwood Press.

Chapter 11
Is the Employer Entitled to Survey Employee's Internet Communications in the Workplace? Case of "Barbulescu v. Romania"

Jaime Cabeza Pereiro and Emma Rodríguez Rodríguez

Abstract The ECHR considers that the interpretation of national courts on the legality of the disciplinary action taken by the employer that justifies the surveillance of Internet communications of the employee in the workplace is in compliance with law. This judgement includes a separate dissenting opinion.

Keywords Company management powers · Right to privacy of the employee · Disciplinary action

11.1 Facts

The case analyzed—Judgment of the European Court of Human Rights, 12 January 2016, Case of "Barbulescu v. Romania"—derives from the appeal filed by a dismissed employee against the judgment of the national court (Romania), which upheld the sanction of terminating his contract imposed by the company.

As it happened, the employee, at his employer's request, created a Yahoo Messenger account for strictly professional purposes. On 1 August 2007, the employer informed the employee of its decision to terminate his contract for breach of the company's internal regulations that prohibited the use of company technological resources for personal purposes.

Specifically, in response to the employee's denial of the violation, the company presented a transcript of personal messages that the employee had exchanged with his brother and his fiancée, as well as messages from his Yahoo personal account. In view of these facts, the employee sued the company and claimed that the dis-

J. Cabeza Pereiro (✉) · E. Rodríguez Rodríguez
University of Vigo (UVigo), Vigo, Spain
e-mail: jcabeza@uvigo.es

E. Rodríguez Rodríguez
e-mail: emmarodriguez@uvigo.es

© Springer International Publishing Switzerland 2017
F.C. Freire et al. (eds.), *Media and Metamedia Management*,
Advances in Intelligent Systems and Computing 503,
DOI 10.1007/978-3-319-46068-0_11

missal had been null and void due to the fact that the company had violated his right to correspondence protected by the Romanian Constitution and the Criminal Code.

The court of first instance dismissed his complaint on the grounds that the employee had been sufficiently informed of the prohibition to use company resources (including Internet) for personal purposes. In the appeal, the employee claimed the violation of art. 8 of the European Convention on Human Rights, in addition to a procedural issue. The Court dismissed it and based its argument on EU Directive 95/46/EC, of 24 October 1995, on the protection of individuals with regard to the processing of personal data an on the free movement of such data. According to its reasoning, the employer had exercised in a reasonable manner its management and control powers with the means at its disposal in order to verify the breach.

11.2 Analysis of the Grounds of Annulment

It must be highlighted that this judgment has been delivered by Sect. 4th of the Court. In other words, it is not a case submitted to the knowledge of the Grand Chamber. Therefore, the Court is not assessing that the application creates the necessity of establishing a new perspective of case law. The Section only needs to follow the path of previous judgments and to apply its doctrine to the current application (Senden 2011: 18 ff). The Barbulescu affair is not developing broad principles about art 8 of the Convention but only solving an individual application. For this reason, its undoubted interest must be put in the context of the functioning of the ECHR. Actually the dissenting opinion develops very interesting criteria about company policies in internet surveillance which would need the involvement of the Grand Chamber for the purposes of broad discussion. We mean, the consideration of the arguments expressed in it would need much more than only a Section of the Court.

Taking for granted that the judgment was going to follow known paths opened by previous case-law it is easy to understand that the reasoning of the majority is brief in terms of extension and intensity. Therefore the outcome seems weak and exposed to criticism and the dissenting opinion appears more robust and well-based. Both the majority and the dissenting opinion will be discussed in the following pages briefly. Anyway, it is important to take into account that only one judge signed the dissenting opinion against the other six in favor of the decision that denied the infringement of article 8 of the Convention.

The Court's assessment recognizes the wide notion of private life included in art. 8 of the Rome Convention. Its doctrine has built a very broad concept which covers, among other particular situations, telephone calls made from business premises. Moreover, it concedes that workers have a reasonable expectation to the privacy of calls made from a work telephone. In similar way the company's computer and internet resources might be subject to the same understanding. In this framework, the public authorities are bound by positive obligations to ensure the effective respect for private life. They have to create the adequate conditions to enjoy this sort of rights although keeping certain margin of appreciation.

However, employers can develop internal regulations prohibiting or limiting the use of e-resources for non-professional purposes. In this case there was a clear rule forbidding that use. So the complainant couldn't trust in any sort of tolerance by the company. The employer had checked the Yahoo Message account of the worker in the belief that all the messages contained in it had a commercial purpose and were related with his professional activities. From the point of view of the domestic court the only consideration consisted in that some messages were not work-related. So the employee had misused the company tools in breach of its well known policy in this field.

It must be remarked that the content of the private messages sent by the complainant was irrelevant for the domestic Court. It gave attention to it only to the extent that it proved that the complainant had misused the Yahoo Messenger account he had created at the employer's request. Nonetheless, the Judgment has to balance two different approaches that appeared in its previous jurisprudence: on one side, it was established that the worker hadn't caused any sort of damage to the employer. So the affectation of his private life was not justified by a legitimate aim from the point of view of the employer's interest. On the other side, it is true that the employer had ordered the employee to use the account only for professional issues. So, it had a reasonable expectation not to find private messages and also a justified reason to verify that the worker was complying with his duties during his time of work.

Moreover the Court takes into account that the only data examined by the employer were the messages of his Messenger account but not other data stored in his computer. Therefore the affectation of his privacy was limited and proportionate. It could said that the checking practiced by the company was the less possibly to verify the correct use of the communication tools during the working time.

Furthermore, it was found that the worker failed to justify convincingly why he had behaved against the express orders of the employer and had used the Messenger for personal purposes during his working time. That lack of explanation implies a plane recognition of his fault and of the breach of his duties.

Taking into account these legal grounds, the Court reaches the conclusion that the domestic tribunals had balanced fairly within their margin of appreciation the applicant's right to respect for his private life covered by art. 8 of the Convention and the employer's interest to survey the adequate compromise of the employee in his duties.

Some of the critical aspects of this majority opinion will be deepened in contrast with the dissenting vote. It departs from one not discussed fact: that there was a clear and undoubted rule that forbade the use of the Messenger for personal communications. This assumption drives immediately to conclude that the worker broke an express rule in breach of his obligations. At the same time, makes the difference with other judgments where there was not a so clear prohibition. The Court pointed out this question and built the decision in it. It is not under discussion that the employer owns the right to limit the access to internet communications and social nets as an expression of his position in the employment contract.

Nevertheless, that position must be balanced with the right to privacy that enjoys the worker as citizen. It is true that this right may be subject to some limitations in the interest of the company but always departing from the principle that the employee has a reasonable expectation to keep his private communications free of scrutiny from the employer. Both, the domestic Court and the ECHR, recognize the respective position of the two parts of the employment contract. But the domestic Court emphasizes that the employer had opened the employee's account in the belief that it only contained professional messages. This is an interesting but controverted point: In fact there was an unresolved dispute about if the company had warned its employees about the possibility of monitoring their communications. The ECHR admits this dispute, but shows its satisfaction with the position of the domestic Court.

It is apparent that the blanket prohibition is not enough as company policy. It might be accompanied with the provision of a channel to supervise the irregular use of the communications and to discuss the controversies about this supposed misuse. That question will be underlined with the analysis of the dissenting opinion. At this respect the majority position of the Court seems rather careless. Some obvious questions arise from the decision of the Tribunal. For instance how is possible to deny without more consideration the issue of tolerance. If there was not any warning about the power that the company reserved itself to monitor the communications, simply the time elapsed from the put into force of the rule that forbade the private use could create the appearance that there was some degree of tolerance.

We must take into account that the ECHR limits its analysis to verify if the decision of the domestic Court is coherent with article 8 of the Convention. However, unfortunately its brief judgment leaves some grey areas where the right to privacy is questioned and where the employer can enjoy some degree of discretion. The worst message is that companies can take advantage of a position of not developing rules in this field to give security enough to their workers. On the other hand its approach to the proportionality principle doesn't suffice. It reaches only the question if the employer knew too much private information of the worker, but not to the doubt if the conduct of the employee was serious enough to justify a dismissal. Again the dissenting opinion deals with this interesting question.

The comparison between this Judgment and the case-law of the Spanish Constitutional Court shows some interesting points. The modern position of this Court is strengthening the power of companies and weakening the right to privacy of workers. In this scenario the Barbulescu judgment has outstanding interest because it shows some inconsistencies between the jurisprudence of both bodies. First the ECHR underlines that the domestic Court hadn't give any attention to the content of the messages of the employee. Nothing in the reasoning of the ECHR allows to conclude that its judgment is opening the possibility of an implied resignation of the worker to his right to privacy. The content of the messages is absolutely irrelevant for the European Tribunal. The disciplinary action taken by employer is not related at all with the meaning of the messages submitted by the worker. In this sense there is a huge difference between the Barbulescu affair and the Decision of the Spanish Constitutional Court 241/2012, of 17th September (a very sharp critique of this Decision in Cardona 2012: 169 ff).

The question of the duty of the employer to inform about the monitoring tools used to verify the adequate use of the internet and communication resources has also been omitted by the Spanish Constitutional Court in a Decision that creates very doubtful jurisprudence—Constitutional Court Decision 170/2013 of 7th October. The facts were related to a contract of employment whose collective agreement of application expressly forbade the use of e-mail for private purposes. With this only provision, the Tribunal submits the supervision of the employee's account to the monitoring power of the employer who can supervise the content of the messages. So that the worker can't have any reasonable expectation to keep his communications private. The prohibition automatically implies, from the point of view of the Constitutional Court, the power of the company to accede to the contents of the mailing. Until this point this jurisprudence fits well with the case-law of the ECHR, although it is important to remark that the prohibition had been made in a national collective agreement, not by an express order or instruction of the company.

But the Spanish Court goes further: admitted that the meddling in the privacy of the employee was justified, necessary and proportionate, the contents of some communications—which discovered reserved information of the company to third entities—were serious enough to support a justified dismissal. Therefore—and this point makes the difference—the main consideration is not the breach of a duty not to use the e-mail for personal purposes, but the disclosure of information through this forbidden tool. In this decision the question was the information disclosed and in the 241/2012 Decision, the gossips and critiques to one working mate. But in both cases the Court supports the disciplinary action of the company based in the contents of the messages. Unfortunately in none of them the Court discussed if previously it had been provided a clear, secure and fair proceeding to supervise the messages sent and received by employees.

For this considerations it is obliged to conclude that the domestic case-law of the Spanish Constitutional Court is not clearly consistent with the case-law of the ECHR, even taking into account the majority position of the case Barbulescu.

11.3 Analysis of the Partly Dissenting Opinion

The ECHR considers that the interpretation made by national courts, under which the employer exercised control and monitoring in a proportionate manner within the limits of its management power, is in compliance with law. Specifically, it does not deem that the employer's conduct constitutes the violation claimed by the employee concerning the right to privacy of correspondence (protected by art. 8 of the European Convention on Human Rights). Therefore, the penalty for the contractual breach was imposed in due process of law.

Nonetheless, this judgment includes a separate dissenting opinion, issued by judge Paulo Pinto. It is an extensive text, with numerous case-law and normative references, within both the international and the European scope. The judge reflects on the importance and the use of new communication technologies in the workplace and on the necessary guarantees for their proper use.

The dissenting opinion is structured in twenty-three points, grouped into six sections. It can be stressed that the main idea of the text is the importance of determining exactly the extent of the employee's knowledge concerning the prohibition or the limitations in the use of electronic instruments for personal purposes; that is to say, the importance of providing legal certainty, by means of the principle of characterization, in the employee's alleged breaching conduct and the resulting disciplinary consequences in the context of the working relationship.

In summary, it seeks to clarify the conflict arisen between the right to personal privacy, which is evident in the privacy of communications on the Internet, on the one hand, and the right to exercise management power, which comprises control and monitoring functions by the employer, on the other hand.

The dissenting opinion begins by reminding judgments of the European Court of Human Rights that recognized the importance of the Internet as a platform to exercise the right to freedom of expression (case of "Delfi AS. v. Estonia", 16 June 2015, among others) and emphasizes that this has been recognized by national courts (such as the French Constitutional Council, in its decision no. 2009/580 DC, 10 June).

In this sense, art. 10 of the Convention protects the right to freedom of expression in the different audiovisual media. And, even though it does not refer explicitly to the Internet, it must be understood that it is included within that list of protected means of communication. Moreover, this provision states that there should be no interferences of any kind ("public authority") in the exercise of that right, unless they are justified in order to protect the best interest of a democratic state—"national security, territorial integrity or public safety, for the prevention of disorder or crime, for the protection of health or morals, for the protection of the reputation or rights of others, for preventing the disclosure of information received in confidence, or for maintaining the authority and impartiality of the judiciary".

This right is complemented with the provisions of art. 8 of the Convention on the right to respect for private and family life, which specifically provides for the protection of correspondence. The limits to this right bear a great similarity to those listed in art. 10, but provisions are added in relation to "the economic well-being of the country", as well as a more open clause concerning "the protection of the rights and freedoms of others".

With these normative premises, the dissenting opinion focuses on reasoning that there is no absolute power of the employer regarding the control of the electronic communications of its employees, even during working hours and with working tools. The doctrine is unanimous on the limits of corporate management power, so that control of Internet communications can not be discretionary or disproportionate.

As noted by other decisions cited by the dissenting opinion (and ignored by the judgment), the corporate conduct should be guided by the adequate protection of the right to ensure the smooth running of the company and to ensure that the information to which the employee has access is suitably used. However, none of these rights is explicitly expressed in arts. 8 and 10 of the Convention, whose restrictions always require proper justification (case of "Palomo Sánchez and Others v. Spain", 12 September 2011; case of "Halford v. the United Kingdom", 25 June 1997 (1997-III); case of "Copland v. the United Kingdom", no. 62617/00, ECHR 2007-I).

Through a comprehensive analysis of the legal texts, from both the European Union and the ILO (some of which are not even mentioned in the judgment; for example, the Recommendation 2015 (5) of the Committee of Ministers of the Council of Europe, adopted on 1 April 2015, on the protection of personal data used for employment purposes), the dissenting opinion emphasizes that control measures must be justified and least intrusive. It also stresses the importance of the employee having an adequate knowledge of the prohibitions or limitations that the employer can impose.

According to the dissenting vote, this has not been proven. As it is a possible violation of fundamental rights, the burden of proof is on the employer. It concludes that there is insufficient supporting documentation showing that the employee knew the limits and restrictions on the use of technological means for personal purposes. Lacking this essential element, the penalty and the whole disciplinary proceedings would not be in compliance with art. 8 of the Convention.

In conclusion, this judge considers that the minimum legal requirements to impose such restrictions on the right to privacy of communications and the right to privacy of personal data, also applicable to labor relations, have not been met. Therefore, according to his opinion, neither the Romanian court nor the ECHR have resolved the breach of the right to privacy of the employee.

The issue discussed is very controversial in the field of current labor relations, in which the electronic means are predominantly used. The majority ruling puts workers in a situation of helplessness and special vulnerability to the exercise of corporate management power. This judgment may be appealed to the Grand Chamber. It is possible, then, that the proposal in the dissenting opinion will be taken up and that the disproportionate use of the privacy of the employee that seems to be allowed by the Romanian court and upheld by the ECHR will be amended.

References

Cardona R.M.B. (2012). Reinterpretación de los derechos de intimidad y secreto de las comunicaciones en el modelo constitucional de relaciones laborales: un paso atrás. Comentario a la STC 241/2012, de 17 de septiembre. *Revista de Derecho Social, 60,* 169–180.

Senden, H. (2011). *Interpretation of fundamental rights in a multilevel legal system. An analysis of the European Court of Human Rights and the Court of Justice of the European Union.* Cambridge: Intersentia.

Part III
Journalism and Cyberjournalism

Chapter 12
Press Photography and Right to Privacy

Esperanza Pouso Torres

Abstract The present paper offers an analysis of the visual coverage of the deaths of former Libyan leader Muammar Gaddafi and Spanish police inspector Jesus Garcia given by three Spanish newspapers: *ABC*, *El Mundo* and *La Vanguardia*. The study will be the starting point from which additional ideas will be provided to infer conclusions regarding the respect for the victims and their legal protection, as well as to how a predominantly visual culture driven by the expansion and consolidation of the Internet is fostering the increase in violations of the right to privacy by the press.

Keywords Privacy · Public image · Journalism · Internet · Law

12.1 Introduction

On 19 January 2000, police inspector Jesus Garcia died of a heart attack while testifying at the hearing previous to the trial of the Lasa and Zabala case. Years later, on 20 October 2011, Muammar Gaddafi, Libya's de facto leader for more than four decades, was killed by rebel forces.

The number of ethical conflicts arising from the clash between the right to information and the right to privacy is increasingly growing in present journalism—particularly in connection to the use of pictures depicting private areas of life. Thus, when newspapers show on their pages such intimate events, as somebody's death, the debate on the ethical code of journalism is re-opened.

This is in partly due to the fact that in the last decades the expansion and consolidation of the Internet helped to develop a predominantly visual culture on a global scale. This situation posed a significant challenge to the press, conditioned by a society saturated with pictures and readers with changing habits.

E. Pouso Torres (✉)
University of Vigo (UVigo), Vigo, Spain
e-mail: epouso@uvigo.es

© Springer International Publishing Switzerland 2017   97
F.C. Freire et al. (eds.), *Media and Metamedia Management*,
Advances in Intelligent Systems and Computing 503,
DOI 10.1007/978-3-319-46068-0_12

12.2 Theoretical Basis

12.2.1 Privacy as an Inherent and Fundamental Right a Person Has

The Spanish Constitution (SC henceforth) recognises in its Article 20.1 d) the right to freely communicate or receive accurate information as a fundamental right, i.e. as an inalienable right of Spanish citizens, especially protected from further legislative action. However, despite sheltering this freedom, the very same article, states on its fourth section the following: "These freedoms are limited by respect for the rights recognised in this Title, by the legal provisions implementing it, and especially by the right to honour, to privacy, to personal reputation and to the protection of youth and childhood" (1978: 13).

Freedom of press is therefore limited by the rest of fundamental rights, and particularly by the so-called personality rights, enshrined in SC's Article 18.1: "The right to honour, to personal and family privacy and to the own image is guaranteed" (1978: 12).

Therefore, the right to information has to be perfectly compatible with the protection of individual privacy, which is one of the clearest limitations to freedom of speech. Displaying photographs of somebody's corpse violates human dignity, regardless of whether the deceased is classified as an ordinary or popular person.

12.2.2 Legislative Development of the Right to Privacy

Organic Law 1/1982 (5 May), which was issued to develop Article 18.1 of the SC, aims at providing civil protection of the right to honour, personal and family privacy and identity. Such protection is dispensed against unlawful interference of third parties in the field of fundamental rights belonging to the category of personality rights.

Remarkably, in its seventh Article this law provides a detailed enumeration of possible illegal interferences to the right to privacy. Amongst them, it lists the act of recording, exploiting or publishing somebody's identity by means of a photographic, filming or any other device, regardless of whether the occasion is private or not, except in the cases provided for in Article eight, point two of the same law.

12.2.3 Self-regulatory Documents

Most newspapers currently follow their own style guide. These books include standards for document design and writing, information on linguistic and professional issues, and sections devoted to ethical and deontological aspects. Style guides by Vocento, *ABC* and *El Mundo*, among others, include references to both photographic information and the invasion of privacy.

Likewise, Spanish journalist associations, such as the Federation of Associations of Journalists of Spain's General Assembly and the Union of Journalists of Madrid state in their guides the principles and deontological rules of the journalistic task. These works aim at getting information journalists and communicators committed to society and thus achieve—both at an individual and collective level—an impeccable conduct able to connect ethics to information. All these manuals make a specific reference to the individual right to privacy, especially when somebody is in a distressing or a painful situation.

12.2.4 The Rise of Photograph in the Printed Media

Digital press arrived as a new competitor forcing printed photojournalism into a serious transformation of the way in which information was displayed and how visuals were exploited until then. As a consequence, every newspaper is nowadays designed not only to be read but also to be seen.

That is the reason why photography gradually ceased to be a mere ornament used in order to avoid the monotony of the printed pages, to become an essential. As Keene (1993: 193) points out: Newspapers use pictures to illustrate stories, as items of interest in their own right, and as design features.

There is a widespread logic argumentation putting forward that most readers call for lurid images, which, in turn, justifies their use. Sontag claims that many of us have "the wish to see something gruesome" (2003: 75). Sadly, "modern life consists of a diet of horrors by which we are corrupted and to which we gradually become habituated" (Sontag 2003: 82). Mass media feed us with daily reports of atrocities from around the world, which often include disgusting images; either filmed ones on TV or photographed on the written press. This informative overflow prevents viewers and readers from telling the important news from the anecdotic ones, and this way, the news consumers are progressively rendered insensitive.

12.3 Methodology and Subject of Research

To carry out the aforementioned analysis, this essay will profit from a methodology typical of the social sciences: content analysis. In this case, it will be obviously focused on photographic content. In so doing, six basic parameters will be taken as reference: three connected to content and three to form. The main elements to take into account when analysing the content of a given picture are the main figure in it, its setting and the elements it shows—that is, its pictorial essence. A detailed examination the picture will be attempted in order to depict the context in which it was taken as clearly, simply and concisely as possible. As to the formal analysis, colour and size of the photo as well as its location within the newspaper (section and page) will be analysed. This thorough study will reveal the importance that the publication gives to that particular picture.

This model of analysis arises from the collection of methodological approaches done by Alonso and Matilla (1990) and Barrett (1990). In order to classify the photographic message as information or opinion and, also, for the analysis of the photograph and its surroundings, Abreu (2004) mentions the ideas of the previous authors in an article on the qualitative analysis of photography in the press.

As it has already been mentioned, the amount of lurid pictures we are exposed to on a daily basis is enormous. It becomes then necessary to narrow the field of study. This essay will focus on three national newspapers: *El Mundo*, *ABC* and *La Vanguardia*; and research will be done taking copies of the printed editions of each of them issued from the 15th to the 17th of January 2001 and form 21st to 23rd October 2011. Front and back pages, as well as the full articles dealing with these pieces of news—a total of 18 pages—will be carefully analysed so as to see how the deaths of Gaddafi and Jesus Garcia were represented photographically by the media. From this analysis conclusions will be drawn, paying special attention to whether the victims are respectfully treated.

12.4 Photographic Analysis and Results

Words addressing issues such as the ones here discussed are usually simple, clear and concise. However, after analysing the coverage these deaths were given, it can be said that nothing compares to the straightforwardness of pictures on these newspapers' front pages showing Gaddafi's lifeless body and Jesus Garcia's heart attack.

On 20 October 2011 all three newspapers placed dead Gaddafi on their opening pages. *La Vanguardia*'s photo shows the hands of several unidentified Libyans taking pictures of the dictator's corpse, whose torso is bare and has traces of blood. It is a four-column-wide American shot of Gaddafi, which is placed at the bottom of the front page. *El Mundo* chose the same snapshot, but in this case, only Gaddafi can be seen; again, it is a colour image that now only shows a medium shot of the tyrant. It is a column and a half wide and located at the bottom, too. On the other hand, *ABC* decided to use a picture of the dictator's bloodied corpse of the after being shot; it is a medium shot, in colour, a column and a half wide, placed once again at the bottom of the cover.

In short, all three newspapers issued the same or very similar pictures of the event, which proves that there is a journalistic routine in the criterion for image selection. Moreover, having into account the fact that these photos are extremely disturbing, it can be deduced that such criterion is based on the hypothesis that the crueller and the more lurid the picture is, the bigger impact on readers is achieved. This way, all three newspapers also chose to cover the event with pictures of Gaddafi's last moments of life after being found by the rebels. Wounded, his face and shirt stained with blood, surrounded by soldiers of the new government or with a rebel pointing a gun to the head of the Libyan ex-leader. Similarly, one of the most striking images is the lifeless body of the dictator displayed in a warehouse in Misrata—included by all the three newspapers—sometimes more than once, as it is the case of *El Mundo* and *ABC*.

As for the photographs depicting the death of Spanish police inspector Jesús García, the most dramatic one shows the moment in which he loses consciousness because of a heart attack. There is an equally though snapshot where he is being revived by somebody who partially covers Garcia's face. Both photographs were published by *El Mundo* and *La Vanguardia*; the former on their front pages and the latter in their inner sections. *El Mundo* issued a three-column-wide photo on its cover, whereas *La Vanguardia* showed a smaller one (two-column wide). In both cases, it is a colour medium shot set at the top of the page. The photos on the inside are black-and-white three-column-wide medium shots of Garcia, and are displayed at the top of the page. On the other hand, *ABC* chose not to publish this image on its front page, making referent to the event with a simple text line. Nevertheless, it included the appalling heart attack on the inside pages; in a black-and-white, four-column-wide and medium shot format at the top half of the page. The reason for this choice is explained in the caption: "Due to the brutality of the photo, we chose to display it on this page and not as the main photograph on our opening page" (translation mine). In addition to these photographs, *El Mundo* and *ABC* published in one of their following issues an image of the inspector Garcia's burial.

12.5 Conclusions

Once the analysis that of the treatment given to the images illustrating the reports of Muammar Gaddafi's and Jesus Garcia's deaths is completed, several conclusions can be drawn. In the first place, *ABC*, *La Vanguardia* and *El Mundo* included in their aforementioned issues photos of the deceased, violating the right to privacy of the victims at a time when their image was especially vulnerable. Ethically and deontological speaking, there are some pieces of reality that cannot be publically released. It could be argued that the right to privacy must be observed even beyond the person's death.

Through these images readers "felt" and "lived" Gaddafi's and García's sorrow and pain. This notwithstanding, it has to be considered that these are personal experiences, and belong to the private life of the sufferer. Thus, an ethical conflict arises between the picture publishing and the right to privacy.

In this situation, it can also be argued that the event is newsworthy because of its what originated it, the person it shows, its context, the consequences it had, etc. But the journalistic interest that a piece of news may have does not entail that the image of a lifeless body—be it in a store in Misrata or on a chair at the High Court—may have the same interest. Some pictures simply should be kept in a drawer and never see the light.

As far as legislation is concerned, it is true that the photos under analysis show people with different degrees of popularity: ordinary citizens and people with a high rank status are not *sub especie iuris* in the same situation. Nevertheless, the fact of being a public figure, even when voluntarily, cannot mean a complete renunciation of privacy. Spanish legal order cannot allow, not even when the subject spontaneously decides so, their renunciation or free disposition of the personality rights. Nobody questions the existence of limits posed by the very right to privacy.

Differences in treatment of the images here analysed are obvious, as in Gadafi's case the amount of them is bigger and the coverage of the topic more detailed. This is partly because he was more popular than the police inspector, but also because of the time elapsed between the two pieces of news: Garcia's death one took place in 2000 and Gaddafi's 11 years later. With the rise of the Internet and the emergence of new media–and, therefore, new rivalry for success—the press was forced to modernize its formats and to use more visuals than before. Over the years, photography has become an essential for the printed media: a powerful element that transmits information straightforwardly thanks to its inherent characteristics.

Disgraceful images can sometimes hurt people's feelings, although their growing number in the mass media can also result in the trivialisation of the feeling, generating the opposite effect on a public already saturated by terrible pictures.

After this study, it can be concluded that citizens have the right right to be objectively informed, just like the victims have the right to privacy. And the connection between both rights is the role played by the media, which must act responsibly and humanly when dealing with such delicate situations as the ones discussed in this essay.

References

Books

Alonso, M. Y., & Matilla, L. (1990). *Imágenes en acción*. Madrid: Ediciones Akal.

Barret, T. (1990). *Criticizing photographs. An introduction to understanding images*. United States: The Ohio State University, Mayfield Publishing Company.

Keene, M. (1993). *Practical photojournalism, a professional guide*. Waltham, Massachusetts (USA): Focal Press.

Martínez De Sousa, J. (2003). *Libro de Estilo Vocento*. Gijón: Trea.

Sontag, S. (2003). *Regarding the pain of others*. New York: Picador.

Unidad Editorial. (1996). *Libro de Estilo*. Madrid: Temas de Hoy.

Vigara Tauste, A. M. (2001). *Libro de Estilo de ABC*. Barcelona: Ariel

Electronic Media

Abreu, C. (2004). El análisis cualitativo de la foto de prensa. *Revista Latina de Comunicación Social, 57(8)* (September, 2015). Retrieved from: http://www.ull.es/publicaciones/latina/20040757abreu.htm

Colegio de Periodistas de la Región de Murcia. *Código Deontológico.* Retrieved October 5, 2015 from: http://periodistasrm.es/codigo-deontologico/

Col·legi de Periodistes de Catalunya. *Declaración de principios de la profesión periodística en Catalunya.* Retrieved from: http://www.xornalistas.com/imxd/noticias/doc/1229538920codigo collegi.pdf. Accessed at October 5, 2015.

Colexio Profesional de Xornalistas de Galicia. *Código Deontolóxico do Xornalismo Galego.* Retrieved from: http://www.xornalistas.com/colexio/interior.php?txt=m_codigo&lg=gal. Accessed at October 5, 2015

Federación de Asociaciones de Periodistas de España. Código *Deontológico.* Retrieved from: http://fape.es/home/codigo-deontologico/. Accessed at October 5, 2015.

Organic Law 1/1982, of 5 May, Civil Protection of the right to honor, personal and family privacy and self-image. BOE, May 14, 1982, no. 115, p. 12546 to 12548. Retrieved from: http://www.boe.es/buscar/doc.php?id=BOE-A-1982-11196. Accessed at October 3, 2015

Sindicato de Periodistas de Madrid. *Código Deontológico.* Retrieved from: http://www.xornalistas.com/imxd/noticias/doc/1229539030codigosindimadrid.pdf. Accessed at October 5, 2015.

Spanish Constitution. (1978). BOE. Retrieved from: https://www.boe.es/legislacion/documentos/ConstitucionINGLES.pdf. Accessed at February 13, 2016.

Chapter 13
Internet and Social Media in the Prevention Journalism Discourse. A Theoretical Proposal and Main Magnitudes

Silvia Alende Castro and Aurora García González

Abstract In the current metamedia context, there is a need of reflection on journalistic thoughts that display content and discourses for social change. In this regard, Prevention Journalism (PJ) is a trend that, in a pioneering way, has the leading role in a Spanish research on the risk management from the communication field. This study introduces a theoretical basis of this journalistic model and quantifies its presence in the daily press and its main magnitudes. Methodologically, this research —in which were reviewed more than 30.000 informative units extracted from 406 copies of a ample of 12 newspapers—employs a mixed methodology: the content analysis and the HJ Biplot multivariate analysis. Among the key conclusions, we can confirm that the risks linked to new technologies, internet and social media are not still part of the Prevention Journalism discourse.

Keywords Journalism · Prevention journalism · Internet · Anticipation · Recent communicative movement

13.1 Introduction

As a journalistic movement linked to useful communication for the citizen day-to-day, Prevention Journalism[1] (PJ) finds itself under the Social Action Theory, since the object is understood as an action and, consequently, the communication

---

[1]This paper comes from the doctoral thesis Alende (2015). *Periodismo de Prevención en Galicia. El concepto de comunicación útil en la prensa diaria,* directed by Ph. D. Aurora García González and Ph. D. José David Urchaga Litago. University of Vigo.

S. Alende Castro (✉) · A. García González
University of Vigo (UVigo), Vigo, Spain
e-mail: silvia.alende@gmail.com

A. García González
e-mail: auroragg@uvigo.es

© Springer International Publishing Switzerland 2017
F.C. Freire et al. (eds.), *Media and Metamedia Management,*
Advances in Intelligent Systems and Computing 503,
DOI 10.1007/978-3-319-46068-0_13

science is integrated automatically in the field of the Social Action General Science (Román 2000: 120). The Social Action Theory has among its referents the Austrian sociologist and philosopher Alfred Schütz, studied by Martín Algarra (1993). The shift from traditional to modern societies has been analysed by different movements and authors. However, since the 80s of the past century, a sociological perspective focused on modernity has arouse (Alfie and Méndez 2000: 174). As Beck notes (1998: 14), "in the developed modernity [...] a new assigned use for *danger* appears, from which there is no way to escape" (1998: 12); in other words, it is the so-called risk society. In this context, the role of the media has been a continuous concern. The social aim of the media can be summarized as the responsibility of these companies to look for the social benefit rather than its own economic profit (Tallón 1992: 19). Accomplishing this social purpose is precisely where emerges the obligation of the media to act in its own surroundings, fulfilling control and security functions through the utility information (Diezhandino 1994: 24). Among the different ways to inform appeared in the 20th century (López 2012), there is the Prevention Journalism, linked to the Preventive Journalism and International Analysis Institute (PJIAI) in Spain (Bernabé 2007). It is necessary to take into account that, unlike the PJIAI proposals, which are centred in the international field, the PJ is focused on the local environment.

13.1.1   A Theoretical Proposal on the Prevention Journalism

From an ethical point of view, helping as possible to solve potential unpleasant circumstances becomes an informative activity with evident repercussions (Diezhandino et al. 2002: 9). Prevention Journalism aims at helping citizens to anticipate potential dangers, conflicts and unsteadiness situations in their close surroundings. That could be understood as one of the current challenges in journalism (Diezhandino 2012: 49). The author sustains that its added value links with the capacity to guide, orient and do surveys. Secondly, the anticipation implies taking into account those events that, although they seem isolated or unconnected, can trigger a conflict. Therefore, the most important risk management is that made in advance (Obregón et al. 2010: 110). This is in relation with the forward-looking ability of the media, formulated by Wright (1972: 19), which implies addressing issues related to the informative approach, since the PJ demands reviewing classical journalistic concepts, such as 'currently'. PJ is in relation with which Alberdi, Armentia, Caminos and Marín define as permanent present (2002: 61). Definitely, PJ could be defined as the journalistic practice that, through a conscious attitude taken by the media and journalists, looks for approaching readers´ reality, disseminating useful information that would enable a suitable management of risks in their close surroundings.

13.2 Methodology

This study seeks to know whether Galician daily press practises the Prevention Journalism. As a second purpose, this paper looks for highlighting the features, functions, and peculiarities of this journalistic trend. Thus, three hypotheses and a series of exploratory questions were considered.

Hypothesis 1: Prevention Journalism has a lack of representation in Galician daily media. This section emphasises the quantitative part of this paper. We try to study the presence of this journalistic perspective, as well as how many inclusion criteria are published.

Hypothesis 2: Thematically, new technologies, the Internet and social media are apart from prevention journalistic speech. In this case, we look for analysing the thematic content of the PJ, that is to say, the elements that compose its speech.

Hypothesis 3: Prevention Journalism is more associated to the agenda setting than to the media preventive attitude. This third assumption stresses the need of stablishing the reasons why this information is disclosed, particularly the interest that media shows to this type of news, analysing the motivation of publications.

To tackle these research questions, we chose the content analysis, a "specific" research approach "which is frequently employed in all the areas related with the media" (Wimmer and Dominick 1996: 169). The following newspapers conform the study sample: *El Correo Gallego, La Voz de Galicia, Faro de Vigo, La Región, El Progreso, Diario de Pontevedra, Diario de Arousa, Diario de Ferrol, El Ideal Gallego-Diario de Bergantiños, Atlántico Diario* and *La Opinión de A Coruña.* In relation with the timeframe, these newspapers were analysed between November 2012 and December 2013. Once we determined the material of analysis, the research gave way to the identification of those informative units that could link with Prevention Journalism and, as a result of cataloguing these units, it was composed the definitive study universe to which the content analysis was applied. To differentiate in each newspaper what was PJ from what was not, we took into account specific identification criteria. In this sense, PJ are those contents that: (A) They derived from a clear intention of the journalist of taking position regarding to the audience welfare. (B) They influenced using a survival or control perspective, in front of risk elements. (C) They valued the existence of protocols and its suitability. (D) They attracted attention on issues that used to cause risk situations. (E) In recent news, content is not reduced to tell facts, but rather helps to understand the situation proposing useful solutions. (F) Guides: they offered indications. (G) They looked for, in a clear and unambiguous way, anticipating, warning, alerting or positioning the reader. Once these criteria were applied in the 406 copies reviewed—implying 30.000 informative units, there were identified 150 informative units that matched the selected criteria. In order to study in detail this sample, we employed the content analysis, developing a code-book composed by the following variables: inclusion, thematic and motivation criteria.

13.3   Results

Results derived from this research are presented attending to the order of hypothesis posted on the methodology.

Hypothesis 1: Prevention Journalism has a lack of representation in Galician daily media. To test this hypothesis, the PJ units (150) have been compared with the total informative units. Altogether there were 31.545 informative units, so that the PJ units that we found were very low (150) compared with the total (31.545) or with the selected copies (406). Therefore, this hypothesis is confirmed. Also, it was established as a research problem the study of inclusion criteria for each unit. Among them, the B criteria (Normality monitoring) was the most present requirement (35.3 %) in the analysed units, followed by the G criteria (anticipate and position the reader) with 25.8 %, and finally the C criteria (protocol compliance rating: 17.2 %). Those were the three more usual criteria in the prevention informative units, in which there were others in lesser extent, as contributing solutions to a concrete fact (E: 6.6 %), warning of possible accidents (D: 6.1 %), using as a guide (F: 5 %) or these news were results of the evident spirit of the journalist for warning (A: 4 %).

Hypothesis 2: Thematically, new technologies, the Internet and social media are apart from prevention journalistic speech. This second section draws special attention to the topic of publications. It is noted that, despite its penetration among the Spanish society—76.2 % of people are internet users (INE 2015)—the new technologies, the Internet and social media are not part of the prevention journalistic discourse, as any prevention informative unit was found (Table 13.1).

Hypothesis 3: The Prevention Journalism practice is more associated to the agenda setting than to the media preventive attitude. This third hypothesis is about the reasons why this information is published. We could check that the present time is the criteria that determined the inclusion in 86 % of the cases. Therefore, it confirms the hypothesis that PJ arises more from *hot news* than from the media or journalist intention for warning. In this point, it seemed interesting, as a methodological contribution, the application of a multivariate statistical procedure which allowed showing simultaneously the relations between the studied variables on a chart (Chart 13.1). For this, we employed a HJ BIPLOT thanks to the Multbiplot Program (Vicente 2010). The following variables were selected: Section (local and society), Addressees (collective), Genre (news and interview), Page (cover), Headline, Signature, Sources, Usefulness (third degree) and used criteria. The resulting *biplot* shows the link between the different variables, and represents graphically the Prevention Journalism reality in the Galician daily press.

Table 13.1 Thematic: frequency and percentage

Thematic	Frequency (n)	Percentage (%)
Health	55	36.7
Personal security/job	32	21.3
Infrastructures/public services	22	14.7
Environmental	14	9.3
Economic/financial	8	5.3
Domestic environment/familiar	6	4.0
Physical environment/weather	5	3.3
Maritime	5	3.3
Agricultural/stock	3	2.0
Sports	0	0.0
New technologies	0	0.0
Another	0	0.0
Total (Σ)	150	100

Source Own elaboration

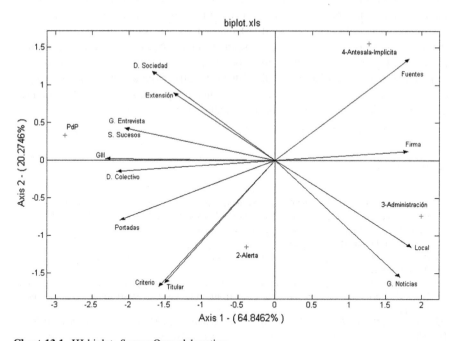

Chart 13.1 HJ biplot. *Source* Own elaboration

13.4 Discussion and Conclusions

Once we corroborated the proposed hypotheses, we could confirm that the PJ still has little presence in Galician daily press. Finding informative units that answer to prevention criteria in these newspapers has been a complex task that required reading almost three copies (271 copies exactly) to encounter this kind of content. Likewise, the few PJ cases found follow a limited number of prevention criteria. At the same time, these units answer more to general parameters focused in surveillance or control, than to a conscious aim of helping the reader to anticipate in risk situations or with guides. Therefore, we noticed that PJ attends to a generic purpose of the surrounding control and warning the reader. However, the PJ is out of seeking solutions among currently facts–in the case of events, for example, of proposing guides or offering information about practice guidelines to the user.

From a thematic point of view, we observe that, despite of their importance and novelty in the current society, new technologies, the Internet and social media are not part of the prevention journalistic speech. It is a type of journalism focused, fundamentally, on health and personal and labour security issues. Besides, the lacking PJ practiced in Galicia arises from currently reasons. They are exceptional the prevention informative units derived from a conscious attitude of the media or the journalist to anticipate potential risks situations. Therefore, the PJ still suffers from the conscious attitude of professionals and the media necessary for this journalism tendency from a theoretical point of view, which demands anticipation to potential risks as a defining attribute itself. Finally, from a methodological point of view, the quantitative and qualitative combination perspective as well as the use of the HJ Biplot show its utility in the content analysis.

References

Alende Castro, S. (2015). Periodismo de Prevención en Galicia. El concepto de comunicación útil en la prensa diaria. Tesis Doctoral. Dirigida por la Dra. Aurora García González y el Dr. José David Urchaga Litago. Universidad de Vigo. Available from http://www.investigo.biblioteca.uvigo.es/xmlui/handle/11093/350

Alberdi, A., Armentia, J. I., Caminos, J. M., & Marín, F. (2002). *El diario de servicios en España*. Oviedo: Septem.

Alfie, M., & Mendez, L. H. (2000). La sociedad del riesgo: amenaza y promesa. *Sociológica, 43*, 173–201. Recovered on January 11th, 2015 from http://www.revistasociologica.com.mx/pdf/4308.pdf

Beck, U. (1998/Trad.). *La sociedad del Riesgo. Hacia una nueva Modernidad*. Barcelona: Paidós.

Bernabé, J. (2007). *Periodismo preventivo. Otra manera de informar sobre las crisis y los conflictos internacionales*. Madrid: Los Libros de la Catarata.

Diezhandino, P. (2012-coord.). *El periodista en la encrucijada*. Madrid: Fundación Telefónica. Recovered on February 18th, 2015 from http://goo.gl/YvVVsQ

Diezhandino, P. (1994). *El Quehacer informativo: el "arte de escribir" un texto periodístico: algunas nociones válidas para periodistas*. Bilbao: Servicio Editorial Universidad del País Vasco.

Diezhandino, P., Marinas, J. M., & Watt, N. (2002). *Ética de la comunicación, problemas y recursos*. Madrid: Edipo.

INE (2015). Población que usa Internet (en los últimos tres meses). Recovered on January 15th, 2015 from http://goo.gl/ikN2D6

López, X. (2012). *Movimientos periodísticos. Las múltiples iniciativas profesionales y ciudadanas para salvar los elementos básicos del periodismo en la era digital*. Salamanca: Comunicación Social, Ediciones y Publicaciones.

Martín Algarra, M. (1993). *La comunicación en la vida cotidiana*. Eunsa: La fenomenología de Alfred Schütz. Pamplona.

Obregón, R., Arroyave, J., & Barrios, M. M. (2010). Periodismo y comunicación para la gestión de riesgo en la subregión andina: discursos periodísticos y perspectivas para un enfoque prospectivo y preventivo. *Folios, 23*, 105–135. Recovered on August 8th, 2012 from http://goo.gl/CG9FI8

Román, M. (2000). Aspectos metodológicos de la historia de la comunicación. *Ámbitos, 5*, 119–128. Recovered on September 13th, 2014 from http://goo.gl/BkGWTs

Tallón, J. (1992). *Lecciones de empresa informativa*. Madrid: Ediciones Ciencias Sociales.

Vicente Villardón, J. L. (2010). *MULTBIPLOT: A package for Multivariate Analysis using Biplots. Departamento de Estadística*. Universidad de Salamanca. Retrieved on December 11th, 2014 from http://biplot.usal.es/ClassicalBiplot/index.html

Wimmer, R., & Dominik, J. R. (1996). *La investigación científica de los medios de comunicación. Una introducción a sus métodos*. Barcelona: Bosch.

Wright, R. (1972). *Comunicación de masas*. Buenos Aires: Paidós.

Chapter 14
Immersive Journalism: From Audience to First-Person Experience of News

Sara Pérez Seijo

Abstract This research is focused on checking the theory about how the application of virtual reality techniques in the audiovisual pieces alters the spectator level of implication with regard to information. To demonstrate, the author revises the academic theory and analyses several audiovisual pieces made with virtual reality. According to the study, journalism trends point to the audiovisual landscape, and they do it with the new technologies on the market. Innovation laboratories of the main European public televisions want to apply virtual reality techniques to their own audiovisual pieces, in order to make information closer to the spectator. Therefore, these new products allow the viewers being an actor of the information or recreated reality thanks to the high immersion level.

Keywords Immersive journalism · Immersion · Virtual reality · Empathy · Recreated scenarios

14.1 Theoretical Framework

The application of VR techniques and the implementation of new immersive formats in audiovisual products has become an increasing trend in the media market. The methods initially tested and used in video games to offer users an ever-closer experience to reality are being introduced in the media. The public demands for greater objectivity and a closer touch about the reality of what has happened, and producers seek to fulfill that demand by providing closer contents about the events they report. The aim is immersion, to take the user to the scene, the exact time, the event itself. And for that, immersive and virtual tools are used which make possible to offer the audience a real experience.

Currently Nonny de la Peña is considered to be the godmother of VR and is recognized for her work in the field of immersive journalism. However, the

S. Pérez Seijo (✉)
University of Santiago de Compostela (USC), Santiago, Spain
e-mail: sara.entienza@hotmail.com

© Springer International Publishing Switzerland 2017
F.C. Freire et al. (eds.), *Media and Metamedia Management*,
Advances in Intelligent Systems and Computing 503,
DOI 10.1007/978-3-319-46068-0_14

potential of immersive narratives in the news scene has been the subject of study since the nineties, especially in the American academia. The evolving experiences and possibilities that VR put on the table at that time raised the curiosity of many researchers that decided to focus their efforts and time to study the potential impact in journalism. In response to this new emerging technology, universities such as Columbia, through the Center for New Media, tried to carry out the first hands-on experiments. One of these trials took place in 1997 during the celebration of St. Patrick's Day in New York City. The students used 360° video to capture the demonstrations of gay community organizations and their subsequent police detention. These experiments were not only based on current events at that moment, but Columbia University also experimented with the recreation of past events. One of these tests was the reproduction of the student revolt of 1968 in the university campus, through augmented reality technologies.

Faced with these new techniques that seem to allow a higher degree of engagement in the stories, the University of Minnesota in 2000 and 2001 organized two seminars through the Institute for New Media Studies. The first one was Playing the news: Journalism, interactive and narrative games, an event focused on lecturing about the possible influence of the video games logic in the creation of press reports. One year later, the University of Minnesota held Sensing the news: What Could the new technologies mean for journalism in order to remark the potential of the latest technology and its ability to offer an sensory immersion experience never-before-seen.

Virtual and augmented reality provide new possibilities to journalism. However, the implementation has been slow. The problem is hold against the high costs of such technologies and the resulting required supplies as well as the need for specialized professionals in handling, creation, production and implementation of these tools and techniques in journalism. Despite this, nowadays several media try to incorporate these new possibilities in order to provide the audience with immersive experiences so they can inform them of a more comprehensive and efficient through a first-person interaction way. This is what public television such as the BBC or France TV, or traditional media like The New York Times are trying to strive for.

This is where the researcher De la Peña has tried to join with the leading group at the University of South Carolina, Immersive Journalism. By the implementation of techniques and tools of VR, De la Peña has carried out several immersive projects on real events. The flagship of her work has been the so-called Project Syria, held in 2014. This production is a 360° recording which allows the user immersion by using VR goggles. Probably the user has never been there in the middle of the war, nor have seen any of the people that has been able to observe via that "fourth screen" (Domínguez 2015: 415). However, to some extent how has it felt to be in middle of the conflict. So far the events have reached the audience through news or documentary video, but the 360° recording techniques are able to take the receiver to the main scene like never before. VR, either from the hand of journalism or video games, allows, in the words of the multinational Sony, the opportunity to experience a sense of transforming presence with enough capacity to trick the mind into believing that the virtual environment is real.

14.2 Methodology

This paper is going to study the consequences that the VR techniques have in the degree of the receptors' interaction, based on a theoretical revision of the academic researches about this topic which have been done up to this point, highlighting the more current ones. Both studies about the application of the VR to journalism, and psychological analyses have been taken into consideration in order to investigate users' immersion level. Furthermore, recordings about how people react when testing VR products have been viewed, together with testing experiences of immersion in PC version of *Ferguson* (reproduces the Ferguson gunfire, in Missouri), and *The Suite Life* (immerses the user into luxurious spaces).

14.3 VR in Journalism

According to *Sony*, 2016 has already gained some nicknames, such as the "Year 0" of VR. The technological and audiovisual industry refers to this year as the moment when VR will fully break out into the mass market. Although the video game industry has been testing and experimenting with these techniques in their entertainment products, it seems that for the first time this technology will jump to many more areas. One of them is journalism. European public broadcasters such as the *BBC* or *France Télévisions* group are already implementing on their audiovisual creations the possibilities offered by VR. However, it is still an inexperienced market.

Thanks to *Oculus Rift* glasses—which Facebook bought in March 2014, to the *Samsumg Gear VR* or even the *Google Cardboards*, the VR technology is becoming more accessible to everyone. Little by little it has being incorporated to the market and it is breaking into journalism, a scenario in which it has a lot of sense. Why? According to Chris Milk, musical director and one of the current model of production concerning VR, "journalism tries to move people to an event that they were unable to attend" (Constine, Feb. 1, 2015). Therefore, VR aims to bring much closer than it has been done so far the event that the public is being informed. This is not being done by a multimedia or textual article, but by transporting the user to the event thanks to a reconstruction of the incident and a pair of VR glasses.

Film director Thomas Wallner once said that the application of these techniques to journalism is leading to "a kind of media that has not existed until now" (*Métamedia*, spring-summer 2015), since never before the spectator was so close to a narrated event. Thus has been proven by people like Nonny de la Peña with her productions, for instance *Project Syria* or *Gone Gitmo*. The last mentioned project allows the subject, by means of gloves and VR glasses, 'being in the digital shoes of a Guantanamo's prisoner that is listening to the interrogation to which someone else at the other side of the wall is being submitted' (Domínguez 2015: 415). De la Peña and the crew *Immersive Journalism* have got to reproduce, through the platform *Second Life*, a real scene thanks to the data extracted from diverse public references. *Gone Gitmo* is therefore an immersive journalistic experience that allows the individual to be translated to the facilities of Guantánamo Bay's prison. Thanks to the technology of VR employed in the recreation, their producers assured that 'the participants had the illusion of being in that place' (De la Peña et al. 2010: 297).

If a person is observed while wearing a pair of VR goggles, probably their movements will not be understood. However, for the user they will make sense in that fourth wall because "the actions taken have implications in VR" (De la Peña et al. 2010: 297), such as standing in front of a mirror and being reflected. The participant plunges totally into the story thanks to an immersive experience that, added to "his first-person perspective inside the virtual body together with the synchronization between the movements of the head and his breathing, may lead to a situation in which the participant feels as the owner of the virtual body" (De la Peña et al. 2010: 298).

In this context, the so-called immersive journalism is growing and maturing. De la Peña et al. (2010) has described it as a way of producing news in a format that allows people to get a first-hand experience of certain news events. By saying this, he means that the user has the possibility of interacting in a secondary reality (recreation) that recreates a primary reality (news events) in a digital avatar quality, so the participant will be able to try a fully immersive experience when jumping into the event, thanks to a pair of VR goggles.

14.4 Ethical Issues of Productions with VR

As a result of the VR pieces that recreate certain events, there may be many ethical problems that could question the veracity and honesty with whom they have been made. Kent (2015), a professor at Columbia University, raised this question: Is the journalism produced with VR thought to be an event in itself, to be an artistic concept of the event, or to be something similar to a historical novel "based on real events?". This dilemma lays on the table a variety of ethical issues that can arise with this new form of narration offered by the immersive journalism.

The main problem appears as a result of elaborating productions with VR from 2D pieces. Those images that only have a 2D view are flat, that is, it is totally impossible to discover other views of the object or subject of the scene that have not being captured by the camera at the time when they were photographed—except that they can be shooting again, obviously. These productions have limited angles or points of view of the object or subject, so various ethical dilemmas arise when recreating them. If we do not have other angles, should they have been recreated likewise using the common sense of the imagination? Should we blur those pieces in the 3D representation as a symbol of inaccuracy? When reproducing a 2D piece to change it into a 3D one, producers should make many important decisions such as the ones exposed. Imagining the back of a print vase, the lateral form of a figure, a person's face silhouette or the subject to reenact… a serie of things without all the necessary angles to reproduce. So, is it really ethical to reenact something from whom we do not have all the information?

In order to offer a solution to this difficulty that professionals can face, *The Online News Association* created a section in its webpage called *Build Your own Ethics Code Project*. It is about an space whose purpose is to reach an agreement between the ethic rules of a project that is going to be elaborated with VR techniques and everyone taking part in its production. Although this has been a first step, it is only a temporary solution before the problem of the ethic regulation of 3D recreations, as this technology still has a long way to go until a complete and official ethic code is consolidated.

Apart from the 2D-3D difficulties, there other problems linked to the digital elimination of objects in the scene. *Associated Press* launched in August 2015a, b a VR project whose aim was to explore diverse luxury spaces in first person. *The Suite Life*, available in the *Oculus* Shop for the glasses *Samsung Gear VR*, transfers the user to places that only a few pockets can pay, for instance the luxurious suit *Ty Warner Penthouse* in the New Yorker hotel *Four Seasons*- Every single scene was photographed with 3D cameras from *Matterport* to be recreated 'a posteriori' by means of VR.

The pieces elaborated in *The Suite Life* can boast about an excellent reconstruction quality, but in one of the spaces the camera used to photograph the place has slipped through. This happened in a hotel room full of mirrors, so it is understandable the difficulty that supposed to photograph the place avoiding the appearance of the camera in the images. In this way, in one of these multiple photographs, a camera on its tripod can be observed. Kent (2015) considered the possibility that a number of professionals would opt for deleting the camera digitally, since it is not part of the room and therefore only slows down the vision of the place. However, there is an inconvenient: *Associated Press* forbids the manipulation of its images. The conclusion is, the camera stays where it is.

14.4.1 The Empathy Machine

It is in this discussion about the ethical limits where the goal of the VR pieces emerges: to cause empathy. The VR aspires to come into the land of feelings. It has the intention to outsmart the human mind tricking it into believing that what he is seeing is real, and he can interact the scenario. However, this process is just the result of a recreation that views through the VR glasses that the user is wearing. Therefore, he is not there and he cannot touch what he is watching. Apart from this disadvantage, the VR allows the user to experience a perception of physical presence on a scenario "through an immersive system that allows the participant the access without precedents of the views and the sounds", and "the feelings and emotions" (De la Peña et al. 2010: 291). The feeling of immersion has the purpose of creating empathy in the user, but this is a goal which is away from the merely fact of telling a story (Kent 2015). Achieving that the receiver tests an emotion towards the journalistic story is not easy. Now, how far are we prepared to go in order to create empathy?

14.5 Conclusions

The immersive journalism is materialised in the production of news in a way that allows us to obtain a first person experience of events or situations described in informative reports. In this way, through the immersion in the event, a connexion between the receiver and the new is developed. The user, once he wears the glasses of VR, turns into a digital avatar which allows him to get fully into the story. The hyper-realism of the scenery impedes the mind of the user to realise that what he is seeing is a recreation of the reality (secondary reality), not the reality itself (principal reality). Unlike the named *newsgames*, the immersive journalism does not imply a progress of the receiver on the action. The user is a merely viewer, he has not to take any decision in order to progress in the story. Although since the moment when he can feel emotions and show instinctive corporal reflexes, the subject becomes an actor of the information. The feeling of presence experienced by the user is so strong that the human mind does not realise (distinguish) the fact that he is in a false reality. Consequently, the individual has the necessity of acting as if he was in the principal reality. Although the VR and the immersive journalism offer a wide range of possibilities never experienced until now, its practice causes multiple ethical discussions about how the principal reality is presented to the audience. Many journalistic organizations try to offer a solution to these problems but this format is still being consolidated. Therefore, at the moment it is just developing. The road to be followed is still very long and it is just beginning.

References

Associated Press. (2015). *Exploring "The Suite Life" of virtual reality.* Retrieved September 12, 2015, from http://www.ap.org/content/press-release/2015/exploring-the-suite-life-of-virtual-reality

Associated Press. (2015). *The suite life.* http://interactives.ap.org/2015/suite-life/

Constine, J. (2015). Virtual reality, the empathy machine. *TechCrunch.* Retrieved September 10, 2015, from http://techcrunch.com/2015/02/01/what-it-feels-like/

De La Peña, N., Weil, P., Llobera, J., Giannopoulos, E., Pomés, A., Spaniang, B., et al. (2010). Immersive journalism: Immersive virtual reality for the first-person experience of news. *Presence: Teleoperators and virtual enviroments, XIX*(4), 291–301.

Domínguez, E. (2015). Periodismo inmersivo o cómo la realidad virtual y el videojuego influyen en la interfaz e interactividad del relato de actualidad. *El profesional de la información, 24*(4), 413–423. Retrieved September 5, 2015, from http://www.elprofesionaldelainformacion.com/contenidos/2015/jul/08.html

Kent, T. (2015). *An ethical reality check for virtual reality journalism.* Retrieved September 6, 2015, from https://medium.com/@tjrkent/an-ethical-reality-check-for-virtual-reality-journalism-8e5230673507

Méta-media. (2015). La TV demain: 10 enjeux de transformation. Retrieved August 30, 2015, from http://meta-media.fr/meta-media/files/2015/06/MetaMediaFTV9_SCREEN-2.pdf

Chapter 15
Web-Native Media in Galicia. Trends and Characteristics of a Booming Model

María Cruz Negreira Rey and Xosé López García

Abstract The social and economic reality of recent years has led to a series of changes in the media, evolving into new organizational, productive and distributive models. In this new reality, the presence of online media is growing in relation to traditional media, prompting its emergence as an increasingly important site of communicative influence and citizen participation. In Galicia, where the first digital newspaper appeared two decades ago, it is not difficult to perceive a continuous growth of web-native media. Such media has been born as a space that seeks greater diversity and freedom of information, whether in terms of geographical space, editorial models, specialization or language. Moreover, it is in this type of online media where citizen journalism is most present, offering spaces for participation and allowing citizens to be the drivers of many digital web-natives. The objective of this paper is to study this media reality, the trends of its development and its particular characteristics. The research is based on an exploratory study involving the identification and location of web-native media. The relevance of such media throughout the Galician media ecosystem is analyzed by applying both quantitative and qualitative techniques, so as to identify the representative characteristics of web-natives.

Keywords Web-native media · Journalism · Communications · Online media · Galicia

M.C. Negreira Rey (✉) · X. López García
University of Santiago de Compostela (USC), Santiago
de Compostela, Spain
e-mail: cruz.nr@hotmail.com

X. López García
e-mail: xose.lopez.garcia@usc.es

© Springer International Publishing Switzerland 2017
F.C. Freire et al. (eds.), *Media and Metamedia Management*,
Advances in Intelligent Systems and Computing 503,
DOI 10.1007/978-3-319-46068-0_15

15.1 Introduction and Theoretical Approaches

15.1.1 Two Decades of Online Journalism and Media Transformation in Galicia

Social and economic changes over the past two decades have transformed the media industry in Spain. The popularization of the Internet and new ICT technologies has been key to the creation of the current journalistic ecosystem, whose actors and media are increasingly present in the online environment. In Galicia, the set of phenomena that build the current media reality entail certain particularities. In 1995, two years after the arrival of the internet, *El Correo Gallego* (elcorreogallego.es) inaugurated its online edition and became the first online media outlet of the autonomous region. Gradually, more traditional newspapers began to publish digital editions, becoming a general trend from 2000 onwards.

Since 2007, coinciding with the outset of the global financial crisis, Galician online media has entered a more intense phase of experimentation. By 2010, all general daily newspapers published digital editions, forming an ecosystem of online media that is evolving toward more innovative models and exploiting the possibilities offered by the web for more participatory media. This entails greater independence and editorial diversity, a greater number of titles in Galician, more local information and service journalism, and new business models and sources of funding.

15.1.2 Web-Natives, the (Old) New Online Media

All these forms of innovation in online media are more visibly known as web-natives, which are the object of study of this research paper. According to Miel and Faris (2008: 3), web-native media comprise "media formats that exist only on the internet and media entities whose first distribution channel is the internet". In Galicia, the first web-native appeared in 1996 with the emergence of *Vieiros* (vieiros.com), defined as the Galician "online neighbourhood".

The unique presence of web-natives on the Internet can be considered a positive circumstance for greater freedom in terms of information management and production. However, a lack of support by multimedia groups for the establishment of budgetary synergies or cross promotion problematizes the supply of content and prompts an income statement that relies heavily on advertising revenue (García Avilés and González Esteban 2012). In addition, web-natives must look for new business models that, in general, correspond with: advertising-based funding; total or partial payment by content providers; electronic commerce; donations and crowdfunding (García Avilés and González Esteban 2012).

The highly unstable context in which web-natives operate requires a great capacity for business adaptation, given the smaller and more flexible structures that affect traditional practices of professional journalism in this context. New business and organizational forms follow global trends experienced by creative industries towards less hierarchical models, and ever more decentralized and collaborative networking (Castells 2010; Deuze and Witschge 2015). This entails more flexible but also more precarious employment conditions, wherein freelance work and temporary contracts are increasingly frequent. This phenomenon also produces a significant increase of web-natives promoted by 'journalist-entrepreneurs' (Deuze and Witschge 2015: 29).

Most web-natives are created with the aim of covering specific spaces and giving voice to social actors neglected by traditional media and corporations (Miel and Faris 2008). As such, many of them opt for specialization, which is thematic and particularly significant in terms of geographical areas. This hyperlocal dimension (Metzgar et al. 2011) represents a clear commitment to the development of new forms of journalism and business models, with an increasingly interest in bringing a more personalized information-offer to readers (Miel and Faris 2008).

The participation of the public in creating content and engaging in online media management is another distinguishing feature of web-natives. Citizen journalism becomes especially important here, allowing readers to adopt different roles (Firmstone and Coleman 2015: 124): producers, contributors, sources or participants. Moreover, social networks become spaces of reference for participation and for the positioning of media among audiences (López and Alonso 2013; Martínez Fernández et al. 2015).

15.2 Objectives and Methodology

The main objective of this research is to develop an initial map of web-native media in Galicia, so as to identify key actors and offer a starting point for future research. More specifically, it aims to outline the most important characteristics and basic tendencies of the development of web-natives.

Our research to construct a media map draws on official directories and agendas, as well as academic works which offer relevant data. Local specificity and the short period during which some online media outlets have been active has demanded more thorough investigation, conducted through major search engines (Google and Bing) by introducing key phrases (such as "journal" and "location") to identify the maximum number of active media outlets. To study the characteristics of web-natives, a general analysis card was applied, by data has been collected through non-participant observation between September and December 2015, and January 2016.

In terms of general classification, the traditional typology proposed by Nieto and Iglesias (2000) has been followed to differentiate between web-natives as press, radio or audiovisual initiatives. Online media are also distinguished as either general or specialized media (Fernández del Moral and Esteve 1993). Their publication frequency is indicated as being daily, weekly or monthly. In addition, editing languages are taken into account, differentiating online media by their use of Galician, Spanish, both or other languages. In order to identify the degree of proximity of web-natives, they are classified according whether they cover the community, provincial or municipal levels. The differentiation of citizen journalism initiatives follows the definition proposed by Firmstone and Coleman (2015). According to their business model, media are classified by payment model, as either mixed or free (Pereira et al. 2004). Finally, the methodology proposed by Toural Bran et al. (2013) has been applied to the social or 3.0 web, in order to study the active social profiles of each media outlet, its visibility and the ease with which its informative content is shared.

15.3 Results

The investigation conducted to construct a Galician media map identified a total of 164 localized media outlets, of which 62 (38 %) are web-natives. Of such online media that is published and distributed exclusively on the Internet, an overwhelming majority, 91 %, are newspapers, compared to 6 % representing audiovisual media and 3 % representing radio.

Locating media outlets on the map, according to the address of their officially registered offices or the addresses otherwise referenced on the internet, reveals that most web-natives are concentrated in the provinces of A Coruña and Pontevedra, representing 53 and 30 %, respectively. Far behind are Lugo and Ourense, with 12 and 5 % of Galicia's online media. Analysis of the degree of proximity of web-natives in the autonomous community demonstrates that 48 % cover the entire community, 7 % cover the provincial level, 20 % the region level, and 25 % the municipal level.

Regarding business models, all but one of the media outlets identified apply the "free model", demonstrating that the main source of funding for web-natives is advertising, coupled with free access to their content. The only newspaper which employs a "mixed model" is *Galicia Confidencial* (galiciaconfidencial.com), which limits readership of its investigative reporting, available only to subscribers of GC plus. Another particular case is *O Botafumeiro* (botafumeiro.com), a satirical magazine that explores alternative funding with a digital magazine and online store. Readers can purchase the digital magazine for the price they choose, conceived of as a donation, while the magazine's online store, *A Tenda*, sells merchandise featuring their designs. Another example is *Praza Pública* (praza.gal), which seeks profits through contributions from its audience as well as through advertising. A notable example of entrepreneurship is *Disque*, a company that promotes two of the studied web-natives, *Disquecool* (disquecool.com) and *Librópatas* (libropatas. com), and expands its activities with a portfolio of online and offline communication.

When analyzing citizen journalism initiatives, only three (5 %) media outlets were identified. *Véspera de Nada* (vesperadenada.org) is a representative case, geared towards creating bridges between groups working for the elimination of oil as an energy source.

The importance that web-natives accord to their social profiles is reflected in the data collected, as only six (9.6 %) of the media outlets studied have no social profiles. For the rest, the hegemony of Facebook (used by all online media) and Twitter (used by 84 %) is clearly evident, with an average of 4265 and 3243 followers, respectively. After these, majority networks are Google+ (used by 36 %, with an average of 60 followers) and YouTube (used by 31 %, with an average of 632 followers). These are followed by other minority networks, such as Vimeo (used by 9 %, with an average of 24 followers), Flickr (used by 7 % with an average of 2 followers), LinkedIn (used by 5 % with an average of 42 followers) and Instagram (used by 3 %, with an average of 81 followers). In addition, 96 % of web-natives give visibility to, and offer access to, their social profiles on their cover pages. 89 % feature buttons within their content pages which link to their social networks and facilitate the sharing of information.

Of the identified web-natives, only 34 % are specialized media, most frequently pertaining to cultural issues (21 %). Regarding the frequency of updates, mainstream media are updated most often, with 87 % renewing information daily. By contrast, only 29 % of specialized media update their information every day, compared to 33 % which updated their content weekly and 38 % which do so monthly.

As for the languages of publication, most web-natives use Galician as their only language of publication (48 %), while 27 % only use Spanish. The use of both languages is also significant, which are combined in the same edition of digital publications or offered through an automatic translation option by 24 %.

15.4   Conclusion

Online journalism in Galicia has continued to experiment and develop in recent years, giving rise to a diverse media reality in which, within two decades of the emergence of its first digital newspaper, the increasingly important growth of web-native media has become evident.

In this proliferation of web-natives, experimentation and searching for new business models, in which advertising and free access to contents remain predominant, is fundamental. The economic contributions of readers are actively sought through different crowdfunding formulas. In addition, the corporate dimension has been enhanced by expanding promoters' activities to provide related internet management and communications services. In this context, paying for content consumption is still a rarity.

Another trend that our research has confirmed is the close relationship between media and the community, revealing an environment of proximity in which online media and local information predominate. While media promoted by citizens still accounts for a minority of web-natives, it seems that media outlets are searching for a more participatory environment. Specifically through social networks, where almost all media outlets maintain a presence and boast a large community of followers.

It does not appear that specialized content is particularly significant in terms of web-natives, as specialized outlets remain the minority when compared to mainstream media. On the other hand, a greater diversity in terms of the language of publication is detected when web-natives are compared to traditional online media.

This research constitutes an exploratory approach to the reality of web-native media in Galicia, establishing an initial reference map for future, more specific and in-depth studies. It is clear from our research that online journalism is now mature enough to host numerous journalism initiatives able to sustain themselves exclusively on the internet.

References

Castells, M. (2010). *The rise of the network society*. Cambridge: Blackwell.

Deuze, M., & Witschge, T. (2015). Além do Jornalismo. *Leituras do Jornalismo, 2*(4). Available http://www2.faac.unesp.br/ojs/index.php/leiturasdojornalismo/article/view/74/64. October 5, 2015.

Fernández del Moral, J., & Esteve, F. (1993). *Fundamentos de la información periodística especializada*. Madrid: Editorial Síntesis.

Firmstone, J., & Coleman, S. (2015). Rethinking local communicative spaces: Implications of digital media and citizen journalism for the role of local journalism in engaging citizens. In R. Kleis (Ed.), *Local journalism: The decline of newspapers and the rise of digital media* (pp. 117–140). London: I.B. Tauris.

García Avilés, J., & González Esteban, J. L. (2012). Cibemedios nativos españoles: Explorando modelos de rentabilidad. *Trativos, 30*. Available http://www.tripodos.com/index.php/Facultat_Comunicacio_Blanquerna/article/view/50. October 5, 2015.

García, X. L., & Rodríguez, S. A. (2013). Los periódicos gallegos en las redes sociales virtuales: Presencia y posicionamiento en el nuevo escenario comunicativo. *Estudios sobre el Mensaje Periodístico, 19*(2), 1001–1016. Available http://revistas.ucm.es/index.php/ESMP/article/view/43484. October 5, 2015.

Martínez Fernández, V. A., Juanatey Boga, O., & Crespo Pereira, V. (2015). Cibermedios nativos digitales y redes sociales: Presencia y actividad de medios españoles en Facebook y Twitter. *10th Conference on Information Systems and Technologies, CISTI 2015*. Available http://ieeexplore.ieee.org/xpl/articleDetails.jsp?arnumber=7170577. October 5, 2015.

Metzgar, E., Kurpius, D., & Rowley, K. (2011). Defining hyperlocal media: Proposing a framework for discussion. *New Media & Society, 13*(5), pp. 772–787. Available http://nms.sagepub.com/content/early/2011/05/26/1461444810385095.abstract#cited-by. October 5, 2015.

Miel, P., & Faris, R. (2008). *News and information as digital media come of age*. Cambridge, MA: Berkman Center for Internet & Society.

Nieto, A., & Iglesias, F. (2000). *Empresa informativa*. Barcelona: Ariel.

Pereira, J., et al. (2004). El impacto de Internet en los medios de comunicación en España. Aproximación metodológica y primeros resultados. *Comunicação apresentada no II Congresso Ibérico de Comunicación*, 23–24. Available http://bocc.ubi.pt/pag/pereira-gago-lopez-salaverria-noci-meso-cabrera-palomo-impacto-internet-medios-comunicacion-espana.pdf. October 5, 2015.

Toural Bran, C., Limia Fernández, M., & López García, X. (2013). Interactividad y participación en los cibermedios: Una propuesta metodológica para la elaboración, registro y análisis de datos. *Investigar la Comunicación hoy, 1*, 187–204. Available http://dialnet.unirioja.es/servlet/articulo?codigo=4228721. October 5, 2015.

Chapter 16
Interactive Feature: A Journalistic Genre for Digital Media

Jorge Vázquez Herrero and Xosé López García

Abstract Main media are seeking new ways of storytelling with important bets on interactive features, an interpretive in-depth genre, with audiovisual content, interactivity and participation. Within the field of interactive non-fiction digital narratives, we have carried out a research through journalistic main media and informative innovation awards to develop a list of significant features from 2014 and 2015. In a global analysis, we identified a heterogeneous production where features made from a template coexist with more innovative and tailor-made features. In these cases, there is a greater transfer of control to the user and a more sophisticated and surrounding design, which provides more immersive and personalized experiences.

Keywords Digital storytelling · Interactive non-fiction · Interactive feature · Digital media · Journalism

16.1 Introduction

The adaptation of media to digital scenario entails the development of new narrative proposals. The boundaries of the traditional classification according to the format are blurring due to the process of convergence. This transition has been accelerated by an increasingly large and active audience on the network, with more complex needs and their attention on multiple screens.

Interactive feature emerges in this context, associated to media and particularly to the major international newspapers. This journalistic genre goes in depth with the information besides the use of multimedia, hypertext, interactivity and participation.

J. Vázquez Herrero (✉) · X. López García
University of Santiago de Compostela (USC), Santiago de Compostela, Spain
e-mail: jorgevazquezh@gmail.com

X. López García
e-mail: xose.lopez.garcia@usc.es

© Springer International Publishing Switzerland 2017   127
F.C. Freire et al. (eds.), *Media and Metamedia Management*,
Advances in Intelligent Systems and Computing 503,
DOI 10.1007/978-3-319-46068-0_16

16.2 Theoretical Framework

The non-fiction stories find a fertile ground for growth in interactive media. Non-fiction includes several forms of expression such as documentary and journalistic genres, so it is shown as a "macro-genre" (Gifreu 2015).

The convergence of media moved to interactive non-fiction digital narratives with the appearance of subgenres that combine characteristics of different industries in the past and breaking with an inherent linearity from the previous format.

At the confluence of interactive media with cinema appears interactive documentary, an issue developed in detail by Gifreu (2013, 2015) and Gaudenzi (2013). Interactive documentary is defined as an interactive non-fiction genre that aims to reflect reality in a story built on different modes of representing reality and it is characterized by the degree of interaction and participation, provided by different interaction and navigation modalities.

Moreover, cyberjournalism supposes the adaptation of genres, paying special attention to the hypertext, use of multimedia and interactivity. The evolution of online media has been studied before, as have their new ramifications such as transmedia journalism (Porto and Flores 2012) and immersive journalism (Domínguez 2013). It reveals the importance of a fundamental sector for Society in the XXI century.

Different stages of development define the multimedia feature (Salaverría 2005), the hypermedia feature (Larrondo 2009) and the interactive feature (Gifreu 2013; Uricchio 2015). The most widespread concept in the academic tradition is the multimedia feature, which emphasizes on multimedia. This characteristic is ideally understood as an integration of different media in a communicative unit, against the most common juxtaposition of elements.

As Larrondo (2009) points out, "the name of 'feature' is exceeded on the Internet by other as 'special', 'in-depth' or 'report'". The same author defends the term "hypermedia feature", characterized by the possibilities of hypertext, with multimedia and interactive applications, as a "container genre".

Interactive feature comes to claim a stronger development of interactivity, participatory and selective, favoring non-linearity and multimedia integration. It is also giving the user more control and allowing deeper experiences, which can be reached by more audiovisual, playable or immersive representations.

16.3 Methodology

Nowadays, we consider that feature has an important role in the online media. Its development is reaching more useful hypertextual, multimedia and interactive opportunities, making more complex and diverse products. In order to know the current state of the genre, the universe of study was delimited to multimedia features published between January 2014 and July 2015, focusing on the major international media—The New York Times, The Washington Post, The Wall Street Journal or The Guardian–, with attention to Spanish media also—El País, El Mundo or RTVE–, referring to digital design information awards—Online News Association and Society for News Design—or Spanish digital journalism awards—Agency EFE, Ortega y Gasset Prize–.

After this search, a sample of 150 articles were studied based on a analysis sheet. The core of the analysis focused on coding hypertextuality, use of multimedia, interactivity, design and innovation in the degree of development that the feature showed.

The resulting reports were published in the *Digital Storytelling Index* website, of Novos Medios research group at the University of Santiago de Compostela, where each product was added with a gallery of screenshots and a link to it. Thus, the website works as an archive of the current status of the multimedia feature.

16.4 Results

In the selected sample we have identified a heterogeneous production of features that combine characteristics with different levels of development. Concerning the terminology used by media, there is no consensus, making search and location more difficult through archives.

The exploration of the current state allows to speak about interactive feature as one that involves a higher level of development in terms of interactivity, enabling the user immersion, losing some author control and breaking with conventional informative structures to more immersive and participatory spaces.

English-speaking media are the most prolific regarding production, led by The New York Times with 36 features analyzed between January 2014 and July 2015. The Washington Post (21 articles), The Wall Street Journal (13) and The Guardian (12) also show interest in multimedia and interactive stories. Other media are also active in this field: National Geographic, National Public Radio, The Globe and Mail.

In addition, news agencies such as Associated Press or Reuters, and even other kind of organizations, such as ProPublica, work on this genre which is adapted to the network. In Spain, El País and El Mundo have certain activity with 10 and 7 analysis, respectively. The public television, RTVE, from its Laboratory of Audiovisual Innovation, and eldiario.es also publish interactive features.

We have noticed that social topics are the most common (40 % of analyzed cases) and sports and international issues are placed below (19.33 and 16 %, respectively) (Table 16.1).

The main characteristics of journalistic genres in the online media are hypertextuality, multimediality and interactivity, according to the researchers mentioned before. In our analysis, design and innovation were added to describe each product.

Hypertext facilitates the connection between documents and the navigation. In this way, we have identified features that allow an extensive deepening in the

Table 16.1 Distribution of analyzed features by section

Section	Analyzed features	%
Society	60	40.00
Sports	29	19.33
International	24	16.00
Politics	16	10.67
Science/culture	15	10.00
Other	6	4

Source Prepared by the authors

subject through attached elements—related news, linked graphics, scanned documents, external links, content downloads, etc.–. Furthermore, there are permanent and temporary menus, also fragmented content on the same page or several, for instance as chapters, and active layers.

The use of multimedia is one of the strengths of the journalistic feature in the digital environment. The most common case is the inclusion of images or photo galleries, however it can happen at different levels until the full integration of the image with the text, in a highly visual interface. Incorporating video, as photography, provides a different point of view and in some articles represents the main format of the information. In most cases, video is presented in a small window beside the text, although we have identified several examples in which the video is more integrated into the interface and it occupies more space. Other elements that are incorporated are interactive graphics for data visualization and maps, as well as audio files that open a space for first-person testimony and ambient sound.

Within interactivity we distinguish selective and participatory. The first one allows the user to decide what content to view or break the linearity. There is a remarkable development of selective interactivity. Nevertheless, the participatory has a lower appearance. It gives the user the ability to change the story or contribute to its expansion. It is usually reduced to common mechanisms for discussing and, in infrequent cases, facilitating contact with the author. Navigation and interaction are one of the most potential areas where we have identified examples of linear stories, choices and free exploration. In the most interactive features, the user can make an itinerary and build his own message and go deeper into what is more relevant to him.

Design is one of the strongest characteristics of these articles and it allows greater experimentation against classical presentation, inherited from paper. Consequently, we have identified projects in which the true innovation is the design, sophisticated and surrounding. There are also full screen features with great audiovisual content or integrated text with other elements that create a nice and continuous presentation. Although designs are very different, some media such as The Washington Post or The Wall Street Journal employ a base template adapted to each issue. In these situations, multimedia integration and design are more limited, but it has the advantage of faster production and more inexpensive development.

Innovation, in our analysis, is based on the degree of novelty compared to the global production of multimedia features, as well as differentiation from the conventional format of inherited traditional media and evaluating each project experimentation. Design and multimedia are the most exploited characteristics in the most innovative features. *Chasing Bayla* (The Boston Globe), *Walking New York* (The New York Times) or *Taming the Wild Tuna* (The Wall Street Journal) stands out because of the integration of different visual elements and the content navigation. On the other hand, interactivity is greater than conventional in articles such as *Losing Ground* (ProPublica), *Vivir en negro* (El Mundo) or *Die narbe der stadt* (Berliner Morgenpost). In them, the user has more control over what is shown, being required or moving freely through the content. Interactive video is another area of interest and the role it plays should be highlighted in *First World War* and *Where I Went Right* (The Guardian).

16.5 Discussion and Conclusions

The adaptation of journalistic genres in the online media has a focus of interest in interactive non-fiction digital narratives, which have progressed in the fields of documentary and interactive feature, as a result of the transformation of the network and the users.

Despite the diversity of terms that we have observed from the point of view of producers, we choose the concept of interactive feature to identify those projects that emphasize the multimedia integration, the interactivity that gives the user control during his itinerary by the story and hypertext that allows to navigate through extensive articles and it also opens new paths to go in depth. It is the adaptation of the journalistic genre to interactive medium, which is gradually approaching the "immersive rhetoric" defined by Domínguez (2013).

Although most of the articles remain categorized as multimedia features, a significant number of projects involve differentiation and add value in the main media. These projects are the result of the integrated combination of online products characteristics, furthermore, the association of media or the advanced development of only one functionality is not sufficient.

The production is heterogeneous, it comes largely from English-spoken countries and it has different levels of development. The New York Times remains at the head of the media that publish interactive features, followed by The Washington Post, The Wall Street Journal and The Guardian. However, we have found examples in 36 different media, which suggests the relevance of this genre in the network. The subjects are broad and the importance of features in the Society section should be noted.

The current state of interactive feature is particularly characterized by the visual power and the integration of elements on creative designs that favors the experience, besides the use of hypertext and the partial transfer of control to the user.

The strength of the multimedia aspect of the design in interactive feature contrasts with limited participatory interactivity. The most common use is for comments, with the same mechanism as the rest of the website, however there are a few examples of active participation with opinion, decisions or contribution to the narrative expansion.

We are facing a journalistic genre, named interactive feature, which remains in constant transformation and adaptation to interactive media. The relevant production located between January 2014 and July 2015 reveals the role of interactive feature in online media. It also predicts a potential growth that comes together with technological and audience changes. All types of media around the world come together now to develop interactive features, one of the most representative genres of convergence and adaptation of media in the digital context.

References

Domínguez, E. (2013). *Periodismo inmersivo: fundamentos para una forma periodística basada en la interfaz y la acción* (Doctoral thesis, Universitat Ramon Llull). http://www.academia.edu/3206171. Accessed October 14, 2015.

Gaudenzi, S. (2013). *The living documentary: From representing reality to co-creating reality in digital interactive documentary* (Doctoral thesis, University of London). http://research.gold.ac.uk/7997/1/Cultural_thesis_Gaudenzi.pdf. Accessed October 14, 2015.

Gifreu, A. (2013). *El documental interactivo como nuevo género audiovisual. Estudio de la aparición del nuevo género, aproximación a su definición y propuesta de taxonomía y de modelo de análisis a efectos de evaluación, diseño y producción* (Doctoral thesis, Universitat Pompeu Fabra). http://agifreu.com/interactive_documentary/TesisArnauGifreu2012.pdf. Accessed October 14, 2015.

Gifreu, A. (February, 2015). Evolución del concepto de no ficción. Aproximación a tres formas de expresión narrativa. *Obra Digital, 8,* 14–39. http://revistesdigitals.uvic.cat/index.php/obradigital/article/download/54/51. Accessed October 14, 2015.

Larrondo, A. (2009). La metamorfosis del reportaje en el ciberperiodismo: Concepto y caracterización de un nuevo modelo narrativo. *Comunicación y Sociedad, 22*(2), 59–88. http://www.unav.es/fcom/communication-society/es/resumen.php?art_id=317. Accessed October 14, 2015.

Porto, D., & Flores, J. (2012). *Periodismo transmedia: Reflexiones y técnicas para el ciberperiodista desde los laboratorios de medios interactivos.* Madrid: Fragua.

Salaverría, R. (2005). *Redacción periodística en Internet.* Navarra: EUNSA.

Uricchio, W. (2015). *Mapping the intersection of two cultures: Interactive documentary and digital journalism.* United States of America: MIT Open Documentary Lab.

Chapter 17
A Communication Law Feared and Discussed by the Press: The Case of Ecuador

Jenny J. Yaguache, Hernán Yaguana and Abel Suing

Abstract Much academic research is focused on the current situation of Ecuador with regard to journalism and media management after the application of the Communications Law. The media lynching and the subsequent liability, censorship, and policies to the professional practice, are the main topics of discussion. Grievances, complaints and the closure of media have been the trends over the past two years. In addition to this situation, in December 2015, the Telecommunications Law was reformed, including the possibility for Internet users and social networks to be punished. The research analyzes the criteria of the promoters of media companies according to national, regional, and local study cases, compared to the regulatory processes running on Ecuador, which are altering the journalistic routines in traditional and digital editions.

Keywords Newspapers · Regulation · Communication policies · Social media

17.1 Theoretical Framework

17.1.1 Media Regulation in Latin America

The regulations in South America revolve around two variants. One grouping of countries—Venezuela, Bolivia, Ecuador and Argentina–, where there have been imminent changes with the creation of laws and constitutions involving audiovisual and print media. This first variant indicates that here has been higher incidence where progressive governments have been creating laws that completely change

J.J. Yaguache (✉) · H. Yaguana · A. Suing
Universidad Técnica Particular de Loja (UTPL), Loja, Ecuador
e-mail: jjyaguache@utpl.edu.ec

H. Yaguana
e-mail: hayaguana@utpl.edu.ec

A. Suing
e-mail: arsuing@utpl.edu.ec

© Springer International Publishing Switzerland 2017
F.C. Freire et al. (eds.), *Media and Metamedia Management*,
Advances in Intelligent Systems and Computing 503,
DOI 10.1007/978-3-319-46068-0_17

133

codes set forth above, such as the Communications Law 2013 in Ecuador and the Media Law 2009 in Argentina. It is also understood that in these countries there was a greater media concentration dominating entire scene and that the state interference was almost null.

On the other hand, there are countries—Uruguay, Chile, Paraguay and Brazil–, where although there was a greater control and laws protecting communications, some changes and updates, like the Chile case, where there were discussing reforms in public TV, the support for independent production and community media law, but certainly one of the most controversial proposals is the new law on digital media, which in case of approval, it could make those owning a website or a social network with four or more posts a week, to be considered as a means of communication.

17.1.2 Regulation of Communication and Social Networks

In June 2013, after more than four years of debate, the Communications Law was approved by the National Assembly of Ecuador (2013). One of the most analyzed sections by the Ecuadorian press and society in general, is Article 26, concerning the media lynching, contained in Chapter II of the Act. This article is considered by ruling as "curb alleged abuses of the press", and by media organizations as a "gag rule" (Duarte 2013). In several opinion articles, national journalists (Montufar 2013; Cornejo 2014; Villaruel 2013, among others) expressed their concern regarding this article because the criteria do not determine what information is considered harmful to an individual's public credibility.

The first trial for "media lynching" in Ecuador starts on April 8, 2014, when the Superintendency of Information and Communication (Supercom) declared admissible the complaint presented by the former Minister of Education, Government of Abdala Bucaram, Sandra Correa, against the journalist of Vision Radio Diego Oquendo, for alleged "media lynching" (Fundamedios 2014). The matter instantly becomes subject of discussion in social networks, especially Twitter, where the Law and its impact on freedom of expression were challenged, as well as restrictions on criticism of the government officials, in many cases through the hashtag #DiegoOquendo.

In December 2015, the General Regulations of the Telecommunications Law were issued, which should come into force since February 2016. Article 22 of this Law, sets the user's Internet access to a content in the network and the service provider provides the conditions for this to take place. However, the standard also includes an exception: "Those cases where the client, customer or user request so prior its express decision reducing or blocking of content, applications, developments or services available, or by order of competent authority. Providers can implement technical actions they deem necessary for the proper administration of the network in the exclusive scope of activities that were authorized for the purposes of ensuring the service."

17.1.3   Digital Ecosystem of Media Companies

The changes raised in the industry of media communication, telecommunications and information have been so dramatic that a new approach in frameworks and analytical tools is required in order to generate a better understanding of them (Katz 2015: 5).

The economist Françoise Benhamou states that new technologies do not only bring new products, but transform the production processes and content, so that once encoded images can be modified, manipulated and transmitted in the same way as any other digital information. In this way, the dematerialization of the work (replaced by digital files) affects the status of the authors, the modes of production and modes of use and purchase of cultural property (Benhamou 2011: 66). Therefore, production process in media companies seek alternatives of convergence, some with the aim of making efficient the processes and others in a greater scale, enabling them to interface with users.

This digital ecosystem of collaborative environment which media companies are facing—described in example (Campos 2015)—positions online social networks as key elements of the ecosystem. All these processes framed in news routines are fully linked to the communication regulations.

17.2   Methodology

The research aimed to know the legal framework in which the Press Company operates in Ecuador according to the regulatory environment of the country. This analysis was done through the criteria of the managers of four print media outlets in Ecuador. The selection of companies was made by the method of multiple case studies. A newspaper of national, regional, provincial, and local cut was identified. The diversity of the sample yielded important information as to the criteria that managers of media companies have facing the Communications Law.

From this approach to the universe, following the method of multiple case study, the selection of four companies was performed—local, regional, provincial and national—, considered a sample of territoriality. These publishers are El Universo of Guayaquil, La Hora of Quito, El Diario of Portoviejo and Centinela of Loja.

The methodology is complemented by an interview with the executive director of the Ecuadorian Association of Newspaper Publishers (AEDEP) and official statements from organizations like Fundamedios, in order to know the official position of the members in relation to the situation facing the company and the practice of journalism in the country.

17.3 Results

The communications Law has some articles focused directly on the work that the press performs. One of the most discussed was the aforementioned media lynching, considering that it opens the possibility of interpretations, because it does not indicate what type of information is prohibited and especially, if the news is repeated twice, then it can be interpreted as it was reiterated. For Guillermo Navarro (2014), this article "creates conditions for all citizens, including socially disqualified or judicially sanctioned benefiting from the right to proclaim himself the subject of a media lynching, which is socially unacceptable."

For Mena (2014), Ecuadorian journalist, what is happening in Ecuador under the context of the Communications Law, decreases the chances that the media bet on investigative journalism. Since mid-2013, the economic situation of companies in Ecuador newspapers became complicated. According to the president of the Ecuadorian Association of Newspaper Publishers of Ecuador, Diego Cornejo, in an interview for this research, since 2013, "newspapers are billing progressively less and less". In his option, one of the reasons "is the political aggression by the government in government advertising topic and the pressure that has been put into private advertisers, a fact that is reflected in the billing."

The Communications Law (LOC), Article 95 and 96, refer to the investment of public and private advertising and both items the criterion of equal opportunities for media requests.

> There are small printed publications in which a reverse process happens because there is precisely one stimulus in the side of government advertising. That is, it is used as a political instrument of government advertising, which is a violation of the Constitution and even the new Communications Law, which provides that there is an equitable allocation process of advertising designation (Diego Cornejo, Executive Director of the AEDEP, personal interview, July 8, 2014).

Regarding the Telecommunications Law, the journalists' associations initiated some processes of rejection, believing that it violates the constitutional rights. Fundamedios (Andean Foundation for Media Observation and Study), on its website states: "Regulations by extending its scope to users of telecommunications services, opens the possibility for all who use the Internet and social networks could be punished, while content or applications can be blocked by mere administrative decision of the authority." On the other hand, official bodies claim that the law regulates "the telecommunications networks and not to the social ones", however, analysts say that it leaves a loophole to regulate, restrict, interfere with, limit and even to punish users of Internet.

Definitely, the managers of companies believe that the main threat the newspaper company in Ecuador face, is the legal situation that has been submitted by the Communications Law. "Excess laws, excessive controls and a little persecution that is made to companies in all sectors, but the media it has more obviously emphasis and dedication" (Vivanco Arroyo, personal interview, August 19, 2014).

I do not consider that the digitization of the media is such a serious threat. It is an opportunity to migrate to new audiences that is to be discussed. The real threat is the context in which we live, political and now legal with which the press operates, let's say that is the greatest threat we live (Zambrano Lapentti, personal interview, 26 July 2014).

17.4 Discussion and Conclusions

According to those interviewed, the law favors local media and harms national ones. According to the first group, it fosters them in the sense that it provides equal proportions of advertising to big and small media. Therefore, they are the little ones who will have advertising that previously could not obtain, but it is also detrimental to local newspapers that hold critical positions because they must keep vigilant to manage various actions to defend them fairly in the market.

Despite the analysis done in relation to the Telecommunications Law, there are still several issues to solve, starting even with the definition of digital media in the current Communication Law does not appear. Some Latin American countries start their discussions on the regulation of digital communication. Chile, is one of the cases. The purpose of this country is considered as a means of communication to any blog, or Web page that has more than four posts a day.

Definitely, managers of provincial and regional companies believe that the main threat to Ecuadorian newspapers is the legal situation that has been submitted by the Communications Law, as well as the new control reforms created for the digital environment.

References

Benhamou, F. (2011). *L'économie de la cultura*. Paris: La Decouverte.

Campos, F. (2015). Adaptación de los medios tradicionales a la innovación de los metamedios. *El Profesional de la Información, 24*, 441–450.

Cornejo, D. (2014, 8 of July). Personal Interview with the Excecutive Director Ejecutivo of the AEDEP.

Duarte, J. (2013, June 25). En vigor ley comunicación en Ecuador, que castiga el "linchamiento mediático". *Metro Hoy*. Retrieved from: http://tinyurl.com/pfke4yd. Accessed at February 28, 2016.

Fundamedios. (2014). Primer proceso por "linchamiento mediático" se inicia en contra de periodista de radio. Retrieved from: http://www.fundamedios.org/alertas/primer-proceso-por-linchamiento-mediatico-se-inicia-en-contra-de-periodista-de-radio/. Accessed at February 3, 2016.

Katz, R. (2015). *El sistema y la economía digital en América Latina*. Madrid: Fundación Telefónica.

Mena, P. (2014, February 7). Periodismo de datos en un ambiente de pugna Gobierno-medios: ¿Por dónde empezar? Caso ecuatoriano. *Investigative Journalism Education Consortium*. Retrieved from: http://ijec.org/2014/02/07/data-journalism-in-an-environment-of-competing-government-media-where-to-begin-ecuador-case-spanish/. Accessed at February 20, 2016.

Montúfar, C. (2013, July 29). El Consejo de Regulación. *El Comercio*. Retrieved from: http://www.elcomercio.com/opinion/consejo-regulacion.html.. Accessed at February 28, 2016.

Navarro, G. (2014, January 25). ¿Procede "el Linchamiento Mediático"? *Observatorio OLAC*. Retrieved from: http://www.goo.gl/AiZpo6. Accessed at February 28, 2016.

Villaruel, M. (2013, June 28). Ley de medios, análisis crítico de dos periodistas. *La Hora*. Retrieved from: http://www.goo.gl/U54YlP.. Accessed at February 28, 2016.

Vivanco, A. (2014, August 19). Personal interview.

Zambrano, L. (2014, July 26). Personal interview.

Chapter 18
The Influence of Printed and Online Diaries in the Attitudes of Their Readers Towards the 2010 and 2012 Spanish General Strikes

Sergio Álvarez Sánchez

Abstract We pretend to find out if printed and digital media—particularly, *El País* and *El Mundo* newspapers and their counterparts on the Internet, *Elpais.com* and *Elmundo.es*—are able to modify their readers' attitudes towards the calls for a general strike. We focused our research on the last three general strikes performed in the country—september 29th 2010, march 29th 2012 and november 14th 2012–. Our approach involves political communication—given that the unions comply with the characteristics of a single-issue movement—, social psychology—with the theory of reasoned action as our main orientation—and emotional communication —assuming Scherer's propositions. Thanks to the hypertextuality and interactivity of the 2.0 Web, readers are able to answer to the information published by online media, therefore influencing at the same time the attitudes of other readers. We organized focus groups to identify the readers' discourses, at the same time that we interviewed experts in laboral topics—journalists, academics and politicians. The results differ from one newspaper to the other, but show how the power of written press over their readers' attitudes is very limited and can not compare to the potential of online media. Trade unions would do better if they focused on other channels or in their own ones.

Keywords Trade unions · General strikes · 2.0 Web · El País · El Mundo

S. Álvarez Sánchez (✉)
Complutense University of Madrid (UCM), Madrid, Spain
e-mail: sergioalvarezsanchez@ucm.es

© Springer International Publishing Switzerland 2017
F.C. Freire et al. (eds.), *Media and Metamedia Management*,
Advances in Intelligent Systems and Computing 503,
DOI 10.1007/978-3-319-46068-0_18

18.1   Theoretical Framework

18.1.1   The Role of Trade Unions in the Theory of Political Communication

The discipline of political communication has its own theories about the roles of those actors of the political scene who do not compete for the institutional power as political parties do, but attempt to have an influence over wide sectors of the public opinion. For this exploratory study we will assume that unions can be classified as single-issue movements, a term suggested by the italian professor of political communication Gianpietro Mazzoleni (1998). According to Mazzoleni, the term may be applied to every organization which promotes action regarding a certain topic without directly fighting for the political power.

There is also no doubt about the pertinence of including the unions in Mazzoleni's definition of a political system: "a set of political institutions which constitute the skeleton of a country's political life" (1998: 28).[1] His mediatic model puts the media as the public space in itself, where each of the actors and citizens construct their own meanings.

Public relations for unions have been described "as a system of skills which can be put to use in improving the union's institutional position, and in specific circumstances […] as a tool to strengthen the union's position against those who stand for a contrary objective" (J. Barbash, cited in Pomper 1959: 484), in order to make the most of said public space.

18.1.2   The Pertinence of Attitudinal Research

Attitudes, according to Eiser (1986) are defined as "subjective experiences of some issue or object in terms of an evaluative dimension" (1986: 11). They are considered as an enduring feeling, be it positive or negative, with a public reference.

Cognitive consistency is a fundamental idea if we want to understand how people organize their own attitudes. This phenomenon was described by Rosenberg and Abelson (1960) and formulated as an homeostatic process: if the attitudes of an individual are consistent, he will not be motivated to modify them; however, when some type of dissonance arouses—usually when he receives new information—, he will tend to look for consistency and finally find it in the long term.

As Leon Festinger reminded us with his theory of cognitive dissonance, "when two cognitive elements exist in a dissonant relation, psychological tension or discomfort will motivate the person to reduce the dissonance and achieve consonance. The only way to completely eliminate the existing dissonance is to change one of

---

[1]"[…] l'insieme delle istituzioni politiche che costituiscono l'ossatura della vita politica di un paese: […]".

the two elements involved" (L. Festinger, cited in Fishbein and Ajzen 1975: 40–41). Therefore, the individual may try to reduce the dissonance by searching new information supportive of their previously established attitudes.

However, the most complete model for the purpose of analyzing the impact of the unions' communication is the theory of reasoned action (Fishbein and Ajzen 1975). According to this theory, intention is the sum of the attitude towards performing a behaviour—given that the individual will have his own beliefs about doing it—and subjective norm—which evaluation will other people make of the behaviour and how much each group's view matter to the individual. In addition to the statements of Fishbein and Ajzen, we suggest the pertinence of working with an additional subjective norm specifically orientated to the attitude: how others will evaluate the individual's evaluation about performing the behaviour, without necessarily arriving to the point of performing it.

We also detect parallelisms between the theory of reasoned action and Scherer's theory of sequential checking. That is the reason why we have chosen it as the cognitive theory of emotion that we will follow. The monitoring subsystem is related to the subjective norm, given that it is the supervisor of an action to make sure it complies with the internal norms of the individual (Scherer 2001).

18.1.3   The Implications of the 2.0 Web

Never before the arousal of online media, had the reader enjoyed the level of bilateral communication that the tools nowadays at his disposal offer to him.

Hypertextuality implies that online media constitute "a model of communication as close as possible to the structure of human mind, which is not an enclosed structure, with a beginning and an end, but it is based in the constant relation between ideas and information via their complex, and apparently aleatory, structures" (López García 2005: 49).[2] To put it another way, thanks to hypertexts the reader choose what contents he exposes to and in what order. If we are able to determine which contents are more relevant to the readers, we can find what aspects of a general strike are interesting to them and which of these are more sensible to different influences.

The possibility of posting one's own comments constitutes another big contribution of online informative media. The interactivity provided by social networks and the spaces for commenting translate into a degree of interactivity where the user/reader has its own say and can show it to the rest of people interested in certain information. This wide range of emmisors has the potential to generate new sets of normative beliefs about general strikes, with differing degrees of motivation to

---

[2]"[…] un modelo de comunicación lo más cercano posible a la estructura de la mente humana, que no es una estructura cerrada, con un principio y un final, sino que está basada en la continua relación entre ideas e información a través de estructuras complejas y, en apariencia, aleatorias".

comply for each individual. However, anonimity may also play a key role by activating distrust among receptors (López García 2005).

This interactivity, which in some way has allowed users to become contributors to the contents, became the foundational point of 2.0 Web, characterized by "collaboration, with users being able to publish, express their opinions and perform a great number of activities [...]" (Becerril-Isidro et al. 2012: 23).[3] This reality has forced the publisher to a "proggresive adaptation of the written press to the characteristics of online media" (Cabrera González 2001: 74).[4]

18.2 Methodology

We part from the hypothesis of a real change of attitudes among readers of *El País* and *El Mundo* diaries, thanks to the informations appeared in both their printed and digital editions. We also aim to find if the readers of *El País* and *Elpais.com* point out different aspects of general strikes compared to those who read *El Mundo* and *Elmundo.es*, particularly in their evaluations of union leaders' statements, the general strike pertinence, and the so called rights to strike and to work.

The findings have been based on the opinions expressed by selected readers, reunited in two focus groups, each of them for one of the two diaries. We combined those focus groups with interviews to journalists—Manuel V. Gómez, de El País; Fernando Lázaro, de El Mundo-; researchers—Antonio Antón Morón (Universidad Autónoma de Madrid); Carlos Prieto Rodríguez (Universidad Complutense)-; and politicians responsible for labor politics in Madrid's regional parliament when this investigation was conducted, representing the then main parties—Bartolomé González (PP); Mª Encarnación Moya Nieto (PSOE)[5]; Joaquín Sanz (IU), and Juan Luis Fabo Ordóñez (UpyD).

18.3 Results

18.3.1 The Reach of Each Diary's Influence

The readers of *El País* did not express a coincidence with the treatment applied by their newspaper and site of preference to the news about general strikes. Their reinforced previous attitudes did not coincide with the points of view they identified

---

[3]"La web 2.0 se caracteriza por la colaboración, los usuarios son capaces de publicar, opinar y realizar un gran número de actividades y se considera que el mundo se encuentra en esta etapa".

[4]"[...] la adaptación paulatina de la prensa impresa a las características del medio on line".

[5]Due to availability issues, we finally interviewed the Deputy Spokesperson of this party, Josefa Navarro.

that El País attempted to submit to them. One participant perceived "hostility" from *El País* towards trade unions. Another participant expressed: "I'm positioned more to the left wing than El País, but I read it because is the leftiest diary on offer", while also recognizing that nowadays he prefers *Elmundo.es* to remain well informed: "It's not the same, and I can say it because I've read a lot of written press".

We were able to prove an attitude change in some participants of *El Mundo* focus group, but it was not the norm: "The fact of being readers of El Mundo does not mean we think same as them about the strikes". Fernando Lázaro remembered that most of the received responses "supported what we were saying".

18.3.2   The Topics of Interest for Each Group of Readers

The six participants in *El País* focus group believed that all three general strikes were appropriate, while fully supporting the right to strike: "Said 'right to work' is not written anywhere, nor the 'right to transportation'", stated an actuant. Some of the readers of El Mundo dissented, and defended the individual right to take the decision of endorsing a general strike.

Both groups of readers conceded a great importance to the role played by the digital editions for the coverage of general strikes in real time. The readers of *El País* manifested being more interested in the success of demonstrations than in the endorsement by not going to work that day. Manuel V. Gómez explains this order of concerns: "It is not the same to call for endorsement to a general strike in a country with two million of unemployed people, than in one with six million of unemployed, because this segment of population cannot go to strike. You can't measure the impact of a strike just looking at how many people did not go to work".

18.3.3   The Fields Where the Readers Trust Their Chosen Diaries

Participants in *El País* focus group supported unanimously the comments made by the trade unions' leaders and published by the newspaper. Measuring this for El Mundo readers is a harder task, as this quote from Lázaro may explain: "when a general strike is performed, the unions say things like it being a success because of the stoppage of a high percentage of transportation. Another interested party will tell you that this is not true, that is lower and therefore they achieved that percentage thanks to illegal activities. The number of arrestations is among the key data from the initial hours, as well as the political and international resonance". With this need of immediate information, there can be no question about why so many participants in the study had migrated from printed to digital editions.

Professor Antonio Antón describes as "ambivalent" the position maintained by *El País* towards the last two general strikes. However, "this has not made their readers change their views from September 29th 2010 to march 29th 2011". Antón points out that both strikes registered similar levels of participation. Manuel V. Gómez does not think it is fair to accuse *El País* of ambivalence: "I remember to have asked a certain union leader when they were going to call to strike, and he answered: 'you always ask the same question'. 'In fact I can not ask any other one!', I replied, 'because if you don't call to strike you won't have credibility'".

Not all *El Mundo* readers accepted what the newspaper and digital edition stated about general strikes, but some of them shared the views of the newspaper: "I have not been for the strikes. I have accepted the social cutbacks as something that we have to assume, although they are tough for us". A reasoning very similar to that of Fernando Lázaro's: "*El Mundo* has never denied that we have suffered a rights cutback What we have defended is that it was necessary to start creating employment".

There were people who questioned the efficacy of strikes as an instrument of pressure in both focus groups: "we have to find an alternative", said an *El País* reader. But the *Elpais.com* visitors criticised the protagonism conceded to the pickets on the web. Among *El Mundo* readers, there were people angry with the pickets' actions, expressing opinions similar to Lázaro's words, who thinks that a general strike "has a very clear component of public order". They "attempt to bring coercion to the streets".

18.4 Discussion and Conclusions

A common characteristic for all the participants in our focus groups was their high need for cognition. The readers of press and digital media are always looking for different sources of information, a point in favour of exposing themselves to the dialogue generated via social media and news' comments.

The experts we interviewed conceded a very limited power for modifying attitudes to the written press while, according to them, the digital media has a greater potential: "I think that, in what concerns opinion about general strikes, the greater impact comes from audiovisual, not written media", said professor Antón. For a given individual, the opinions found among his group of belonging are more relevant than those spread by media, be them written or digital. And, if the readers who comment the news published in a website or in social media can be identified as part of said group of belonging, their opinions will be particularly valued.

To consider the media as the main public space remains a valid theory, but we found that the most influential representations are performed on online media. It is a reality not only stated by Antonio Antón, but clearly showed by the great number of actuants who admitted having migrated from the print edition to the digital one.

By applying the theory of reasoned action to the results, we can conclude that the subjective norm is the most influential component when the individual is delucidating if he endorses a general strike. But we also acknowledge that the individual will feel more motivation to comply with the expectancies of people he knows and trusts, above those of his preferred newspaper.

In addition, the readers of both diaries tended to reject the information which was inconsistent with their own beliefs. However, the *El Mundo* and *Elmundo.es* readers regarded the september 29th 2010 strike as an action performed by the unions due to a felt external pressure; and described as "very weak" the level of endorsement of workers to the march 29th 2012 general strike. Both were opinions aligned with what the diary wanted to express according to their journalist Fernando Lázaro. To sum it up, *El País* and *Elpais.com* readers had to filter a greater amount of information, so that it did not cause them a feeling os dissonance with their own attitudes.

Both groups of readers wanted to be informed about what the union leaders said, as well as about the response given to the general strike by the population. But the greatest divergences between readers of each diary appeared in the discussion about right to strike versus the so called 'right to work'. The *El País* participants had no doubts about the right to strike being the fundamental one, and the publication can do little to change their positions. The *El Mundo* actuants were more diverse in their opinions when this topic arised. Once again, a result from this study reflects the limited capacity of written and digital press for attitude changing.

When a strike approaches, those responsible for the communication of trade unions would do better by exerting an influence over journalists who work in online diaries, and who dispose of the audiovisual tools to promote attitudinal changes. However, the shortest way would be to get to the readers via their groups of belonging, using for it the union's own tools, without having to accept a third party as a mediator of the message. This method is easier nowadays thanks to the 2.0 Web possibilites; it can multiply the effect of attitude formation and, consequently, the probabilities of transforming the intention of the individuals.

References

Becerril-Isidro, J., Vallejo-Lassard, A. P., Lumbreras-Sotomayor, A., Chavez-Ojeda, G. A., Duk-Sanchez, A. R., & Torres-Parra, R. (2012). La web 2.0: Un análisis de su impacto en lo social, político, cultural y económico. *Investigación Universitaria Multidisciplinaria: Revista de Investigación de la Universidad Simón Bolívar, 11*, 23–34. Retrieved from: dialnet.unirioja.es/descarga/articulo/4281033.pdf. Accessed at January 16, 2016.

Cabrera González, M. A. (2001). Convivencia de la prensa escrita y la prensa "on line" en su transición hacia el modelo de comunicación multimedia. *Estudios sobre el Mensaje Periodístico, 7*, 71–78.

Eiser, J. R. (1986). *Social psychology: Attitudes, cognition and social behaviour.* Cambridge: Cambridge University Press.

Fishbein, M., & Ajzen, I. (1975). *Belief, attitude, intention and behavior: An introduction to theory and research.* London: Addison-Wesley.

López García, G. (2005). Características de la comunicación en red. In G. López García (Ed.), *Modelos de comunicación en internet* (pp. 37–62). Valencia: Tirant Lo Blanch.

Mazzoleni, G. (1998). *La comunicazione politica.* Bolonia: Il Mulino.

Pomper, G. (1959). The public relations of organized labor. *Public Opinion Quarterly, 23*(4), 483–494.

Rosenberg, M. J., & Abelson, A. P. (1960). An analisis of cognitive balancing. In M. J. Rosenberg, C. I. Hovland, W. J. McGuire, R. P. Abelson, & J. W. Brehm (Eds.), *Attitude organization and change: An analysis of consistency among attitude components* (pp. 112–163). New Haven and London: Yale University Press.

Scherer, K. R. (2001). Appraisal considered as a process of multilevel sequential checking. In K. R. Scherer, A. Schorr, & T. Johnstone (Eds.), *Appraisal processes in emotion: Theory, methods, research* (pp. 91–120). New York and Oxford: Oxford University Press.

Part IV
Audiovisual Sector and Media Economy

Chapter 19
Public Service Media on Social Networks: The European Case

Andrea Valencia-Bermúdez and Tania Fernández Lombao

Abstract The aim of this research is to show the evolution of public service media (PSM) on social networks -especially Facebook and Twitter-, as a response to the challenges imposed by the new media market. Constant technological changes and the audience fragmentation have put PSM at the crossroads, so they should design long-term strategies and change management models in order to justify their existence, funding, indispensability and value. Within the frame of governance on the Internet, and characterized by horizontal relationships and dialogue, it seems that PSM have faced the challenge, opening up the path to more active publics. Results obtained show, however, that PSM in Europe have a strong presence on Facebook and Twitter, although less than 10 % of them are supported by social media guidelines. The number of followers is constantly increasing, but it does not result in an improvement of users' engagement.

Keywords PSM · Social networks · Digitisation · Public value · Public service media

19.1 Theoretical Framework

The legitimacy of Public Service Media is compromised across Europe. On the one hand, management and control systems have progressively lost weight because of internal and external pressures; on the other side, PSM's stakeholders do not trust anymore their media, opting for other alternatives (Jakubowicz 2007).

Public criticism on this point led commercial companies to add more arguments against public service media: (1) the traditional "unfair competition"; the "mediamorphosis" conflict (Fidler 1998); the service digitization (Moe 2010); the

A. Valencia-Bermúdez (✉) · T. Fernández Lombao
University of Santiago de Compostela (USC), Santiago de Compostela, Spain
e-mail: andrea.v.bermudez@gmail.com

T. Fernández Lombao
e-mail: t.lombao@gmail.com

© Springer International Publishing Switzerland 2017   149
F.C. Freire et al. (eds.), *Media and Metamedia Management*,
Advances in Intelligent Systems and Computing 503,
DOI 10.1007/978-3-319-46068-0_19

unsustainability of the more stable funding source (Campos 2013); the abundance of contents and services, together with the choice enabled by technological innovation, which empower individuals to obtain whatever they need; and the offer is not distinctive enough and sometimes beyond the PSM remit, hindering private initiative (EBU 2015).

But, while there are still major obstacles in the way, PSM can "play a critical role in sustaining a vibrant public sphere while making good use of the affordances of digital media" (Burri 2015). In fact, the launch of innovation departments–known as MediaLabs of InnovationLabs- might be seen as a sign, while they aim at strengthening the relationship between the organization and the digital environment, serving as experimentation laboratories, and increasing stakeholders' engagement. Experimentation tools -Periscope, Instagram- (Saikalis 2015) are joining to traditional social networks (Facebook and Twitter), on which large part of public media have implemented strategies—especially in that related with the life cycle of the news.

As Hujanen (2004) notes, if PSM are to survive in the new media scene, they have to concentrate efforts on their role as content and service providers, while the EBU (2015) states that PSM have to change their positioning from content to service providers. PSM have to adapt to the media environment and the society they sit within, extending their public value into the new areas and services their audience demand and have right to expect (IBEC 2011).

At present, audience are assuming a more active and prosumer role (Jenkins 2006) thanks to social media, but, while this is undoubtedly positive, user engagement did not just evolve spontaneously, but turned out to be the result of mediated interaction in a growingly complex ecosystem of connective media.

It should be noted that the concept of "public service broadcaster" is not used anymore, since the new media arena requires a paradigm change towards "public service media". Once these services have started to offer contents in various digital platforms, the term "broadcasting" seems to be inappropriate to designate all the services and platforms offered by these media, which go beyond radio and television.

19.2   Methodology

The work is based on the analysis of the use of social media—Facebook and Twitter—in the 28 European State PSM, as well as the strategies for implementing in the digital arena. Results were obtained in three periods: October 2014, March 2015, and July 2015. Principal accounts of public broadcasters have been chosen as subject of study. Variables analysed are related to the number of followers, followed, daily posts, and kind of interaction (tweet and retweet, in the case of Twitter). It should be noted that this analysis serves as a general setting of PSM activity on social networks. Cultural, social, and economic values are not analysed, as well as how deeply social networks have been adopted by television producers and audiences across Europe.

19.3   Results

It is not only a requirement of their mission, but also a key for their strategic future. Public service media should not be absent in new digital platforms and social networking sites, in which they directly communicate and interact with their audience. Social interaction networks have four relevant inputs: connections, cooperation, relationships, and virtuality (Boyd and Ellison 2007). These could be supplemented by viral propagation and operational functionality for automated management of contents and objects through digital networks.

Table 19.1 show the evolution of the activity of PSM on the two analysed social networks. As regards Facebook, it may be observed a positive trend when it comes to followers, while the activity and posts have remained in the last two years. Any public service media in Europe do interact with audience. Once in a blue moon, the Swedish SVT conducted social talks with its audience, but this was stopped in 2015.

It is worth stressing the BBC case as regards the number of followers, as it manages to achieve 17 million on its own. As for the rest, there are significant differences: while Cyprus does not reach one thousand followers, Spain, Portugal, Italy and Germany exceed 370 thousand. Data collection from March 2015 revealed followers' increase in all European public media. From a *Like's* average of 507,985 fans, the figured increased in 2015 to 622,326.

The BBC is still the most outstanding, followed by far by the Spanish RTVE, with almost 460 thousand followers, and the German ZDF, with a little more than 400 thousand. Corporations with less "social supporters" are still the CyBC from Cyprus, the VRT from Belgium, the LTV from Latvia, and the ERR from Estonia.

Facebook is widely used as a portal for publishing news, provide further information, and announce channel schedules. No significant changes on data collections between October 2014 and March 2015 were noted. Finland and France make a corporative use of their profiles, which would explain the small number of daily posts. In October 2014, the most active PSM were the PBS from Malta, the RAI from Italy, the RTVE from Spain, and the HRT from Croatia, with 34, 18, 16 and 15 daily publications on average, respectively. In 2015, first place in the ranking was still held by the PBS, followed by the RAI and the Danish DR.

As regards Twitter, the EU28 have corporative profiles, while data from October 2014 showed that VRT from Belgium and PBS from Malta had not an account. The VRT has already created one. As shown in Table 19.2, the average followers were 381,734, and the BBC still leads the ranking, followed by the German and the Spanish corporations. On the other hand, public media from Greece and Cyprus do not reach one thousand followers, while Hungary does not get one hundred. In March 2015, the average was 462,989 followers. The BBC reached 9 million, followed by the RTVE and the ZDF. Public media from Greece, Cyprus and Hungary remain the least active.

When it comes to interaction, i.e., creation and management of media conversation, European PSM have not yet made a big step forward. They confine their

Table 19.1 European PSM on Facebook

Facebook

RTV	Account	Followers			Daily posts			Interactions		
		Oct-14	Mar-15	Jul-15	Oct-14	Mar-15	Jul-15	Oct-14	Mar-15	Jul-15
ZDF	ZDF	376.888	403.525	413.416	10	7	8	No	No	No
ORF	ORF	18.669	24.630	31.009	4	5	4	No	No	No
RTBF	RTBF TV	35.527	40.044	43.062	4	4	4	No	No	No
VRT	VRT	1.288	1.759	2.248	No	No	No	No	No	No
BNT	BNT	40.662	54.796	63.000	8	15	12	No	No	No
CyBC	Cyprus broadcasting corporation	897	967	973	No	No	No	No	No	No
HRT	Novi mediji, Hrvatska radiotelevizija	27.774	30.831	31.972	15	4	4	No	No	No
DR	DR Nyeheder	97.909	121.128	154.303	10	20	17	No	No	No
RTVS	RTVS	17.445	20.082	22.615	1	3	2	No	No	No
RTVSLO	Uradna stran RTV Slovenija	14.526	16.543	17.844	5	6	5	No	No	No
RTVE	RTVE	429.106	459.622	474.497	16	15	15	No	No	No
ERR	Eesti Rahvusringhääling	6.981	7.293	7.346	5/month	No	No	No	No	No
YLE	Yleisradio	46.155	50.667	53.233	2	5	5	No	No	No
FR	France télévision	46.992	51.607	53.127	1	Eventual	Eventual	No	No	No
ERT	ERT	–	–	1.310	–	–	No	No	No	No
MTVA	MTVA	11.160	13.368	15.421	Eventual	Eventual	Eventual	No	No	No
RTÉ	RTÉ news	44.550	63.484	87.844	9	11	10	No	No	No
RAI	RAI.TV	374.840	391.173	391.726	18	22	21	No	No	No
LTV	latvijasTV	1.510	2.303	2.933	Eventual	8	10	No	No	No
LRT	LRT.LT	63.575	71.114	76.781	Eventual	6	8	No	No	No
PBS	TelevisionMalta	21.093	28.352	48.106	34	50	45	No	No	No

(continued)

Table 19.1 (continued)

Facebook

RTV	Account	Followers			Daily posts			Interactions		
		Oct-14	Mar-15	Jul-15	Oct-14	Mar-15	Jul-15	Oct-14	Mar-15	Jul-15
NPO	NPO.nl	6.959	13.322	14.350	1	2	2	No	No	No
TVP	tvppl	47.689	56.645	67.138	1	4	3	No	No	No
RTP	rtp	374.124	389.005	381.885	4	4	4	No	No	No
BBC	BBC News	11.914.878	14.885.573	17.036.010	10	13	12	No	No	No
CT	ceskatelevize	61.972	69.732	74.592	1	1	1	No	No	No
TVR	televiziunea.romana	67.703	69.954	72.608	6	5	6	No	No	No
SVT	SVT	68.007	81.496	92.568	3	5	5	Yes	No	No

Source Prepared by the autors

Table 19.2 European PSM on Twitter

Twitter		Followers			Following			Daily posts			Interactions		
RTV	Account	Oct-14	Mar-15	Jul-15	Oct-14	Mar-15	Jul-15	Oct-14	Mar-15	Jul-15	Oct-14	Mar-15	Jul-15
ZDF	ZDF	503.000	577.000	648.000	220	226	264	24	18	11	Retweet	Retweet	Retweet
ORF	ORF	7.041	9.980	11.300	–	–	–	6	4	3	No	No	No
RTBF	RTBFtv	35.300	37.300	39.600	1.273	1.276	1.265	8	5	5	Retweet	Retweet	Retweet
VRT	VRT	–	10.800	12.200	–	399	246	–	3	2	–	Retweet	Retweet
BNT	BNT_1	5.915	–	8.330	107	–	121	5	–	6	No	–	No
CyBC	cybc2012	874	898	919	47	47	47	No	No	No	–	–	–
HRT	HRTvijesti	–	–	5.198	–	–	191	–	–	20	–	–	Retweet
DR	DR Nyheder	20.800	97.100	128.000	116	13	15	30	Eventual	Eventual	No	No	No
RTVS	RTVS	2.545	3.207	3.897	29	28	29	1	1	1	Retweet	Retweet	Retweet
RTVSLO	RTV_Slovenija	23.000	30.600	35.200	1.546	1.307	1.343	4	7	5	Retweet	Retweet	Retweet
RTVE	RTVE	739.000	795.000	842.000	255	260	267	30	85	42	Retweet	Retweet	Retweet
ERR	err_ee	4.352	4.922	5.255	40	45	48	7	6	8	No	No	No
YLE	Yleisradio	27.300	31.900	38.900	332	344	366	5	6	4	No	NO	No
FR	francetele	57.500	70.300	82.000	110	123	205	4	3	4	No	No	No
ERT	EPT	–	–	47.000	–	–	560	–	–	75	No	–	No
MTVA	–	–	–	–	–	–	–	–	–	–	–	–	–
RTÉ	rte	91.600	164.000	206.000	20.900	21.600	22.200	23	20	30	No	Retweet	No
RAI	rai.TV	274.000	328.000	368.000	247	268	304	23	21	19	Retweet	Retweet	Retweet
LTV	latvijasTV	4.152	5.469	6.408	413	564	577	4	4	5	Retweet	Retweet	Retweet
LRT	LRTinklas	3.994	4.493	4.825	78	68	78	11	4	5	No	No	No
PBS	TelevisionMalt	1.190	2.062	3.053	142	700	863	31	40	53	No	No	No
NPO	PubliekeOmro	15.000	19.500	21.400	506	505	470	8	10	6	Retweet	Retweet	Retweet
TVP	TVP Info	–	269.000	348.000	–	870	932	–	50	50	–	Retweet	Retweet

(continued)

Table 19.2 (continued)

Twitter

RTV	Account	Followers			Following			Daily posts			Interactions		
		Oct-14	Mar-15	Jul-15	Oct-14	Mar-15	Jul-15	Oct-14	Mar-15	Jul-15	Oct-14	Mar-15	Jul-15
RTP	rtppt	230.000	274.000	310.000	10.430	10.500	10.500	14	12	9	No	NO	Retweet
BBC	BBCWorld	7800.000	9180.000	10600.000	61	61	61	40	50	50	Retweet	Retweet	Retweet
CT	CzechTV	47.300	83.700	106.000	8	10	10	1	1	1	No	Retweet	Retweet
TVR	_TVR	1.548	1.797	2.063	170	170	170	7	4	5	No	No	No
SVT	svt	21.200	25.500	29.300	6.786	6.800	6.766	4	6	Eventual	Retweet	Retweet	Retweet

Source Prepared by the autors

actions to retweets, without answering messages and opening dialogue with audiences. Also, more than a dozen corporations do not engage in any activity in this regard.

19.4 Conclusions

Results obtained from this work, which belongs to a deeper research, could be summarized as follows:

1. PSM are aware of how the important are social media in the new digital, convergent and increasingly complex scenario.
2. However, the high presence of PSM on social networks does not result in an increase of interactivity, which denotes an under-utilisation of social media's potential.
3. Also, social media are generally used as platforms for pasting contents broadcasted and published on their official websites, leaving behind the new forms of audience's engagement.

References

Boyd, D., & Ellison, N. (2007). Social network sites: Definition, history, and scholarship. *Journal of Computer-Mediated Communication, 13*(1), 210–230. Blackwell.

Burri, M. (2015). *Public Service Broadcasting 3.0. Legal design for the digital present.* London: Routledge.

Campos, F. (2013). El futuro de la TV europea es híbrido, convergente y cada vez menos público. *Revista Latina de Comunicación Social.*

EBU. (2015). *Contribution to society. Keynote by Roberto Suárez Candel on the EBU knowledge exchange.* Switzerland: Genève.

Fidler, R. (1998). *Mediamorfosis: comprender los nuevos medios.* Barcelona: Granica.

IBEC. (2011). *Online business models and the content industry.* Dublin: Audiovisual Federation in conjunction with Enterprise Ireland.

Hujanen, T. (2004). *Content production as the new identity of public service broadcasting: Lessons of a digital television.* Copenhagen.

Jakubowicz, K. (2007). Public service broadcasting in the 21st century: What chance for a new beginning? In G. F. Lowe & J. Bardoel (Eds.), *From public service broadcasting to public service media* (pp. 29–49). Göteborg: Nordicom, Göteborgsuniversitet.

Jenkins, H. (2006). *Convergence culture: Where old and new media Collide.* New York: NYU Press.

Moe, H. (2010). Governing public service broadcasting: "Public valuetests" in different national contexts. *Communication, Culture & Critique, 3*(2), 207–223.

Saikalis, M. (2015). *Discussion on the Spanish innovation observatory from RTVE.* Barcelona: Sant Cugat.

Chapter 20
The Behaviour of Ecuadorian TV Micro-Enterprises

Ana Cecilia Vaca Tapia and Mónica López-Golán

Abstract This study analyses the structure of television in Ecuador by reviewing 50 limited companies, 25 of which are microenterprises, 14 small enterprises, 9 medium-size enterprises an 2 large enterprises. Microenterprises are in a dominant position, and stock companies (35) prevail over limited liability companies (15). The data from every microenterprise was gathered, tabulated and fixed to stablish like-for-like comparisons of the financial situation of each category and a broad picture for the years 2012–13. The results point out that the financial situation of television companies in Ecuador is uneven and weak.

Keywords Structure · Television · Microenterprise

20.1 Theoretical Framework

Television is a cultural expression, a place of social encountering and aesthetic possibilities. It also represents the world and local culture, personal faces and identity. Therefore, these media are the most legitimate scenario for culture recognition as they are different, give voice to many different people, and work in order to learn how to be tolerant to others (Rincón and Estrella 2001).

Besides, television is considered as the classic communication method because of its attractive, potential and social performance. Also, they have been always been in the subject of social, political and educative discussions. From a critical perspective, these media are advertising with interleaved shows; they are business, industries, but a public service as well. "No one makes television to lose money" (Rincón and Estrella 2001).

A.C. Vaca Tapia (✉)
University of Santiago de Compostela (USC), Santiago de Compostela, Spain
e-mail: acvaca@pucesi.edu.ec

A.C. Vaca Tapia · M. López-Golán
Pontifical Catholic University of Ecuador (PUCE-SI), Ibarra, Ecuador
e-mail: molopez@pucesi.edu.ec

© Springer International Publishing Switzerland 2017   157
F.C. Freire et al. (eds.), *Media and Metamedia Management*,
Advances in Intelligent Systems and Computing 503,
DOI 10.1007/978-3-319-46068-0_20

Several authors (Collins et al. 1986; Kopp 1990) highlight this media as a "public asset", free of access and not destroyed by its consumption. Thus it is an inexhaustible service. Because of that, even though its fixed costs are high, the marginal cost added in order to reach a new consumer is almost insignificant.

In the last few years, there has been a high increase in costs due to competitiveness, as well as a rise in the value of producing information using new technologies. This increase in expenses requires an increase in revenue that the many funding sources cannot cover, leading to a serious economic crisis in channels (Cebrián Herreros 2003).

The financial situation of television companies in Ecuador is uneven and weak. Most of these companies can cover their short-term liabilities, but they have to take care not to accumulate account receivables and to monitor debt ratio, since they are high whereas profits are generally low.

In March 2012, Superintendence of communications (SUPERTEL) published the number of open television station. As Table 20.1 indicates, there were a total of 501.

In 2012–2013 there was a total of 63 companies registered in the Superintendence of Companies, Securities and Insurance, whose corporate purpose was the television economic activity. However, this study focuses on 50, since the other 13 did not provided any data in the two financial years taken for comparison.

In Ecuador, the Internal Revenue Service (IRS) and SUPERTEL require that all the television channels fulfil with certain legal and tax requirements legally regulated. According to Coronel (2012), in 2012 100 % of the Ecuadorian television sector fulfilled these requirements.

Table 20.2 provides a classification of SMEs, obtained from RESOLUCIÓN No. SC-INPA-UA-G-10-005 (2010), Artículo 1. DE LA CLASIFICACIÓN DE LAS COMPAÑÍAS, following the regulation implanted by the Andean Community of Nations in its Resolution 1260 and the existing domestic law.

The studied structure of television in Ecuador consists of 50 television limited companies, 25 of which are microenterprises, 14 small, 9 medium-size and two large. Microenterprises occupy a dominant position and stock companies prevail over limited liability companies.

All the television companies registered in the Superintendence of Companies, Securities and Insurance with this corporate purpose were taken as a starting point.

Table 20.1 Open TV stations in Ecuador

Category	No. of stations	Percentage (%)
Private commercial	355	71
Public service	146	29
Regional	0	0
Total	501	100

Source Superintendence of communications (SUPERTEL). Broadcasting stations and open televisión by category until March 2012 (Coronel 2012)

Table 20.2 Classification of companies

Variables	Microenterprise	Small enterprise	Medium-size enterprise	Large enterprisess
Personnel employed	1–9	10–49	50–199	More than 200
Gross value of annual sales	Less than 100,00,000	100,00,100–1,000,000.00	1,000,00,100–5,000,000.00	More than 5,000,00,000
Amount of assets	To 100,00,000	100,00,100–750,00,000	750,00,100–3,999,99,900	More than 4,000,00,000

Source Resolución No. SC-INPA-UA-G-10-005 (2010)

From there, the data from every microenterprise is gathered, tabulated and fixed to stablish like-for-like comparisons of the financial situation.

It is worth mentioning that the companies ECUADOR INFORMATION TECHNOLOGY ECUAIT CIA. LTDA., EFIELEC S.A., GRANDECONSTRU S. A., KEIMBROCKS MULTI NEGOCIOS COMPAÑIA LIMITADA, ORPCO S.A. and TROPISERVI S.A perform activities of operation, maintenance, or facilitation of access to voice, data, text, sound and video services, by using satellite communication systems. Conversely, the other companies develop the full program for a channel in a TV station. The full program can be broadcasted by production units or by third party distributors.

In this study, the two available years in the Superintendence of Companies—2012 and 2013—are reviewed.

20.2   Methodology

Descriptive research is one of the methodologies used in this study, since "descriptive studies seek to specify the important properties of people, groups, communities or any other phenomenon subjected to analysis" (Hernández et al. 2010:80).

Likewise, a quantitative methodology is used to analyse the financial state of each television by applying horizontal trend analysis. Simple ratios are applied in the areas of study of liquidity and asset, debt and profitability management.

Arias (2012) defines the horizontal method as a tool to evaluate the performance of a company over time. It makes possible to distinguish the advance or regression and, based on the administration or the owner's preferences, to take steps to meet short-, medium- and long-term goals.

Horizontal methods analyse the organization by comparing each of the elements contained in the main financial statements in the period under evaluation.

Also, it is used the trend analysis, which uses several accounting periods of the same Company, such as absolute values and rates. Changes in the companies do not occur in a certain point in time; therefore, by observing the past behaviour with a view to the future, tendencies can be rectified so that they are in line with the company's goals.

According to Arias (2012), financial ratios are used to highlight the features that require closer attention due to their inefficiency. With the data of the main financial states (balance sheet and income statement) the accounting information is analysed, by relating the different accounts in both states. It aims to establish the financial situation of the company in one or more accounting periods. Ratios can be:

- **Simple**: a tool for conducting analysis of the company and detecting areas of improvement. It is done by relating a line or a group of lines with others.
- **Standard**: a tool for measuring the efficiency to detect areas of improvement by comparison with real or current date. They are obtained by symmetric distribution.

The areas of study in the simple ratio method are (Arias 2012):

- **Liquidity**: it proves the capacity of the company to cover expirations of obligations (loans, interests, taxes) in the short-term. The easier the conversion from assets to cash, the more liquidity. Within this parameter, the *current ratio* or *credit rating* is applied to determine the capacity of the company to cover its close commitments or short-terms liabilities with all its current assets.
- **Assets or activity management**: a number of ratios used to determine the effectiveness which with a company administrates its assets. The *total assets turnover* or *return on total investment* are used to point out the effectiveness of the company's investments, also known as assets to generate sales.
- **Debt, indebtedness or leverage management**: this group of ratios measures the level of indebtedness of the company, if it is still owned by the owners and if they are managing effectively those funds for which they have to pay interest. Moreover, this indicator helps to decide the maximum of debt depending on the preferences of the owners towards the risk. The bigger the risk, the higher the performance.
- **Profitability or earnings**: group of ratios that show the combined effects of liquidity, asset and debt management on the results of operation. Their main objective is to determine whether the enterprise has the capacity of generating profit through investment usage and determining the profit percentage of the sales and capital. If an enterprise is not profitable, it will not be able to effectively make their operations and will perish in the long term.

20.3 Results

For the financial analysis through the horizontal tends method, the following table is used as a base, corresponding to the average of TV micro-enterprises. If we analyse the comparison of the four reference indicators (liquidity, asset turnover, indebtedness and profit margin), the average evolution of the Ecuadorian television micro-enterprises between 2012 and 2013 was negative. Even if they register a positive liquidity average, their asset turnover, indebtedness and profit margin are negative in 2013 are negative compared to 2012 (Table 20.3).

The reason why some micro-enterprise do not register data (N.A.) is that they do not have information on the current assets, total assets, current liabilities, total liabilities, and sales.

20.4 Discussion and Conclusions

The average of the analysis of financial ratios in microenterprises reflects that their ability to cover their immediate obligations with all its current assets has significantly increased from 2012 to 2013. They cover their short-term liabilities with 3.44 dollars per dollar of debt in 2012 and 15.36 in 2013.

It is observed that the least capable company of covering its liabilities in 2012 is SATCONTV SATELITE CONECCION S.A. with 0.38 dollars. In 2013 it is TELEFICAZ S.A. with 0.14 dollars. On the other hand, the most capable trading company of paying its dues in 2012 is I.Q.PROYECTOS S.A. with 26.16 dollars. In 2013 it is BLUEDIGITAL S.A. with 26.37 dollars, proving they are in a very liquid position.

Concerning the effectivity of asset usage in order to generate sales, it is observed that for each inverted dollar the asset rotates 2.04 times in 2012, 1.81 times in 2013 tanks to sales.

It can be concluded that the enterprise with the least asset turnover in 2012 is TROPISERVI S.A. with 0.29 times turnover. In 2013 it is WHILAN S.A. with 0.16 times turnover, revealing a poor administration of the inversion in the total asset. On the contrary, the company with highest asset turnover in 2012 is WHILAN S.A., with 12.53 times turnover, which indicates that the sales level is high and allows the recovery of the total investment.

Regarding the total investment obtained through debt, it is observed that in 2012 there was an indebtedness of 69 and 56 % in 2013. The smallest indebtedness percentage in 2012 is represented by I.Q.PROYECTOS S.A. with 3 %. In 2013 it is ECUADOR INFORMATION TECHNOLOGY ECUAIT CIA. LTDA with 1 %. This data reflects that there is no inversion from capital providers and its total asset is low.

During 2012 the highest indebtedness percentage corresponds to the enterprise BLUEDIGITAL S.A. with 118 % and in 2013 to IMPORTADORA EQUIPMENT MODERNITY CIA. LTDA., with 135 %.

Finally, it is observed that the net profit margin in 2012 is 7 %, while in 2013 is 2 %. The limited companies with less cents earned on each invested dollar in 2012 is SATCONTV SATELITE CONECCION S.A. with −61 % and in 2013 it is COMPAÑIA TELEVISION SATELITAL ZAMORA CIA. LTDA with −36 %, indicating that they obtained a loss in the financial year.

Table 20.3 Average profit margin of the national and local TV micro-enterprises of Ecuador

Company name	Liquidity ratio—current ratio		Asset management ratio—total assets turnover		Debt management ratio—indebtedness ratio		Profitability ratio—net profits margin	
	Year 2012	Year 2013	Year 2012	Year 2013	Year 2012	Year 2013	Year 2012	Year 2013
BLUEDIGITAL S.A.	5.64	26.4	1.33	5.04	1.18	0.04	0.07	0.08
COMPAÑIA TELEVISION SATELITAL ZAMORA CIA. LTDA.	0.40	0.64	2.05	2.14	0.92	0.79	0.02	−0.36
ECUADOR INFORMATION TECHNOLOGY ECUAIT CIA. LTDA.	N.D.	166	N.D.	N.D.	N.D.	0.01	N.D.	N.D.
EFIELEC S.A.	N.D.	N.D.	N.D.	N.D.	N.D.	N.D.	N.D.	N.D.
EUFORICORP S.A.	1.32	2.01	2.26	1.2	0.76	0.5	0.91	0.28
GRANDECONSTRU S.A.	N.D.	N.D.	N.D.	N.D.	N.D.	N.D.	N.D.	N.D.
I.Q.PROYECTOS S.A.	26.2	22	N.D.	N.D.	0.03	0.04	N.D.	N.D.
IMPORTADORA EQUIPMENT MODERNITY CIA. LTDA.	N.D.	0.45	N.D.	7.06	N.D.	1.35	N.D.	−0.05
KEIMBROCKS MULTI NEGOCIOS COMPAÑIA LIMITADA	0.97	1.32	0.93	1.06	0.57	0.35	0.01	0.01
MULTICANAL CIA. LTDA	N.D.	N.D.	N.D.	N.D.	N.D.	N.D.	N.D.	N.D.
NATAS PRODUCCIONES S.A. NAPRODNE	N.D.	N.D.	N.D.	0.57	N.D.	N.D.	N.D.	0.93
ORPCO S.A.	N.D.	N.D.	1.11	N.D.	0.43	N.D.	−0.39	N.D.
PROFILMS S.A.	N.D.	1.17	N.D.	1.39	N.D.	0.6	N.D.	0.04
REDVISUAL CIA.LTDA.	0.46	0.71	0.42	0.48	0.92	0.83	0.14	0.15
SATCONTV SATELITE CONECCION S.A.	0.38	1.41	1.12	2.68	0.73	0.35	−0.61	0.14
SISTEMA DE TELEVISION UHF S.A. SISTEUSA	0.84	0.6	1.65	N.D.	0.91	0.86	−0.02	N.D.
SYSTEM ONE DEL ECUADOR S.A.	1.1	N.D.	N.D.	N.D.	0.99	0.9	N.D.	N.D.

(continued)

Table 20.3 (continued)

Company name	Liquidity ratio—current ratio		Asset management ratio—total assets turnover		Debt management ratio—indebtedness ratio		Profitability ratio—net profits margin	
	Year 2012	Year 2013	Year 2012	Year 2013	Year 2012	Year 2013	Year 2012	Year 2013
TELEFICAZ S.A.	0.96	0.14	0.95	1.05	0.13	0.28	0.09	−0.22
TELEGAME S.A.	3.98	0.61	0.56	0.26	0.99	1.23	0.65	−0.9
TELESAT S.A.	N.D.	N.D.	N.D.	N.D.	N.D.	N.D.	N.D.	N.D.
TENAVISION T.V. CIA. LTDA.	1.88	2.66	N.D.	N.D.	0.23	0.33	N.D.	N.D.
TROPISERVI S.A.	6.81	N.D.	0.29	0.37	0.21	0.05	−0.01	0
WEBTEC COMUNICACION CORPORATIVA CIA. LTDA.	1.99	1.62	2.7	0.55	0.84	0.74	0.01	0.16
WHILAN S.A.	0.81	0.33	12.5	0.17	1.19	0.98	0.02	−0.02
ZAIGOVER S.A.	1.29	2.07	0.59	3.1	0.69	0.48	0.04	0.04
Average	3.44	13.56	2.04	1.81	0.69	0.56	0.07	0.02

Source Prepared by the authors

After making the comparative study of the financial state of the national and local TV micro-enterprises of Ecuador, it can be concluded that:

- The average of the analysis of the profit margin reflects that their capacity to cover their dues with the total of their floating assets is increased in 2013 compared to 2012, reaching to cover their short term liabilities for each indebted dollar.
- The average of the analysis of the asset management ratios of the micro-enterprises diminishes in 2013 regarding the prior year, showing that they are in a very liquid position.
- The average of the analysis of the debt management ratios of the micro-enterprises diminishes from 2012 to 2013, but the percentages show that the resources come from the capital agents, being a risky position as they could not stand more debt and they would be accountable to the lender.
- The analysis of the profitability ratios shows that micro-enterprises obtain low earnings on each invested dollar or euro in 2012 and 2013.
- The comparative analysis between the selected cases of the liquidity, asset management, debt management and profitability ratios allows to deduct that micro-enterprises must correct their administration as they register losses in both analysed years.

References

Arias Anaya, R. M. (2012). *Análisis e interpretación de los estados financieros*. México: Ed. Trillas.

Cebrián Herreros, M. (2003). *Información televisiva. Mediaciones, contenidos, expresión y programación*. Madrid: Ed. Síntesis.

Collins, R., Garnham, N., & Locksley, G. (1986). *The economics of UK television.The London centre for information and communication*. Policystudies, texto mimeografiado.

Coronel Salas, G. (2012). *Anuario de las empresas de comunicación de Ecuador 2011-2012*. Loja, Ecuador: Universidad Técnica Particular de Loja.

Hernández, R., Fernández, C., & Baptista, P. (2010). *Metodología de la Investigación* (pp. 4–80). Mc Graw Hill. Quinta edición.

Kopp, P. (1990). *Televisiones en concurrence*. París: Anthropos.

Rincón, O., & Estrella, M. (2001). *Televisión. Pantalla e identidad*. Quito: Editorial el conejo.

Superintendencia De Compañías, & Valores Y Seguros (2010). RESOLUCIÓN No. SC-INPA-UA-G-10-005. Retrieved from http://www.russellbedford.com.ec/images/Boletines2010/12. ResolucionSUPERCIASPYMES-SC-INPA-UA-G-10-005.pdf. (Accessed August 15, 2015).

Chapter 21
The Weaknesses of Spanish Communications Groups and the Challenges They Face in the Wake of the Economic Crisis

José Vicente García Santamaría, María José Pérez Serrano and Lidia Maestro Espínola

Abstract The Spanish multimedia groups that have survived or emerged from the economic, media and advertising crisis of 2007–2014 are quite different from those that thrived before its onset, which had either been forged through horizontal integration processes and had accumulated large press holdings (Unidad Editorial, Godo, Vocento and Zeta) or after convoluted processes of vertical integration had achieved a presence throughout the communications value chain but maintained a significant stake in the television sector (Prisa Mediapro, Atresmedia or Telecinco). The protracted crisis has wrought a number of significant changes within the Spanish communications sector. A prime example is the Prisa group, Spain's "national champion", whose total annual revenues in the wake of a long string of divestments have shrunk to levels similar to those the company reported in the early nineties. Within the "mutating oligopoly" that has always dominated the sphere of Spanish communications, the Telefónica group has re-entered the media landscape as a television broadcasting company and the void left by the retrenchment of Prisa has been filled by Planeta and Mediapro. The research presented here was undertaken to provide an analytical overview of the current status of Spain's major communications groups, the significant changes the communications sector has undergone in recent years and the threats it faces going forward.

Keywords Communication groups · Media groups · Telecommunications companies · Multimedia content

J.V. García Santamaría (✉)
Carlos III University of de Madrid (UC3M), Madrid, Spain
e-mail: jvgsanta@hum.uc3m.es

M.J. Pérez Serrano
Complutense University of Madrid (UCM), Madrid, Spain
e-mail: mariajoseperezserrano@pdi.ucm.es

L. Maestro Espínola
International University of La Rioja (UNIR), Logroño, Spain
e-mail: lidia.maestro@unir.net

© Springer International Publishing Switzerland 2017   165
F.C. Freire et al. (eds.), *Media and Metamedia Management*,
Advances in Intelligent Systems and Computing 503,
DOI 10.1007/978-3-319-46068-0_21

21.1 Introduction

The beginnings of most of the major communications groups operating in Spain date back to the early 1980s. The debut of commercial, non-public broadcasting in 1989 and the entrance of foreign investors paved the way for the emergence of multimedia enterprises that would provide a fresh injection of leadership in the Spanish media sector, which up to that point had been dominated by a press sector disinclined to pursue the vertical integration strategies necessary for expansion into multimedia. The fact that none these groups has been operative for as long as three decades explains, in part, their past and present weaknesses, which include their lack of internationalisation, elevated levels of debt and their executives' deficient expertise in audiovisual communications (García Santamaría 2013).

The media scene in Spain in the late 1980s was a "mutating oligopoly". Since that point in time, the media sector has been more or less dominated by a cluster of three or four conglomerates that has shifted in composition as some organisations have lost weight in the sector due to the failure of their vertical and horizontal integration strategies or poor management decisions and others have ascended to take their place.

A media enterprise's position within the dominant cluster—once advertising revenues generated by the press sector began to slump—(see Table 21.1) was almost entirely determined by the weight of its holdings in the audiovisual sector. It has been observed on numerous occasions that the only Spanish media corporations capable of generating annual advertising revenues on par with other major European media conglomerates (one billion euros or higher) during the last quarter century have been those with consolidated television broadcasting operations (Table 21.2).

Table 21.1 Advertising investment in conventional media 2007–2013 (in millions of euros)

Sector	2007	2008	2009	2010	2011	2012	2013	2014	% Var. 2007–2013
Newspaper	1894.4	1507.9	1174.1	1128.4	967.0	766.3	662.9	656.3	−65.36
Sunday supplements	133.5	103.9	68.9	72.2	67.1	52.0	38.7	37.7	−71.76
Magazines	721.8	617.3	401.9	397.8	381.1	313.7	253.9	254.2	−64.78
Internet	482.4	610.0	634.1	798.8	899.2	880.5	896.3	956.5	98.28
Radio	678.1	641.9	537.3	548.8	524.9	433.5	403.6	420.2	−38.03
Television	3468.6	3082.4	2237.8	2471.9	2237.2	1815.3	1703.4	1890.4	−45.50
Other	606.4	539.3	416.8	445.2	420.6	348.8	302.2	317.6	−47.63
Total Investment	7985.2	7102.7	5471.3	5863.1	5497.1	4610.1	4261.0	4532.9	−46.64

Source Infoadex (2007–2014)/The autors

Table 21.2 Comparison of the 2008 and 2013 net and operating incomes (in millions of euros) of the seven largest media conglomerates in Spain

Communication group	Net income 2008	Net income 2013	Operating income 2008	Operating income 2013
Prisa	4001	2728	698	−801
Mediaset	982	827	298	70
Atresmedia	724	830	147	63
Vocento	855	530	−22	6
Unedisa	636	372	22	−33
Godó	294	201	−15	4
Zeta	370	208	−56	−5
Total	7862	5696	1072	−696

Note All data above has been drawn from corporate annual report
Source Corporate annual report/Authors

21.2 The Evolution of Spain's Major Media Conglomerates During the 1980s and 1990s

The collective turnover of print media in Spain in 1989 was 2.104 billion euros, more than that generated by television (at a moment when private channels were making their debut) and three times more than radio, a sector that had a collective turnover of 671 million euros the same year. A 1990 Fundesco report noted that the turnover generated by Spanish newspapers had risen tenfold within a decade, from 150.3 million euros in 1980 to 1.4576 million euros in 1989.

By the end of the 1980s, three of these groups (Prisa, Correo and Zeta)—stood out from their colleagues in terms of revenue. In 1990, these groups generated a combined turnover of over 800 million euros and maintained a collective workforce of 5200. However, their strengths at that particular moment would prove to be Achilles' heels going forward. One by one, they would eventually lose their leadership positions: Prisa would be forced to withdraw from the audiovisual sector; Zeta would see its scope reduced once again to print media after selling its stake in Antena 3 and Correo's incursions into radio and television would end in failure. The common denominator of their experiences would be their inability to create or sustain the complicated structure inherent to all multimedia enterprises (Table 21.3).

Telefónica's acquisitions and Planeta's promising entry notwithstanding, the Spanish communications group with the clearest lead in terms of multimedia during the 1990s was Prisa. Proof of this success was the group's net earnings for 1997, which exceeded 411 million euros, which at that time was an impressive figure. Another player that shone throughout this golden decade for print media in Spain was Grupo Zeta, whose turnover topped 300 million euros the same year.

The 1990s was also a period of underlying change in Spanish communications sector. A period of selective asset purchasing took place in the wake of the

Table 21.3 Holdings of main communications groups in Spain at the beginning of the 1990s

Prisa	Bilbao editorial/comecosa	Godó	Zeta	Once	Grupo 16	Prensa española
El País	Print media: *Marca* (25 %) and	*La Vanguardia*	*El Periódico de Catalunya*	Tele 5 (25 %)	*Cambio 16*	*ABC*
Cinco Días	*Expansión* (8 %)	*El Mundo Deportivo*	*La Voz de Asturias*	*Publiespaña* (30 %)	*Marie Claire 16*	*Blanco y Negro*
Mercado	Antena 3 TV	*Historia* and	*El Periódico de Extremadura*	*Diario de Barcelona*	*Motor 16*	
SER	Antena 3 Radio (1 %)	*Vida*	*La Gaceta de los Negocios*	*El Independiente*	*Gente y Viajes 16*	
Canal Plus	*El Correo Español*	LID (55 %)	OTR Noticias	*El Diario de la Bahía de Cádiz*	Radio 16	
Sogetel	*Diario Vasco*	Antena 3 Radio (60 %)		*El Periódico de Guadalete*		
Santillana	*Ya*	Radio 80 (80 %)		Radio Amanecer		
Crisol	*La Verdad*	Antena 3 TV (25 %)		Cadena Rato		
Ihca	*Diario Montañés*					
Publintegral	*Ideal*					
	Hoy					

Source Fundesco (1990)

establishment of private television networks: the *ancient régime* was giving way to more modern concepts of corporate structure.

Taking stock of the fact that Vía Digital (a satellite platform launched by Telefónica in 1996) had accumulated over 600 million euros in losses, the group's next CEO César Alierta brokered a merger between the ailing venture and Canal Satélite Digital, a similar company created in 1997 in which Sogecable (owned by rival Prisa) was a major stakeholder. Although the two companies announced their merger plans in 2002, the deal that lead to the creation of Digital+ was not legally formalised until the following year (Pérez Serrano 2010).

21.3 Spanish Communications Groups at the Beginning of the Twenty-First Centurys

Taking the annual reports released by these companies, it would appear that having undergone a dramatic transformation the Spanish press had reached a state of full "maturity", at least in terms of circulation if not from the viewpoint of profitability, the latter of which expected to improve during the early years of the twenty-first century thanks to rising advertising rates (Aede 2000–2007).

In the words of Díaz Nosty (2001: 135), the Spanish media system at that moment was mired in "a longstanding and antiquated tradition of highly specialised family-run companies that were obsolete technologically as well as in terms of editorial approach, whose intransigence [towards change] was reflected in a mish-mash of formats and poorly organised content". In fact, although three communications groups (Prisa, Recoletos and Correo) controlled 47 % of the newspaper market in Spain, a single competitor (Prisa) with a market share just over 15 % published the country's most successful newspaper in terms of revenue (*El País*), which generated a gross annual revenue of 58 million euros in 1999—more than double the 25 million reported by Recoletos for that same year.

In terms of the total turnover of Spain's largest communications companies in 1999, Prisa led the pack with an operating revenue of over 1.1 billion euros and a consolidated net profit of 91 million euros. Second-place Planeta reported a turnover of more than 750 million euros and a pre-tax profit of 51 million euros.

Following an apparent stabilisation of the sector between 1991 and 1996 (see Table 21.4) the entry of Telefónica created a much more polarised environment that lasted from 1997 to 2000. This landscape was dominated by Telefónica and Prisa, the two highest-ranking companies in the field but far from equals from the perspective of financial resources, management capabilities and size. During this period, Correo, Zeta and Recoletos functioned as lesser satellites jockeying for relative position around one or the other of these two poles of attraction (García Santamaría 2011).

Table 21.4 The annual results of Spain's seven top communications groups for 1997* (Figures in millions of euros)

Communications group	Net income	Operating revenue	Net profit	Employee headcount
Prisa	411.5	53.2	50.2	2331
Zeta	336.4	29.6	18.4	1833
Correo	282.4	51.5	47.1	1720
Recoletos	185.1	not available	not available	852
Prensa Española	164.8	18.1	18.2	987
Prensa Ibérica	123.7	23.1	16.0	1171
Unidad Editorial	141.3	13.2	12.1	685

Source Informe anual de la comunicación (1998)

21.4 Results: The Post-crisis Performance of Spanish Multimedia Groups

Generally speaking, Spain's main communications groups entered the new century with respectable turnovers that can be attributed above all to positive trends in the press sector and advertising revenues generated by generalist television channels. For example, the revenue streams of Spanish dailies practically doubled between 1989 and 1998, rising from 1.175 to 2.3 billion euros.

These rosy perspectives aside, Spanish media organisations entering the new millennium were faced with a number of pending twentieth-century challenges they had yet to successfully address: expansion, consolidation, internationalization, audiovisual diversification and the threat posed by the Internet. The golden era of easy prosperity was drawing to a close.

Planeta assumed Teléfonica's leadership position in the sector when it acquired controlling stake in Antena 3 and assumed control Onda Cero in 2003. Founded and run by the Lara family, Grupo Planeta became one of Spain's most important communications conglomerates. It became patently clear during the first decade of the twenty-first century century that groups without solid audiovisual projects could at best expect a secondary position in the media industry. The same fate awaited those that had accumulated high levels of debt.

Prior to the outset of the most recent financial crisis in the summer of 2007, Prisa stood out as the dominant player in a pack of media groups that included Telecinco, Antena 3, Unidad Editorial and Vocento. The combined 2008 net income of the top seven Spanish communications groups added up to 7.862 billion euros—five times more than 1997. Despite the difficult financial climate of 2013, the combined turnover of the top seven groups in this sector came close to reaching a very respectable 5.7 billion euros.

The operating revenues of these top seven companies seriously deteriorated during the second decade of the new century, falling precipitously from 1.072 billon euros in 2008 to a negative balance of −696 million in 2013. Although this drop was largely due to the delicate financial situation of Prisa, it also reflected poor performances on the part of Godó, Zeta and Vocento.

The multimedia groups that have survived or emerged from the 2007–2014 crisis are therefore quite different from those that thrived at the beginning of century. A prolonged period of sharply falling revenues and a severe contraction in the advertising market have reduced the net incomes of many by as much as 50 %, leaving them, in terms of constant euros, below levels achieved in the 1990s. For example, Unidad Editorial's turnover of 358.1 million euros in 2014 was 43 % lower than the 628.5-million-euro turnover it reported in 2007 (Chart 21.1).

The winners, or quite simply the groups that that emerged from the crisis in the best condition, had little to do with the "nuts and bolts" of traditional journalism. They were more apt to be firmly situated in the worlds of general publishing, television or telecommunications. Putting all one's eggs in the basket of an industry in decline such as press sector had been a risky proposition.

The most remarkable aspect of these major communication groups' behaviour and obsession with expansion was their constant willingness to resort to the financial leveraging common in other sectors. The continual reliance of these organisations on credit paved the way for a process of "financialisation" that has supposed a setback for pluralism in the sector. Although the combined debt of Spain's private communications groups and public television broadcasters exceeded ten billion euros in 2010 (García Santamaría 2013), listed and unlisted companies alike have since taken advantage of the crisis to refinance their debt load and sell off non-strategic assets. Figures for the end of 2014 show (see Table 21.5) that sector debt has been reduced by 50 % and now stands at 5.759 billion euros.

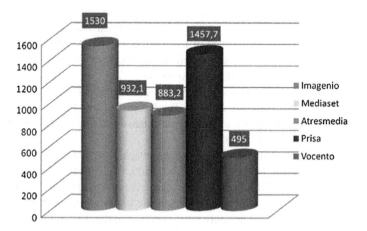

Chart 21.1 Individual turnover figures for Spain's main media groups in 2014 (in millions of euros). *Source* Corporate annual report/Authors

Table 21.5 Debt levels of private communication groups and public television broadcasters at the autonomous community level in Spain in 2014 (in millions of euros)

Communications group	Debt
Prisa	2406
Mediapro	165
Unedisa	1053
Zeta	117
Vocento	126
Antena 3 (Atresmedia)	270
Public TV broadcasters autonomous communities	1622
Total	5759

Source Authors

21.5 Discussion and Conclusions

The research confirms that only those Spanish communications capable of mounting and retaining solid television broadcasting operations have risen to benchmark status in their sector. Telefónica's bid to enter the media sector (1997–2003) created a bipolar media environment that Telefónica and Prisa strove to dominate. In the end, however, a trend towards media concentration in the sector that began in 2010 with Telecinco's takeover of Cuatro and terminated in 2013 with Antena 3's takeover of La Sexta and included Prisa's withdrawal from the arena and the reentry of Telefónica, radically transformed the Spanish media landscape.

In hindsight, one can only reach the same conclusion as Bustamante (2014: 32), which is that the rush towards mergers within the media sector is "proof of an across-the-board failure of Spanish media groups' audiovisual and multimedia strategies that have paved the way for an acceleration in the internationalisation of capital by foreign investors". That said, the Spanish communications oligopoly nevertheless continues to mutate and evolve. Telefónica now holds a dominant position in pay TV and a prime slice of premium content and has become a major sponsor of TV movies. On the other end of the spectrum, Planeta-De Agostini, which has overtaken Prisa's leadership position.

The retrenchment of Prisa (once referred to admiringly as the "national champion") has opened the field for its rival Telefónica but left the sector without a counterweight to the duopolistic control of Spanish television broadcasting by Atresmedia and Mediaset España.

It is abundantly clear that groups focusing intensively or fully on print media such as Unidad Editorial, Prensa Española, Vocento, Godó, Zeta and Prensa Ibérica have not been as succesful as competitors that have consolidated positions in the audiovisual visual sector. As far as the foreseeable future goes, companies desiring to move into multimedia will not only need expertise in the field but also significant capital and/or an excellent credit position—advantages that only large telecommunications and Internet enterprises enjoy today.

The future is unforeseeable, but it is clear we have entered an era of strategic alliances that may be forged either between telecommunications companies and communications groups or the latter and and newer Internet players. In any case, we are convinced that the fifth largest media market in the European Union will not be exempt from these changes.

References

Aede. (2000–2015). *Libro Blanco de la Prensa Diaria.* Madrid: AEDE.
Alférez, A. (1986). *Cuarto poder en España. La prensa desde la Ley Fraga 1966.* Barcelona: Plaza & Janés.
Apm. (2010–2015). *Informe Anual de la profesión periodística.* Madrid: Ediciones APM.
Bustamante, E. (2014). La democratización del sistema cultural y mediático español. Ante una situación de emergencia nacional. In M. Chaparro, *Medios de proximidad: participación social y políticas públicas* (21–34). Girona/Málaga: Imedea/ComAndalucía, Luces de Gálibo.
Díaz Nosty, B. (2001). *Informe Anual de la Comunicación 2000-2001.* Barcelona: Grupo Zeta.
Fundesco. (1989–1996). *Comunicación Social/Tendencias.* Madrid: Informes Anuales de Fundesco.
García Santamaría, J. V. (2011). Reorganización en los grupos multimedia españoles: la nueva cartografía. *Observatorio (OBS*) Journal, 5*(1), 157–174.
García Santamaría, J. V. (2013). Las debilidades estratégicas de los grandes grupos españoles de comunicación y su viabilidad futura. *Global Media Journal México, 10*(19), 85–102.
Pérez Serrano, M. J. (2010). *La concentración de medios en España. Análisis de casos relevantes en radio, prensa y televisión.* Madrid: UCM.

Other Sources (for Tables and Charts)

Atresmedia: *Informe Anual (2007–2014).* Retrieved from: http://www.atresmediacorporacion.com/informe-anual (Accessed October 26, 2013).
Díaz Nosty, B. (1998–2001). *Informe Anual de la Comunicación 1997-2001.* Barcelona: Grupo Zeta.
Fernández-Beaumont, J., & Díaz Nosty, B. (2006) (Dirs.). *Tendencias 06.* Madrid: Fundación Telefónica.
INFOADEX. (2010–2015). *La inversión publicitaria en España.* Retrieved from: http://www.infoadex.es/
Mediaset. (2004–2005). *Informe Anual.* Retrieved from: www.mediaset.es/.../Informe-Actividades_MDSFIL20111228_0031.pdf
PRISA. (2000–2014). *Informe Anual.* Retrieved from: http://www.prisa.com/es/datos/cuentas-anuales/
RCS Mediagroup. (2008–2015). *Informes Anuales.* Retrieved from: http://www.rcsmediagroup.it/pagine/investor-relations/

Chapter 22
The Scientific Structure of Media Management: Strategies for Emancipation

Manuel Goyanes

Abstract The present article problematizes the scientific structure of the relatively recent sub-field scientifically coined as "media management". Concretely, the manuscript is displayed around the critical reflection of what I conceptualize as F.A.T.I., that is, four structural interrelated dimensions that configure the "spiritus" of its form: (1) Field, (2) Autonomy, (3) technepractical domination and (4) Identity. The F.A.T.I, which is normally portrayed as template for good scholarship, is configured in this essay as a comfort academic zone in which a series of dispositions, practices and socialized habitus by the media management consortia obstruct and jeopardize alternative intellectual dispositions and modus of scientific discovery and verification. I propose the concept of transcendence-breaking as strategy of individual and disciplinarian emancipation of the structural ideological statu quo. I suggest three versions: theoretical-breaking, functional-breaking and, more ambitiously, epistemological-breaking.

Keywords Media management · Field · Transcendence · Standardization · Media studies

22.1 Introduction

In general terms, any field is objectively and subjectively oriented by its hierarchical agents to achieve an intellectual maturity through the development of certain autonomy with respect to other fields. This genuine teleology of fields cover up, however, a double intrinsic complexity: on the one hand, maintaining the original "essence" and "spirit" of its creation as autonomous field and, on the other hand, enriching (or impoverishing) its conserved and conservative structure with external influences empirically supported by other fields (what in modern academic jargon is consecrated as crossfertilization). The struggle for the domination of the field and

M. Goyanes (✉)
Carlos III University of Madrid (UC3M), Madrid, Spain
e-mail: mgoyanes@hum.uc3m.es

© Springer International Publishing Switzerland 2017
F.C. Freire et al. (eds.), *Media and Metamedia Management*,
Advances in Intelligent Systems and Computing 503,
DOI 10.1007/978-3-319-46068-0_22

the hierarchization of its agents has been traditionally the quid in social science, intensified along the last three decades as a result of the rocketing specialization in sub-fields.

This is the particular case of media management, a sub-field halfway between the pure management and media and communication studies. A hybrid sub-field that, however, has achieved certain autonomy by means of developing some scientifically structural homologies with other (hard) fields (mainly spaces and institutions) and a relatively sovereign identity. Both the specialization and socialization of and within a particular communitarian scheme of (objectified) objects and scientific habitus have led, as a result, to the collective acknowledgment of "sound research" that neither pure management nor media and communication studies would conquer without the aid of one another. In addition, the collective sense of belonging has reported substantial benefits to their integrants: on the one hand, it has created a scientifically sanctioned and intellectual school and, on the other hand, and perhaps most importantly, it has boosted the productivity of the tribe (mainly in terms of publications in scientific journals).

In this essay, I explore the intrinsic dilemmas of academic stagnation in the comfort zone. Via the critical description of four interrelated structural dimensions (F.A.T.I, initial derived from Field, Autonomy, Domination and Identity) that shapes the "spiritus" of its academic form, I discuss the scientific dispositions that we, as a collective, have socialized as heteronomous agents. Learning about the conventions of each structural dimension maximizes a bucolic, fenced and "at home" academic dispositions in which the productivity and visibility of the tribe jeopardize alternative and radical approaches. I propose transcendence-breaking as an alternative strategy to the current academic schemes developed through the four interrelated structural dimensions.

22.2   The Academic Comfort Zone of Media Management

22.2.1   Field

The first structural dimension that constitutes media management is the structuring of its field. The case of media management represents a hybrid sub-field halfway between the field of pure management and the field of communication and media studies. Each of these fields permeates schemas, habitus and (determined) agents which struggle to dominate the sub-field that, as a whole, constitutes and delimits the boundaries of its relatively scientific autonomy. The theoretical articulation of the original fields, from where media management receives its scientific schemes and dispositions that shape its structure as academic space, are, on the one hand, the field of pure management and, on the other hand, the field of communication and media studies. Finally, the hybridization that characterizes the sub-field of media management.

The fundamental agent to scientifically establish media management as a sub-field is the distinction between field/object and the privilege of the object as field establisher. Media management takes the media as object, but in terms of organization(s) (that is, blindly articulated through three of its fundamental social forms: newspapers, radio, television and, perhaps, cinema and new media), but it does not take them as (mainly) terrains or analysis spaces of social (or not) relations "in" or "from" the media as objects of study of the society and its mutations. The accurate and positive configuration of the object triggers the importation of agents of domination which establishes the matrix field from which it inherits its fundamental scientific dispositions, that is, from the pure management and, on the contrary, the impoverishment of the inherited tradition of the field of communication and media studies as a terrain social analysis through the media.

The domination of the academic and social category of organization as object in studies of media management implies, therefore, a theoretical disposition that privileges a pure management approach and its dominant epistemologies. This intellectual disposition does not lead, however, to exclude multi-disciplinarian theoretical perspectives from other fields such as sociology, psychology, or economics or other approaches to the object from other sub-fields such as journalism studies. On the contrary, the inclusion of these multi-disciplinarian perspectives as well as the assimilation of distinct approaches to the object are temporarily produced, but in such a way that its assimilation tends to reproduce the schemas of domination of pure management. This fact naturally bureaucratizes research orientations in media management, where media means organization and organization field of pure management. This is triggered in such a way that the socialized scientific schemas enhances the apparent progress both of the sub-field and the tribe, already immersed in a higher order network of relations within and out of sub-field. This stops at the same time, the self-questioning of research outcomes and a template of alternative research dispositions.

22.2.2   Autonomy

Similarly to any other field (or sub-field), media management struggles for maintaining its autonomy and boundaries, that is, the articulation of its own laws in the search of a position within the social scientific order. The relative autonomy of media management is articulated through the hybridization of the two fields from which it originally inherits its academic and intellectual dispositions. This hybridization results in positive and negative outcomes for its scientific position: on the one hand, it encourage the symbolic assimilation of different (fundamentally theoretical, methodological, epistemological) approaches naturally and rationally inherited from the original fields (or external ones) that enrich (or impoverish) its schemas of scientificity and; on the other hand, the non-sanctioned contribution of researchers scientifically positioned on the original fields or sub-fields (or other) into the media management sub-field. This configuration that, a priori, seems pretty abstract, could be explained more easily.

The scientific authority that emerges from the specialization(s) has led to theoretical or empirical contributions that, in multiple occasions, have not been sanctioned by the original field. This is the particular ambivalence of media management specialization: the empirical enhance of notable discoveries based on a meticulous analysis of a portiuncula portion of the universe, and the apparent contribution of results based on the specificities that the original field takes naturally for granted but the field to which is linked is totally unaware. This is the barbarity of specialism: its ambivalent capacity to produce ignorant-sages who are expert connoisseur of a portiuncula portion of universe but philistine in everything else.

The relative autonomy of media management as sub-field provides its integrants with a strong power of identification, basically through the establishment of different academic interaction networks and the socialization of schemas and habitus. This autonomy, which is articulated through the hybridization of the intellectual and research schemas of the original fields, also places media management in a position of intellectual maturity that implies the reproduction and conservation of scientific dispositions. Within the tribe, its results are not self-questioned in terms of broad outcomes and audiences.

22.2.3 Identity

Researchers have a sense of belonging to the media management sub-field which permeates the categories that organize the objective and subjective research orientations. With the notion of category I mean the invisible structures that organize perception and determine what we see and don't see, that is, the eyeglasses articulated through our education, scientific experience and habitus. The perceptive categories place researchers in a particular position within the sub-field, they soothe the collective chaos inherent to all social knowledge and provide their integrants with a determined scholastic identity.

The identity framework provides the sub-field and its tribe with an individual and/or collective island of integration, it enhances its autonomy as scientific organization as well as triggers a strong integrative and integral self-definition of objects and subjects (I am a media management researcher and I study multiplatform strategies). The sense of belonging provides, therefore, a template for developing one's research identity, the collective and collegiate creation of patterns/models of identity dispositions providing distinct and distinctive identity templates as well as distinct research profiles as modus of personal brand. In addition, it establishes individual and/or collective and semiautonomous interaction and interactive clusters (congresses, conferences and seminars about different (sub)topics within the sub-field) where the tribe is collectively self-reaffirmed while also reaffirming the autonomous and sovereign identity of the sub-field as a whole.

The collective and collegiate identity acquisition and its external sanction, set the ground for a fertile, fenced, comfortable and islander habitat. The communitarian identity, as well as its socialized and acquired intellectual dispositions, drives the sub-field to a (symbolically) violent reproduction of scientific schemas and habitus that, holistically, enhance the conservative dispositions of the sub-field versus the suspicions of interlopers/outsiders who could disturb its identity as a relatively autonomous sub-field. At home, dispositions take control and jeopardize alternative research notions.

22.2.4 Domination

The sub-field of media management is structured around the object of media in terms of organization. The privilege of the object as main field establisher of the sub-field implies the importation of dominant dispositions from the field of pure management for the analysis of the specificities of media. As a consequence, media management naturally and rationally inherits (and therefore reproduces) its scientific schemas of domination and academic dispositions from pure management, basically rooted in a techne-practical research orientation focused on the use of standardized models of scientific discovery and production of symbolically powerful fields (hard sciences).

Standardized research publications are the products of a sequence of interrelated codified and formulaic practices that involve standardized research, a standardized editorial process, standardized reviewing, and more generally, standardized mind-sets, that is, standardized ways of thinking about what constitutes scholarship (Goyanes 2015). This dominant vision of what is generally considered as "sound research" is articulated through four dimensions that prescribe how scientific production and discovery should be designed in the sub-field: 1) hyper-focalization and tribe, 2) gap-spotting, 3) rationalism and statistical reductionism and 4) stylistic engineering and rhetorical gymnastics (see Goyanes 2015 for more evidence).

The social dominance of standardized schemas of research discovery and verification implies the symbolic acquisition of a rational, natural and taken-for-granted model of knowledge production. Through this socially constructed configuration, structurally homologous to hard sciences dispositions, both media management and its tribe reproduce the formulaic accepted signs of scientificity. As a result, media management enhances its autonomy and political power to protect and promote the sub-field in relation to other fields (or sub-fields) smoothing the path to potentially increase the research output and career prospects of the community. However, non-standardized alternatives struggle to compete for (mostly) theoretical and methodological representations in the sub-field in exchange.

22.3   Transcendence-Breaking

My suggestion, transcendence-breaking, challenges and problematizes the taken-for-granted ideals and expectations of the dominant logic within media management. They are based on enhancing intellectual dispositions aiming to be transformative rather than merely confirmative and incremental. I suggest three research strategies, fundamentally based on the transformation and modification of the object media, the experimentation with other theoretical streams as frame of reference and the collective consideration of imported epistemologies from other intellectual terrains.

22.3.1   Theoretical-Breaking

The theoretical-breaking aims to transform media management through skeptically questioning, problematizing and challenging its dominant theoretical assumptions, fundamentally by drowning upon theoretical inspiration from sources other than the usual suspects. This approach is oriented towards an ideal that emphasize a broad audience based on the dominance of diverse and varied intellectual fields in order to provide an alternative theoretical framework that challenges taken-for-granted assumptions and scientific schemas of media management. Rather than the adaptation to conventional theoretical ideals and sensible automatism of sound and robust theoretical categorizations, the theoretical-breaking reveals and problematizes the hidden scholarship operations, that is, those clear and taken-for-granted theoretical assumptions that do not need to be discussed nor problematized because they are, precisely, beyond any theoretical doubt.

Conversely to the uncritical assimilation of ideals and theoretical conventions of the "school", the theoretical-breaking encourages an outsider vision that problematizes the dominant logic as well as a skeptical relation with the sub-field. Hence, the theoretical-breaking is established through: (1) the challenge and problematization of the assumed theoretical structures of domination and (2) a broad research commitment (and as a consequence a certain level of disloyalty to media management) with other intellectual terrains as the frame of reference.

The main ambition of theoretical-breaking is to enhance an intellectually-open disposition that challenges and uncovers the shortcomings of the dominant sedentary schemas of media management as a fenced and prepackaged sub-field. The immersion in alternative theoretical adventures, the amplification of dominant social relations and rituals (such as conferences) and the self-scientific-production-interrogation, are all vitally important for subvert the traditional statu quo. The theoretical-breaking, therefore, encourages intellectual nomadism that "betrays" its socialized and taken-for-granted assumptions and conventions in the search of a great transcendence and broad audience.

22.3.2 Functional-Breaking

The functional-breaking aims to modify the object media in the accurate and positive terms of an organization and to locate the most pressing scientific priority in taking as one's object the social work of construction of the pre-constructed object (the media). The functional-breaking acknowledges the object media as a privileged observatory terrain for the study of society and its mutations. The object media as "privileged terrain" emphasize the many interrelated dimensions that model the object: managerial, technological, symbolic, ideological, sociological, etc. Media are not objectively objectified as organization(s), but as genuine and relatively autonomous spaces for the study, discussion and reflection about the social mutations and the effects of the mediation of the media in society. The functional-breaking aims, departing from the intrinsic specificities of media, to highlight the theoretical and epistemological space for enhancing an alternative objectified vision of the object of study.

22.3.3 Epistemological-Breaking

Media management naturally and rationally inherits its particular theory of knowledge in search of "truth" from pure management. Essentially, the approach of media management's research towards scientific discoveries and (re)production is rooted in a positive epistemological orientation. This orientation takes place through the cultivation of ultra-rational scientific methods of hard sciences and the (violent) symbolic representation of a standardized discourse in search of a replication and exterior scientific sanction. This source of scientific legitimization is rooted in the acceptance of representations of scientificity and the norms to be respected in practice in order to produce a scientific effect and thereby acquire symbolic efficacy and the social profits associated with conformity to scientific appearances (Bourdieu 1988).

Media management acquires the status of scientific sub-field when an impression of scientificity is produced based on at least apparent conformity to the norms by which science is generally recognized. Hence, researchers, whose exaggerated concern with linguistic finesse might threaten their status as scientific researchers, can resist this, more or less consciously, by rejecting literary elegance and drain themselves in the trappings of scientificity (mathematical formalism, experimentation, structural equations, etc.). This "methodologist" approach is penchant to separate reflection on methods from their actual use in scientific work and to cultivate method for its own sake.

The epistemological-breaking aims the emancipation of media management from the dominant positive regime in social science as quasi-paradigmatic mechanism of scientific discovery and production. The epistemological-breaking is based on a methodological polytheism, although not rooted in the "anything goes" as in the epistemological anarchism of Feyerabend (1975). It is rather based on the implementation of an array of methods oriented to problem at hand, constantly reflecting upon in actu, in the very movement whereby they are deployed to resolve particular questions. This breaking departs from a critical empiricism, based on acknowledging the sensory experience but under the supervision of a higher order theoretical schema grounded in the problematization and critical reflection.

22.4   Discussion and Conclusions

A central ambition of this essay has been to provide an alternative research strategy in contrast to the current scientific schemas being symbolically developed and structured within the media management sub-field. As an antidote to the prevalent state of academic comfort, I have suggested the notion of transcendence-breaking. The transcendence-breaking provides an alternative set of criteria for what can be seen as a good researcher but, opposite to the current dispositions, it tries to capture a more reflective, critic and imaginative orientations. I have also developed a vocabulary for grasping the different ideals and encouraging different ways of thinking. Furthermore, I have outlined theoretical-breaking, functional-breaking and epistemological-breaking.

In conclusion, the central practical implications emanating from the observations made in this essay include: (1) a redefinition of structures that configure our sub-field which drive the media management community to an apparent progress and (2) the need to go in depth in the transcendence of our contributions as element of development of our sub-field.

References

Bourdieu, P. (1988). *Homo Academicus*. Cambridge: Polity Press.
Feyerabend, P. (1975). *Against method: Outline of an anarchist theory of knowledge*. Londres: Humanities Press.
Goyanes, M. (2015). ¿Hacia una Investigación Estandarizada? *Observatorio (OBS*) Journal, 9*(3), 85–99.

Chapter 23
New Advances in Transmedia Storytelling in Spanish Fiction. Case Study of the Television Series 'El Ministerio del Tiempo'

Mª Isabel Rodríguez Fidalgo and Adriana Paíno Ambrosio

Abstract If there is one thing that characterises transmedia storytelling, it is without any shadow of doubt, making way to great creative potential. Essentially, it opens a door to all the possibilities connected with the part the spectator-user plays within the narrative. Fiction television series have not overlooked this new way of telling a story, and there are several examples which can be found both in national and foreign TV channels. The series '*El Ministerio del Tiempo*' (Spanish for 'the Ministry of Time') which is broadcast in Spanish national TV (RTVE), is a clear example of the creation of a transmedia TV series. It offers innovative storytelling elements within a TV fiction series. This article, using the methodological structure of a case study, aims to analyse the narrative structure of this production, where the followers have become the new main characters of this elaborate fictional universe.

Keywords Transmedia · Prosumer · Entertainment · Fiction series · Fan phenomenon

23.1 Theoretical Framework

The number of devices that each person has access to when searching for media has multiplied, which means nowadays audiovisual products are both mobile and multi-screen. In addition, there is 'the multiplication of filing and distribution channels which allow the download and view of contents after the moment of their airing' (Rodríguez-Mateos and Hernández-Pérez 2015: 96). This frees the spectator to choose where and when they want to use said contents. The channel and screen multiplication has forced broadcasters to adapt and find new ways of transmitting the contents to a

M.I. Rodríguez Fidalgo (✉) · A. Paíno Ambrosio
University of Salamanca (USAL), Salamanca, Spain
e-mail: mrfidalgo@usal.es

A. Paíno Ambrosio
e-mail: adriana.paino@usal.es

© Springer International Publishing Switzerland 2017
F.C. Freire et al. (eds.), *Media and Metamedia Management*,
Advances in Intelligent Systems and Computing 503,
DOI 10.1007/978-3-319-46068-0_23

public which is more geographically disperse each day (Canavilhãs 2011). However, not only are technology and the access to contents different, but all this change has created a new kind of spectator, one which wants to participate and interact. In this sense we can start speaking of the term 'prosumer', coined by Toffer (1980).

The 'directed passivity' denounced by Noam Chomsky (Fernández 2014) was predominant in the Web 1.0, however, it is improving as the users become more important and start interacting with the product, creating content themselves. As David Caldevilla points out 'the new version (2.0) of the web offers an environment where you can participate, mixing authors and readers'. He also suggests an active recipient who searches, investigates, connects, hyper-connects, thinks, answers, compares and creates content from others that already exist' (2009: 33).

From now onward, recipients-users will be offered experiences where they will have the opportunity to participate as co-creators of the story. 'They all narrate and they all listen (to each other)' (Costa 2013: 562). We are witnessing formation of a media-centered culture which puts the consumer in a position of prominence, thus reaching a new dimension. Therefore, there is a constant search for new digital material since the public no longer settles for only receiving information and entertainment, they want to participate and interact directly (Rodríguez and Molpeceres 2014; Rodríguez and Sánchez 2014). This brings us to the term 'transmedia storytelling'.

The term was coined by Jenkins (2003), and has been defined by Carlos Scolari as '… a way of telling a fiction where the story spreads through multiple media and communication platforms and in which consumers play an active part in the expansion process' (2013: 46). From this definition, Professors Rodríguez and Molpeceres (2014) have drawn two of the main features of transmedia storytelling. The first one is the idea of dispersing the story across multiple media and channels. During this dispersion each medium will provide typical characteristics of their format, varying and adding new information to the structure of the fictional world. Hand in hand with this expansion idea, is the second main feature: consumers play an active role in this process. However, as Professors Rodríguez and Molpeceres state 'it isn't enough with the active participation of the audience, there must be a creative dialogue with the foundation of the story, being able to change it in the process' (2014: 318).

There are several examples of fiction series, such as *Breaking Bad* (Scolari 2013), *Game of Thrones* (Molpeceres and Rodríguez 2014), *Fringe* (Scolari 2013; Belsunces 2011), or the Spanish series *Isabel* and *Águila Roja* (Molpeceres and Rodríguez 2014; Barrientos-Bueno 2013), all of which have allowed spectators participate creatively in the story.

One of the main Spanish models of transmedia storytelling, as Varona and Lama point out, can be found in the different proposals that have been developed throughout the years in the RTVE.es Laboratory. Therefore, this 'effort to generate different storytelling adds to the production of materials which will accompany television and radio broadcasts, since this is a field where transmedia experiences have had some prominence' (2015: 209) within Spanish public broadcasting channels.

23.2 Methodology

In order to breach the subject of the planned study we have decided to take a qualitative methodology perspective, in particular the case study technique. Its first analysis will show which platforms and channels the *El Ministerio del Tiempo* uses to expand their storytelling and which contents travel through each of them. A second analysis will reflect on the role that the subsequent fan phenomenon has played in the series, since this element is of great importance to determine the profile of the true 'transmedia consumer'.

23.3 Results

El Ministerio del Tiempo is a RTVE fiction series (2015). The combination of adventure and historical content as well as a tinge of humour used by the series had a good reception from the public and the critics since it began, even though it went through some changes in the day and time of broadcasting (which almost meant its cancellation). However, social media is where we can truly see *El Ministerio*'s success and the fan phenomenon that developed around the show.

23.3.1 The Transmedia Experience and the Inclusion of Recipient Discourse in 'El Ministerio del Tiempo'

The creators of the show were aware of its potential from the start; they even built a whole transmedia universe around it, including a 'Making of' which is broadcast every week after the series, profiles in social media (Facebook, Twitter and Instagram), a WhatsApp exclusive group, an online programme for the show's followers ('*La Puerta del Tiempo*' or The Gate of Time) and an Official Forum where users can share their opinions.

The centre of all content is the series official website, where we can access all the episodes aired up to that moment *on demand,* in addition to a great deal of extra contents.

The following can be found on the website:

1. Cover: It's the access point to whole episodes of the first season and includes information about the filming of the second season. There is also an area called 'The transmedia universe of *El Ministerio del Tiempo*'.
2. The series: Here you can find a summary of the series and there is a short introduction of each of the three main characters, who comprise 'the Patrol'. In addition, there are profiles for each cast and crew member.

3. Characters: There are three sections '*La Patrulla*' (The Patrol), '*Los Pilares del ministerio*' (The pillars of the ministry) and '*La experiencia es un grado*' (Experience is a degree).

4. The Gate of Time: section centred in the programme '*La puerta del tiempo*' (the gate of time), which is broadcasted exclusively in the website featuring interviews with the characters of the show. It also includes a section with additional videos called '*los extras de la puerta del tiempo*' (The extras of the gate of time).

5. Complete episodes: here we can find all eight episodes of the first season as well as some additional content.

6. Videos: this link directs to the RTVE '*A la carta*' section devoted to the series.

7. Forum: It has eight big debate sections organised by topic: 'the Ministry's cafe', 'Chapters', 'The characters', 'Ministry's caretaker's office', 'Audiovisual documentation', 'What would have happened if...', 'Ministry's archives' and 'the Ministry's library'.

23.3.1.1 The Dialogue with the Audience in Social Media

'*El Ministerio del Tiempo*' has more than 35,000 followers on Facebook, almost 36,000 on Twitter and near 3000 on Instagram.

The kind of messages they send in social media are informal, that is, besides offering information about the series, they use the official profiles to send gifs, collages, videos of the filming sessions and of course to retweet fans, actors and members of the crew.

After asking Twitter fans how they would like to be known, they answered the '*ministéricos*' (which combines the Spanish words for ministry and hysteria). This started an internet phenomenon, called '*ministeria*'. It seemed to be frequent in foreign fiction, but up to this moment nothing of the sort had happened in any Spanish television production. Nobody questions how active the *ministéricos*'s participation is when taking a look at their creations: There are *Twitter* accounts for the characters of the series, blog entries, podcasts, *Facebook* groups. We would like to highlight two of these: '*Funcionarios del Ministerio del Tiempo*' or Civil servants of the Ministry of Time and '*Ministerio del Tiempo*' or Ministry of Time. As well as this, there is a blog in *Tumblr*, an automatic certificate generator, Amelia Folch's diary on *Tumblr*, rol plays with educational purposes, a *Wikipedia* page about the series, *Youtube* content, *Sims* games, an app dedicated to the Ministry's different doors, wallpapers, comics, gifs, drawings, posters, etc. It is even possible to find subtitles online created by fans so that the series can reach other parts of the world.

However, the transmedia element that stands out the most is 'Basic training for civil servants and rookie patrol members of the 'Ministry of Time'', which is a fictional online course described as 'Basic training course for new members of the civil servant corps of the institution'. It comprises four short videos (no more than one minute long each) which feature characters from the series. It is considered transmedia because it prompted other content from the 'ministéricos', such as the already mentioned certificate generator.

23.4 Discussion and Conclusions

The above analysis allows us to determine that a change is taking place in the handling of television content, particularly in fiction series. For some time we have been appraising the potential of transmedia storytelling in this context, but to be able to find examples we had to turn to foreign fiction. '*El Ministerio del Tiempo*' has been a qualitative leap in terms of 'Spanish transmedia series'. Through the *ministéricos* and the *ministeria* movement, participation has ceased to be only active (fans who follow the content of the narrative universe in its many forms and formats) but it has transformed into 'transmedia participation' (where the fans not only follow the transmedia content designed by the series but they also create content which enriches the very narrative universe of the series). It has been thanks to this fiction's fandom that the series has renewed the second season (with the hash tag #TVERenuevaMdt).

This fact poses an inherent debate, one connected to TV audience measurements; since examples like the one we have been analysing have highlighted the importance of social audience measurements. Moreover, all the great media groups know that the broadcast of audiovisual content has to be through a multiple-screen system; and that the spectators-users demand content they can interact with in social media.

The transmedia project behind the analysed series shows as innovative traits not only that it has been capable of providing its audience with contents in any form and format (from the aired episodes, the series official website with additional content, and different content for social media- mainly designed for phones and tablets, etc.) but also it has generated a social audience like nothing ever seen in Spanish television before. This allows us to speak of a 'before and after' of '*El Ministerio del Tiempo*' in the context of Spanish fiction television series.

References

Barrientos-Bueno, M. (2013). La convergencia y la segunda pantalla televisivas: el caso de Isabel (TVE). In Proceedings of the *I Congreso Internacional Comunicación y Sociedad*. Logroño: UNIR. Retrieved from: http://goo.gl/gc3TKU. Accessed at October 16, 2015.

Belsunces, A. (2011). *Producción, consumo y prácticas culturales en torno a los nuevos media en la cultura de la convergencia: el caso de Fringe como narración transmedia y lúdica*. Universitat Oberta de Catalunya. Retrieved from: http://goo.gl/c2j6S1. Accessed at October 16, 2015.

Canavilhãs, J. (2011). El nuevo ecosistema mediático. *Index.comunicación, 1*. Retrieved from http://goo.gl/Tzcfki. Accessed at October 21, 2015.

Corona, J. M. (2014). El poder de las historias: Los retos para investigar las narrativas transmedia. In E. Rueda y C. P. Martínez (Coord.), *La investigación de la comunicación ante el nuevo marco regulatorio de las telecomunicaciones y la radiodifusión en México* (pp. 392–402). Retrieved from: http://goo.gl/lHH1qv. Accessed at October 22, 2015.

Costa, C. (2013). Narrativas Transmedia Nativas: Ventajas, elementos de la planificación de un proyecto audiovisual transmedia y estudio de caso. *Historia y Comunicación Social, 18.* Retrieved from http://goo.gl/VzSR0Z. Accessed at October 16, 2015.

Fernández, C. (2014). Prácticas transmedia en la era del prosumidor. Hacia una definición del Contenido Generado por el Usuario (CGU). CIC *Cuadernos de Información y Comunicación, 19.* Retrieved from http://goo.gl/jRElrH. Accessed at October 22, 2015.

Jenkins, H. (2003). Transmedia storytelling: Moving characters from books to films to video games can make them stronger and more compelling. *Technology Review.* Retrieved from http://goo.gl/6nkcU2. Accessed at October 13, 2015.

Jenkins, H. (2007). Transmedia storytelling 101. *Confessions of an AcaFan.* Retrieved from http://goo.gl/JZmFZ7. Accessed at October 13, 2015.

Jenkins, H. (2009). The revenge of the Origami Unicorn: Seven principles of transmedia storytelling. *Confessions of an Aca-Fan* (blog), 12 de diciembre. Retrieved from http://goo.gl/HsQSch. Accessed at October 22, 2015.

Jenkins, H. (2010). Transmedia storytelling and entertainment: An annotated syllabus. *Continuum, 24*(6), 943–958. doi:10.1080/10304312.2010.510599.

Jenkins, H. (2014). Transmedia 202: Reflexiones adicionales. *Confessions of an Aca-Fan* (blog), 8 de septiembre. Retrieved from http://goo.gl/Q8ig1w. Accessed at October 22, 2015.

Molpeceres, S. & Rodríguez, M. I. (2014). La inserción del discurso del receptor en la narrativa transmedia: el ejemplo de las series de televisión de ficción. CIC *Cuadernos de Información y Comunicación, 19.* Retrieved from http://goo.gl/1Ju75i. Accessed at January 22, 2015.

Renó, L., & Renó, D. (2013). Narrativa transmedia y mapas interactivos: periodismo contemporáneo. *Razón y palabra, 83.* Retrieved from http://goo.gl/ZLwmFS. Accessed at October 22, 2015.

Rodríguez, M. I., & Gallego, M. C. (2012). Las webs de series de ficción como nuevas experiencias narrativas en el contexto hipermediático. In B. León (Coor.), *La televisión ante el desafío de internet* (pp. 110–121). Salamanca: Comunicación Social ediciones y publicaciones.

Rodríguez, M. I., & Molpeceres, S. (2014). The Inside Experience y la construcción de la narrativa transmedia. Un análisis comunicativo y teórico-literario. *C.I.C Cuadernos de Información y Comunicación, 19*, 315–330. doi:10.5209/rev_CIYC.2014.v19.43918

Rodríguez, M. I., & Sánchez, A. (2014). La interactividad, hipertextualidad y multimedialidad al servicio del género documental. Estudio de caso del webdoc En el reino del plomo de Rtve.es. In B. León (Coord.), *Nuevas miradas al documental* (pp. 82–94). Salamanca: Comunicación Social.

Rodríguez-Mateos, D., & Hernández-Pérez, T. (2015). Televisión social en series de ficción y nuevos roles el documentalista audiovisual: el caso de «El Ministerio del tiempo». *Index.comunicación, 5*(3). Retrieved from http://goo.gl/xovsvv. Accessed at October 16, 2015.

Scolari, C. A. (2013). *Narrativas transmedia: Cuando todos los medios cuentan.* Barcelona: Deusto, S.L.U.

Toffer, A. (1980). *The third wave.* Canadá: Bantan Books.

Tur-Viñes, V., & Rodríguez, R. (2014). Transmedialidad, Series de Ficción y Redes Sociales: El Caso de Pulseras Rojas en el Grupo Oficial de Facebook (Antena 3. España). *Revista Cuadernos. Info, 34*, 115–131. doi:10.7764/cdi.34.549.

Varona, D., & Lara, P. (2015). "Be ministerico, my friend": Diseño de una estrategia transmedia. In C. Cascajosa (Ed.), *Dentro de El Ministerio del Tiempo* (pp. 203–210). España: Léeme.

Chapter 24
Corporate Communication and Social Media. Spanish Companies' Communicative Activity Index on the Audiovisual Social Networks

Bárbara Fontela Baró and Carmen Costa-Sánchez

Abstract The present work studies the presence of 50 companies in main audiovisual social media. These companies hold the top positions of the 2014 Merco (Corporative Reputation Business Monitor) ranking according to corporative reputation. The objective of the study is to determine how Spanish companies are using second generation social media (González Macías in Pinterest. La red social visual y creativa. Editorial UOC, Barcelona, 2014) depending on their proactivity level and the content flow. In addition, the suggestion and validation of an index that allows measuring the aforementioned proactivity establishes an interesting methodological proposal that may be useful for subsequent works.

Keywords Corporate communication · Business communication · Social media · Audiovisual · Strategy

24.1 Theoretical Framework

Corporate communication is recently going through a new challenging stage for its established dynamics, structures and goals. Internet arrival, evolution towards web 2.0 and the successful reception of social media has set out a new scenario for which corporate communication should be prepared (Celaya 2008; Merodio 2012).

The symmetric bidirectional model proposed as an utopia is close to become feasible. "What we metaphorically called conversation is becoming tangible and it is taken to the extreme in social media" (López Font 2011: 23).

B. Fontela Baró (✉) · C. Costa-Sánchez
University of a Coruña (UDC), Corunna, Spain
e-mail: barbara.baro@udc.es

C. Costa-Sánchez
e-mail: carmen.costa@udc.es

© Springer International Publishing Switzerland 2017
F.C. Freire et al. (eds.), *Media and Metamedia Management*,
Advances in Intelligent Systems and Computing 503,
DOI 10.1007/978-3-319-46068-0_24

In this new period, images are the winning contents. We are 'homo videns'. Visual stimuli capture our attention, and they do it more than ever in a multiscreen era.

Continuous and interesting dialogue with users and potential clients requires attractive content, participative initiatives, and experiences that increase loyalty to the brand (Alloza 2010; Martínez 2011). It is necessary to adapt the content to each of the channels, as well as to make the most of the dialogic effect between them. In addition, it should be taken into account that social media strategy it is not just a presence strategy in which the company multiplies its communication channels. In social media it is not enough to be, it is necessary to define where the brand wants to be and why (Macnamara and Zerfass 2012).

Up to now, research about corporate social media use has been mainly theoretical (Kaplan and Haenlein 2010; Kietzmann et al. 2011). This study is innovative and necessary because of two reasons: first, because it analyzes the presence and visibility of the main Spanish companies in audiovisual social media or second generation social media: Youtube, Vine, Vimeo, Instagram, Flickr and Pinterest. Second, because it proposes and validates an activity measuring index, which can be useful for subsequent studies.

24.2 Methodology

The paper analyzes how the 50 top ranked companies in Merco (Corporate Reputation Business Monitor) 2014[1] use audiovisual social media. The study reviews the following social networks: Youtube, Vimeo, Instagram, Vine, Flickr and Pinterest. Linkedin was added taking into account that it can be also used for multimedia communication (Costa-Sánchez and Corbacho-Valencia 2015). The period of the study covers from March to June 2015.

The study checks the number of social media in which each company has an account, the opening date and their proactivity level, which was defined according to a self-elaborated scale.

Regarding the content, the paper also reviews the level of content reduplication in social media by locating the primary source and checking its repetition among other social networks.

The communicative activity index was obtained by relating the number of accounts opened in social media and the proactivity level measured (Table 24.1). The formula applied was the following:

$$Activity\ index = \frac{\sum proactivity\ value}{number\ of\ social\ media\ accounts}$$

---

[1]http://www.merco.info/es/ranking-merco-empresas?edicion=2014.

Table 24.1 Proactivity values for the activity index formula

Proactivity value	Inactive	Passive	Active	Very active
Numerical value	0	1	2	3

Source Self-elaboration

24.3 Results

24.3.1 Presence

The analyzed companies were present in 3 audiovisual social media on average. The mode in the sample was 2 social media, mainly YouTube and Linkedin.

The most popular network is YouTube, since 88 % of the studied companies have an account on it. YouTube is followed by Linkedin (78 %), Flickr (42 %), Instagram (34 %), Vine (32 %) and Vimeo (16 %).

24.3.2 Proactivity Level

Regarding the level of proactivity in social media, Youtube is the network more actively used in business, as well as Vine and Linkedin.

On the contrary, Vimeo and Flickr have high inactivity and passivity levels, which exceed 50 %. They are both social networks focused on segmented audience, like video or photography professionals.

The case of Vine is significant. Despite only 32 % of the sample has Vine profiles, their proactivity is noteworthy: 56 % of the Vine profiles checked are active of very active, which points out that their presence on Vine is not incidental, Vine is consciously used (Table 24.2).

Top ranked companies in Merco (2014) are not the ones with the best activity index, which implies that there is not a relationship between the company's reputation and its communicative leadership in social media (Table 24.3).

24.3.3 Video Social Media

Concerning the video platforms studied, YouTube is the leader, since it has 72 % more companies registered than Vine, as well as more users.

Business presence on Vimeo and Vine is more recent than on YouTube: 40 % of the companies reviewed opened a YouTube account before 2008, while all the Vine and Vimeo profiles were opened after 2012.

It is a significant fact that Banco Santander is the only company in the sample that keeps an active Vimeo account.

Table 24.2 Companies by social network according to their proactivity level

Social network	Very active companies (%)	Active companies (%)	Passive companies (%)	Inactive companies (%)
Vine	50	6	38	6
Linkedin	46	21	18	15
YouTube	39	36	25	0
Instagram	29	12	41	18
Flickr	19	0	76	5
Vimeo	0	25	13	62

Source Self-elaboration

Table 24.3 Sample of companies' activity index

Company name	Activity index	Company name	Activity index
Danone	3.00	ESIC	1.60
Sanitas	3.00	BBVA	1.50
Procter & Gamble	3.00	Caixabank	1.50
Hewlett Packard	3.00	Siemens	1.50
Toyota	2.75	NH Hotel Group	1.50
Leroy Merlin	2.75	Endesa	1.50
Telefónica	2.50	IESE	1.40
IKEA	2.50	ING Direct	1.40
Indra	2.50	Mahou San Miguel	1.25
Accenture	2.50	Abertis	1.25
El Corte Inglés	2.50	Inditex	1.00
ESADE	2.50	Google	1.00
La Fageda	2.40	Once and Once foundation	1.00
Santander	2.20	Mutua Madrileña	1.00
Instituto de Empresa	2.20	Banco Sabadell	1.00
Repsol	2.00	Meliá Hotels International	1.00
Iberdrola	2.00	DKV Seguros	1.00
Acciona	2.00	Mapfre	0.67
L'Oreal	2.00	Gas Natural Fenosa	0.60
Calidad Pascual	2.00	Mercadona	0.50
Garrigues	2.00	Nestlé	0.50
Grupo Siro	1.80	Apple	0.00
EAE Business School	1.75	Microsoft	0.00
IBM	1.75	Novartis	0.00
Gamesa	1.67	Zeltia	0.00

Source Self-elaboration

24.3.4 Photography Social Media

Instagram has unseated Flicker. Nowadays, the companies' favorite photography social network is Instagram. More companies are registered on Flickr than on Instagram. However, 81 % of the Flickr accounts checked have been inactive for years. Companies registered in Flickr are migrating to the new photography network, Instagram, leaving their Flickr profiles abandoned.

24.3.5 Content Reduplication in Social Media

The key social media for content is not yet audiovisual. 80 % of the companies analyzed use Facebook to diffuse their audiovisual content, mainly YouTube videos. Nonetheless, YouTube is the social media for which more original content is produced.

Regarding photography social media, content is reduplicated in all of them. Flickr is often used to publish the complete report rather than single pictures, infographies are a trend in Pinterest and Linkedin is used to post corporate news and job offers.

24.4 Conclusions

Presence does not equal proactivity in social media. The interest in participating in *2.0 boom* makes companies open profiles in social media, however, they do not use them to converse. In order to dialogue and engage, it is necessary to establish a continuous relationship with the audience (Merodio 2012).

Despite the sample is made up of the biggest and most renowned Spanish companies, their use of social media needs improvement. 48 % of them, almost half of the sample companies, has a low Activity Index (equal or below 1.5), which means that they worry more about being in social media than about dialoguing with their audience.

By social network, YouTube and Linkedin have both high presence and activity percentages, while Flickr and Vimeo do not awaken as much interest for communication strategy.

It is possible to conclude that Flickr has become stagnant by noticing its presence percentage (42 %) and its passivity and inactivity percentage, which reaches 81 %.

The study of the level of content originality/reduplication shows that pictures are likely to be reduplicated in social media and Facebook is currently the diffuser for video, while YouTube is the primary source.

Concerning the methodology of the study, the formula applied has been revealed as useful for evaluating the online presence and the online communication strategy. Therefore, this article validates the activity index in social media for subsequent research.

In regards to future projects, it could be interesting to do a comparative study about social media use by sector in order to know the industries in which they are applied. Likewise, it would be convenient to know how communication departments are incorporating this new area in their daily work and which routines are applying for content creating.

References

Alloza, A. (2010). Brand engagement: Marca, experiencia. In P. Capriotti & F. Schulze (Eds.), *Responsabilidad social empresarial*. Barcelona: Executive Business School. Retrieved from http://www.bidireccional.net/Blog/Libro_RSE.pdf

Celaya, J. (2008). *La empresa en la Web 2.0*. Barcelona: Gestión 2000.

Costa-Sánchez, C., & Corbacho-Valencia, J. (2015). Linkedin para seleccionar y captar talento. *Prisma Social, 14*, 187–221.

González Macías, M. (2014). *Pinterest*. Editorial UOC: La red social visual y creativa. Barcelona.

Kaplan, A. M., & Haenlein, M. (2010). Users of the world, unite! The challenges and opportunities of social media. *Business Horizons, 53*(1), 59–68.

Kietzmann, J. H., Hermkens, K., Mccarthy, I. P., & Silvestre, B. S. (2011). Social media? Get serious! Understanding the functional building blocks of social media. *Business Horizons, 54* (3), 241–251.

López Font, L. (2011). Comunicación corporativa y redes sociales: cambiarlo todo para que nada cambie. *adComunica. Revista Científica de Estrategias, Tendencias e Innovación en Comunicación, 11–19*. Retrieved from doi:10.6035/2174-0992.2012.3.2

Macnamara, J., & Zerfass, A. (2012). Social media communication in organizations: The challenges of balancing openness, strategy, and management. *International Journal of Strategic Communication, 6*(4), 287–308.

Martínez, A. C. (2011). ¿Cuánto vale un fan? El reto de la medición de la audiencia en los social media. *Pensar la Publicidad. Revista Internacional de Investigaciones Publicitarias, 4*(2), 89–110.

Merodio, J. (2012). *Todo lo que hay que saber de estrategia empresarial en redes sociales*. Madrid: Wolters Kluwer España.

Chapter 25
Indirect Management and Outsourcing of Contents as an Alternative to the Crisis of the Spanish Public Broadcasters

Ana María López Cepeda

Abstract Autonomous public televisions in Spain are confronting with a major crisis that is challenging the viability of regional public broadcasting services. The maintenance of public audiovisual services is being questioned because of the heavy politicisation and the financial crisis with its respective cutbacks. One of the most discussed measures has been the adoption of the Act 6/2012, which legalises an outsourcing model —and the possibility of indirect management—that had been previously applied by autonomous public broadcasters. Against such a background, the aim of this paper is to analyse the practical effects of changes in the Spanish regional situation.

Keywords Act 6/2012 · Autonomous public televisions · Outsourcing · Privatisation · Indirect management

25.1 Autonomous Public Television in Spain

For the last few years, there is a talk of a relevant crisis in the autonomous public broadcasters in Spain. Notwithstanding, it could be said that it has always been criticism related to management and governance of these services, since they have been politicised on many occasions (López Cepeda 2012, 2015). That has resulted in an extreme loss of credibility, together with expense control policies, downsizing of budgets, and the fall in revenues, allowances (Campos Freire 2015) and audiences (Barlovento Comunicación 2011–2014).

One of the most discussed issues related to regional audiovisual media in Spain has been their management model, especially since the emergence of *Televisión*

A.M. López Cepeda (✉)
University of Castilla La Mancha (UCLM), Albacete, Spain
e-mail: ana.lopezcepeda@uclm.es

© Springer International Publishing Switzerland 2017
F.C. Freire et al. (eds.), *Media and Metamedia Management*,
Advances in Intelligent Systems and Computing 503,
DOI 10.1007/978-3-319-46068-0_25

195

Canaria, which initiated in 1999 a system based on the outsourcing of a large part of its production.[1] This system is characterised by a total or partial transfer of production and certain services to external companies (Sarabia et al. 2012a: 175). Public media in Balearic Islands, Aragon, Extremadura, Murcia and Asturias will follow that patron. Questions have been asked as to whether this model is valid, since it sometimes infringe on the terms of the Act 46/1983 on the Third TV Channel—currently repealed. The approval of the Act 7/2010 on Audiovisual Communication changed that situation, since it repealed the previous regulation and dictate that the only interdiction to the contents outsourcing is to assign to third parties the production and edition of news programmes—article 40.1—(Sarabia et al. 2012a: 175).

However, the real change appears to come with the approval of the Act 6/2012. This act has repealed the previous article—40.1—of the Act 7/2010—, while opening up regional governments the possibility of choosing between various management models for their public media (Lozano 2012): the provision of services through its own bodies, mediums and entities; the assignment to third parties of the indirect management of a service and the production and editing of audiovisual programmes; and the provision of these services through public-private collaboration tools. They may also agree to transform the direct management into indirect management through the sale of the ownership of the entity proving the service. This policy has fostered many normative changes in some autonomous regions, but the point is to find out if changes have been also implemented in practice.

25.2 Methodology

The purpose of this investigation is this to analyse the practical consequences of the main communication policies related to the management of autonomous public broadcasters in the face of the crisis that they are experiencing. Reforms in the regulatory framework, favoured by the Act 6/2012 and some autonomous laws, ought to guarantee practical changes, which foreseeably increase the outsourcing of production in public broadcasters. Public media from Canary Islands, Balearic Islands, Asturias, Aragon, Extremadura and Murcia have already chosen this model. The starting hypothesis is that changes are already underway, albeit very slowly.

25.3 Autonomous Communication Policies in Favour of Outsourcing

The outsourcing of a large part of the production as a way of managing some Autonomous televisions in Spain has existed since 1999, when *Televisión Canaria* emerged. Since then, public media from Extremadura, Aragon, Asturias, Murcia

---

[1]Outsourcing means that public service and its responsibility remain in public ownership against privatisations themselves.

Fig. 25.1 Main awards of production services and content supply of EPRTVIB. *Source* Own elaboration from the data of Europa Press (20/08/2004); Diario de Mallorca (28/11/2008), Contracting party profile of EPRTVIB; Fernández Alonso (2002); Laprovincia.es (16/05/2008), Contracting party profile of RTVC, BOC and BOE

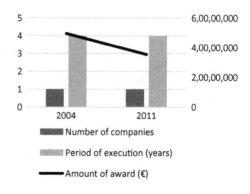

Fig. 25.2 Main awards of program production services of RTVC. *Source* Own elaboration from the data of Europa Press (20/08/2004); Diario de Mallorca (28/11/2008), Contracting party profile of EPRTVIB; Fernández (2002); Laprovincia.es (16/05/2008), Contracting party profile of RTVC, BOC and BOE

and Balearic Islands have also chosen that model.[2] The change of audiovisual policies encouraged the amendment of several rules affecting the Autonomous audiovisual environment. Regulation of Galician and Balearic televisions were modified in 2010 and 2011 with the entry into force of the Act 7/2010 on Audiovisual Communication. Nevertheless, the entry into force of the Act 6/2012 implied more intense changes in Balearic Islands—2013—and Canary Islands—2014—and, to a lesser extent, in Asturias—2014. However, these three public media have hardly changed, even though one could speak of different models.

The system chosen by Balearic Islands and Canarias is the outsourcing of news supply services—Balearic Islands—and the outsourcing of program production—Canary Islands—to an only company for very long periods (Sarabia et al. 2012a). Accordingly, that has not led to a real increase in outsourcing of contents (Figs. 25.1 and 25.2).

On the other hand, Asturias has a different model from the Balearic and Canary one. The Asturian model is characterised by the outsourcing of various services, particularly the production and contents provision (Sarabia et al. 2012a) to various

---

[2]For a detailed analysis of the main audiovisual companies awarded outsourcing contracts of production and content supply in some autonomous televisions, see Sarabia et al. (2012a).

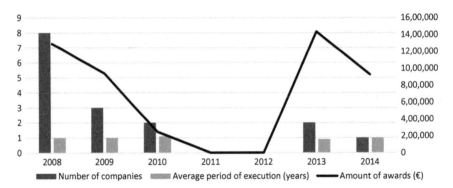

Fig. 25.3 Main awards of production services and contents provison of RTPA. *Source* Contracting party profile of RTPA and BOPA

companies for short periods of time. In short term, nor do there appear to be an outsourcing of contents increase after the national regulatory change—Act 6/2012 —and the regional change in 2014—Contracting party profile of RTPA and BOPA —, probably because it is too recent (Fig. 25.3).

Furthermore, Extremadura and Aragon do increased the outsourcing of contents (Reig et al. 2014), although few new normative changes occurred after the Act 6/2012. Both of these two media have at present a mixed-model, since these two regions have chosen to award these services to various companies for periods of one to two-three years. In the case of Aragon, it is awarding the execution and provision of contents, as well as news production; in the case of Extremadura, it is awarding the provision of audiovisual contents and the service of news coverage (Figs. 25.4 and 25.5).

25.4 Communication Policies in Favour of Indirect Management

Possible privatisation of publicly owned audiovisual entities has always been present in some Spanish regions—Madrid, Valencia, Murcia, and Castile La-Mancha. Despite these attempts, at present only Murcia has chosen an indirect management for its public television, following the entry into force of the Act 6/2012. But these outsourcing of contents is not new for Murcian television. Since 2006 it is awarded to a company the production of news programmes for 7 years (Sarabia et al. 2012b). In August 2012, this television decided to implement a new management model, terminating the outsourcing contract. This brings to an end of emissions of the regional channel, which broadcasts repeated programmes until it has recovered its news service and the programmes, these latter through two awards in 2012 and 2013 (Contracting party profile of RTRM). In 2015 the contract

Fig. 25.4 Main awards of executing, content supply and news production of CARTV. *Source* BOE, Contracting party profile of the Aragon Government and Extradigital. es (09/06/2015); contracting party profile of CEXMA; Official Journal of European Union

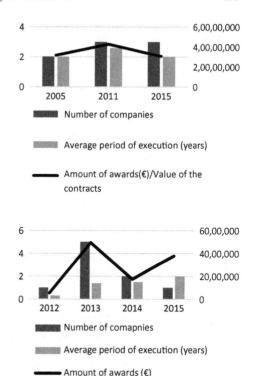

Fig. 25.5 Main awards of direction services, content supply and media coverage service of CEXMA. *Source* BOE, Contracting party profile of the Aragon Government and Extradigital. es (09/06/2015); contracting party profile of CEXMA; Official Journal of European Union

"*Servicio público de comunicación audiovisual televisiva de titularidad de la Comunidad Autónoma de la Región de Murcia*" is resolved in favour of a company (Elpais.com, January 28, 2015 and BOE).

25.5   Conclusions

Autonomous public televisions in Spain are suffering from a major crisis that should be resolved in a very short time. The state government has chosen to make management of regional public televisions more flexible. This measure has resulted in regulatory changes in some autonomous regions, such as Balearic Islands—2013 —, Canary Islands—2014 (Zallo 2015)—and Asturias—2014. These changes have not greatly materialised, while Extremadura and Aragon do increased outsourcings in their autonomous televisions. On the other hand, Murcia has opted for an indirect management of its contents, once allowed by law.

The economic reason is the one that points to the permissiveness of outsourcing to limits that, sometimes, are close to privatisation, but for now figures do not match with that measure. In fact, if we compare the evolution of allowances, advertising income and audiences of traditional public companies and those that have chosen to

Table 25.1 Variation in budgets, allowances and advertising income in the autonomous public broadcasters in Spain (2007–2014)

	Budgets variation 2007–2014 (%)	Allowances variation 2011–2014 (%)	Variation in advertising income 2007–2014 (%)	Variation in audience (2007–2014) (points)
Traditional autonomous regions	−31.0	−15.1	−50.9	− 2.8
Regions that externalise contents	−25.6	−24.4	−35.3	−1.2

Source Own elaboration from the data provided by Campos Freire (2015) and Barlovento Comunicación (2011–2014)

outsource large part of their production, results are not so different, even so there is a minor fall of advertising income and audiences in the second group. A thorough analysis would show that the fall of advertising revenue occurs primarily on Madrid, Castile-La Mancha and Balearic Islands, while the fall of audiences occurs essentially in Andalusia, Madrid, Valencia (before its closure), Canary Islands and Murcia. Therefore one cannot conclude that the outsourcing model is more productive, since it is not in the regions that have chosen it (Table 25.1).

The question at issue is what is the solution to the crisis of the audiovisual public sector in Spain. To find it, it should be assumed that the only problem is not the economic crisis and the solution should not be the simple dismantling of the public sector. The strong politicisation, the fight of political parties for turning television into propaganda, the poor quality of contents and the subordination to private interests has sometimes caused an absolute loss of credibility. None of these problems must be addressed individually as it has been done with some communication policies. The solution is to address them globally and, to do so, there is a need to political will.

References

Barlovento Comunicación. (2011–2014). *Informe de audiencias de televisión.* Retrieved from http://www.barloventocomunicacion.es/publicaciones.html. Accessed at November 3, 2015.

Campos Freire, F. (2015). Financiación e indicadores de gobernanza de la radiotelevisión pública en Europa. In J. Marzal, J. Izquierdo, & A. Casero (Eds.), *La crisis de la televisión pública* (pp. 189–216). Aldea Global.

Fernández Alonso, I. (2002). *La externalización de la producción de los operadores públicos de televisión de ámbito autonómico en España.* Revista Latina de comunicación social: Los casos canario y extremeño. 46.

López Cepeda, A. (2012). Modelos audiovisuales públicos en España. Perfil profesional, empresarial y político de sus principales órganos internos de gestión. *Comunicación y sociedad, XXV*(1), 399–427.

López Cepeda, A. (2015). *Nuevos y viejos paradigmas de la televisión pública. Alternativas a su gobierno y (des) control*. Salamanca: Comunicación Social.

Lozano, B. (2012). Nuevos modelos de gestión y posibilidad de privatización de las televisiones autonómicas (Ley 6/2012). *Análisis*.

Reig, R., Mancinas-Chávez, R., & Nogales-Bocio, A. L. (2014). Un acercamiento en 2014 a la conformación de la estructura audiovisual en España y el caso de Canal Sur TV. *Revista Latina de Comunicación Social, 69*, 571–617.

Sarabia, I., Sánchez, J., & Cano, Á. P. (2012a). La externalización de la producción de los programas informativos en la televisión autonómica pública y su repercusión en el mercado audiovisual regional. In J. C. Miguel de Bustos & M. Á. Casado del Río (Coords.), *Televisiones autonómicas. Evolución y crisis del modelo público de proximidad* (pp. 173–190). Barcelona: Gedisa.

Sarabia, I., Sánchez, J., & Fernández, I. (2012b). La externalización como estrategia del tercer canal. *Trípodos, 29*, 101–115.

Zallo, R. (2015). Análisis de la nueva ley de Radio Televisión Pública de Canarias. *Revista Latina de Comunicación Social, 70*, 322–346.

Press Articles

Diario De Mallorca. (2008, November 28). Los informativos propios costarían a IB3 4,5 millones menos según un estudio del Govern. Retrieved from: http://www.diariodemallorca.es/mallorca/2008/11/28/informativos-propios-costarian-ib3-45-millones-estudio-govern/413906.html. Accessed November 28, 2008.

Elpais.com. (2015, January 28). Murcia privatiza la gestión de su televisión. Retrieved from http://cultura.elpais.com/cultura/2015/01/28/television/1422472763_793633.html. Accessed November 7, 2015.

Europa Press. (2004, August 20). RTVIB adjudica el servicio de noticias de radio y televisión de IB3 a Serveis Balears de Televisió. Retrieved from http://www.vilaweb.cat/ep/ultima-hora/887433/20040820/rtvib-adjudica-serveis-noticies-radio-televisio-dib3-serveis-balears-televisio.html. Accessed September 28, 2015.

Extradigital.es. (2015, June 9). Mediapro vuelve a Aragón TV con la producción de programas. Retrieved from http://www.extradigital.es/mediapro-vuelve-a-aragon-tv-con-la-produccion-de-programas/. Accessed September 29, 2015.

Laprovincia.es. (2008, May 16). Socater gestionó para la TV canaria 40.000 horas de programas con 65 productoras. Retrieved from http://www.laprovincia.es/canarias/2008/05/16/socater-gestiono-tv-canaria-40000-horas-programas-65-productoras/151171.html. Accessed September 28, 2015.

Chapter 26
Consumer-Oriented Business Models in Spanish Cybermedia

Manuel Gago Mariño, Carlos Toural Bran
and Moisés Limia Fernández

Abstract The feasibility of cybermedia and the business models that underpin them are now a classic theme of academic and professional literature within broadcast journalism. We believe it is important to examine revenue strategies aimed at the consumer of Spanish cybermedia in a context of maturity at digital journalism that contrasts with the sharp economic crisis this country is currently going through. The insights of our study highlight the coexistence of two diametrically opposed business models between the most popular and reference cybermedia in Spain and the establishment of 'value-added' strategies in those cybermedia that have established pay per view systems.

Keywords Business models · Online journalism · Convergence

26.1 Theoretical Framework

The convergence process—symptomatic of the 21st century—is a new organizational and production model: the confluence among information and communication technologies. The journalistic convergence, as pointed by Masip et al. (2010: 573) has received attention from researchers on cyberjournalism over the last few years. From an academic perspective, it is possible to identify three differentiated "schools" when studying the convergence processes (Salaverría et al. 2010). Hence, the convergence has been defined as: (1) a product (De Sola Pool 1983); (2) a system (Jenkins 2006); and (3) a process (Lawson-Borders 2003; Dailey et al. 2005).

M. Gago Mariño (✉) · C. Toural Bran
University of Santiago de Compostela (USC), Santiago de Compostela, Spain
e-mail: manuel.gago.marino@usc.es

C. Toural Bran
e-mail: carlos.toural@usc.es

M. Limia Fernández
University of Minho (UMinho), Braga, Portugal
e-mail: mlimia@gmail.com

© Springer International Publishing Switzerland 2017   203
F.C. Freire et al. (eds.), *Media and Metamedia Management*,
Advances in Intelligent Systems and Computing 503,
DOI 10.1007/978-3-319-46068-0_26

Finally, most recent researches at European level have focused on the gradual nature of the process (Erdal 2007; Salaverría and Negredo 2009; García-Avilés et al. 2009). This leads to a large number of case studies on the convergence at international level, such as Moreno's (2009) and López and Pereira's (2010).[1]

The current text aims to determine the consumer-oriented revenue flows established by ten Spanish cyber media involved in convergence processes. It means to analyze the existing business models based on the transaction and user interaction on large traffic and reference media websites in Spain, and verifying differences and similarities among models. We believe it is relevant to carry out this analysis during an acute period of general economic crisis in Spain that has had a strong impact on the fall in advertising revenue in the press, even though its behavior on the Internet as a support continues to grow (13.7 % in 2011 with respect to 2010, more than 900 million euros across Europe.

26.2 Methodology

In order to determine the revenue mechanisms open to the analyzed cybermedia it was decided to carry out a sample drawing from ten media organizations that operate on the Internet and have "reference" roles in the communication system.

The media organizations should reach comparable levels regarding their basic characteristics: they must all operate in the same information market segment –at least in thematic terms and in their territory orientation towards the Spanish sphere. Therefore, these cybermedia must have their head office and newsroom somewhere along the Spanish territory, and publish most of their contents in Spanish. The selected cybermedia born from printed ones were *Elpais.es*, *Elmundo.es*, *20minutos.es*, *ABC.es* and *Lavozdegalicia.es*. All of them, except *20minutos.es*, correspond to mastheads with more than twenty years of existence.

In this study, it is intended to see a comparison of the data for cybermedia born from a masthead of an existing print edition with regards to those new ones designed especially for digital cybermedia and, therefore, without alternative revenue flows or with added benefits for its print edition due to the presence of the digital brand. Hence, we also decided to analyze five reference digital media with different trajectories and chronologies, but featured by innovation and their effective role as actors in the circulation of social, political and dissemination contents in the Spanish Internet audience. They are *Lainformacion.com*, *Eldiario.es*, *Elconfidencial.com*, *Publico.es* and *Elhufftingtonpost.es*.

---

[1]Xosé López and José Pereira are the coordinators of the monograph entitled *Digital Convergence: The reshaping of media in Spain* developed to disclose the results of a macro research at Spanish level on the phenomenon of convergence with the participation of more than 10 universities and around 50 researchers. The project, designated "Evolution of Spanish cybermedia in the framework of convergence", was funded by the Ministry of Science and Technology of the Spanish government (Ref. CSO2009-13713-C05-01,-02,-03,-04).

From our perspective, we chose to carry out an analysis of specific items on the website, accesibles for an ordinary customer, limited to a very restricted time slot: October 2012. It is not intended to check the evolution of business models but to provide a reliable picture of the means of raising money used by Spanish cybermedia in a communicative, social and technological context of particular relevance in this precise conjuncture. All the data were coded systematically.

26.3 Results

26.3.1 Advertising

Advertising is a form of communication for traditional marketing in the media that has been given new flows on the Internet. By analyzing advertising, we try to understand the intensity of the offer of advertising space, the limits and the uses provided and allowed by the media in regards to more exhibitable formats and to more interruptive ones.

26.3.1.1 Banner Formats: More Diversity in Cybermedia Born from Print Editions

The research reveals the widespread use of the conventional banner format in all cybermedia. The research, however, provides an assessment of the use of different formats according to the media and their origin. While the use of conventional banner is shared by web-originated and printed-originated media, we observe an interesting trend towards the diversification in the other formats in the printed-born. The explanation, from our point of view, lies in the power of the brand. They make wider and more complex campaigns and they seem to have a greater ability to attract publicity towards the brand and with a greater user experience.

26.3.1.2 Interstitial and Interactive Formats: The Risk Is for Digital Natives

It is quite interesting the fact that, while cybermedia with a prior printed edition have a greater variation of banner formats, born-digital media are the ones that most frequently invest in more dynamic and interactive advertising formulas. Digital native media are those which invest in more disruptive and interstitial formats, in preference to employing this type of advertising resource in the run-up to the initial loading of the front page. This trend seems to indicate that digital native media accept and are hired by advertisers for campaigns with a higher percentage of technological innovation.

26.3.1.3 Google Adsense, the Great Removal

Although *Google Adsense* advertising program has occupied for many years a large secondary advertising area in cybermedia, the fact is that a 100 % of the analyzed cybermedia no longer use it and have started to develop strategies and tools to offer contextual advertising to their own advertisers. This process of replacement for contextually targeted ad networks of a much more specialized nature was initiated in the summer 2012, founding contextual advertising networks with support from third parties (*El País* through AdMeta technology) or through editorial group associations. The most significant movement is the creation of Premium Audience Network (PAN), behind which are the groups *Unidad Editorial* and *Vocento*, among other representatives. PAN is especially targeted for great advertising accounts and prestigious brands. In contrast to *Google Adsense*, the small advertiser does not participate here in the same channel the large one does.

26.3.2 Paid Content: An Expanding Model

Our analysis allows stating the two flows for digital journalistic business models sustained in Spain at this moment and which can be easily assessed. 4 out of the 5 analyzed major reference media have developed paid-content systems and the only media based on its prime paper edition that remains essentially free is *20minutos.es*, and its paper version is also free. However, the five web-born media built the basis of their digital business through total free access to their contents. We can say that, nowadays, the major Spanish digital press which is based on the analogue world contains methods of payment for content usage, though there are remarkable differences in their business models and sales structure.

26.3.2.1 Pay Per View

All major Spanish cybermedia currently have pay per view services. They all enable the purchase of full daily editions, at the expense of formulas also tested worldwide such as the micro-payment for the downloading of full online articles. Except *El País*, they all provide the purchase of a digital version of one daily edition for 0,79€ (approximately 30 % less than the price of the paper version). Pay per view services, however, are not only limited to the purchase of the full daily edition, but three out of the four also sell additional contents or derived products, such as guidebooks, yearbooks, eBooks or products of services generated from broadcast news workflows.

26.3.2.2 The 'Kiosk' Formula and Additional Contents and Services

Spanish media tried to increase the subscription value. Hence, a distinction can be made between two types of strategies in the offer: the benefits in terms of accessibility (multiplatform access or the consultation of a facsimile edition of the newspaper) and content. In this last case, solutions can be more complex, but we can group them under the concept 'Kiosk'. Editorial groups, or alliances across different groups, offer low access at a same price to a large number of different mastheads, both from general press of national coverage and regional press, as well as to specialized press and magazines (sports or life styles). Multiplatform service combines reading software for facsimile editions with a services club that offers discounts in shows and free access to cultural and entertainment events.

26.3.3 Paid Apps and Free Apps

Different industry or media consultancy reports have highlighted the opportunity posed by tablets and smartphones as a means to reestablish pay channels that user behavior, culture and habits on the web do not permit. Our research allows placing this industry statement in its practical context: we did not find a clear relationship between media that implement payment systems for content usage and specific paid apps (or shared with other media), though payment is not for application download but for inline content download.

26.4 Discussion and Conclusions

It is feasible to depict the existence of two major models that define very different concepts of business flows in Spanish digital communication media.

On one hand, we have traditional mastheads based on their prime print version and that were integrated into the digital world in the nineties in the majority of cases. They combine, on one hand, the symbolic capital and the structure of an editorial office consolidated as social actor for decades in printed editions and for a long time in digital media with their experience in internet operations. These editorial offices strongly invest in a mixed model that combines a more optimized advertising offer of access to free contents, with more ambitious strategies that, in many cases, pass through convergence initiatives. These initiatives are supplemented by marketing strategies of their own brand. In the case of *Kiosks* incurred in alliances by different editorial groups, those imply very interesting value-added cases linked to a technological and viewing platform. Sometimes seems they are suspicious of the media own brand's capacity to generate digital costumers by itself. Media groups are conceiving the brand in an integrated form, though accessibility through these platforms is an obvious value included here. But mastheads follow, at the same time, a horizontal growth strategy of their apps, aimed at defining their own niches of consumption from gratuity models that perfectly fit in horizontal and vertical convergence strategies in the media.

On the other hand, we observe the born-digital media model. They all invest in 100 % free access to their contents through any of the well-established triad of supports (web, cell phone, and tablet). Although they are still far away from consolidated mastheads in terms of volume, digital native media seem to find a niche through unrestricted access to content, but in the light of the data they also hold cultural differences regarding printed-originated media: a larger flexibility in advertising formats and investments to achieve and diversify revenue flows without leaving the gratuity model. Some of them have been in the media scenario for many years now and others have recently been joined, but they are dynamic actors that compensate their relatively low audience shares with visibility in external platforms such as social networks.

It is clear that we are facing a dichotomy that indicates there isn't a single direction in business models and that former or newly formed business cultures play a major role in determining and shaping the business models in Spanish cybermedia.

Acknowledgments This paper was developed with data collected in contextualization and state of the question works of the Research Project "Innovation and development of the cybermedia in Spain. Journalistic interactivity Architecture in multiple devices: information, conversation and services formats—reference CSO2012-38467-C03—, funded by the Ministry of Economy and Competitiveness of Spain.

References

Dailey, L., Demo, L., & Spillman, M. (2005). The convergence continuum: A model for studying collaboration between media newsrooms. *Atlantic Journal of Communication, 13*(3), 150–168.

Díaz Noci, J. (2012). A history of journalism on the internet: A state of the art and some methodological trends. In Comunicación presentada al *XII Congreso de la Asociación de Historiadores de la Comunicación*. Barcelona: Universitat Pompeu Fabra.

Erdal, I.-J. (2007). Researching media convergence and crossmedia news production. Mapping the field. *Nordicom Review, 28*(2), 51–56. Retrieved from http://www.nordicom.gu.se/common/publ_pdf/255_erdal.pdf

Jenkins, H. (2006). *Convergence culture. Where old and new media collide*. New York: New York University Press.

Lawson-Borders, G. (2003). Integrating new media and old media: seven observations of convergence as a strategy for best practices in media organizations. *The international Journal of Media Management, 5*(II), 91–99. Retrieved from http://www.mediajournal.org/ojs/index.php/jmm/article/viewPDFInterstitial/10/3

López, X., & Pereira, X. (Coords.). (2010). *Convergencia digital. Reconfiguración de los medios de comunicación en España*. Santiago de Compostela: Servicio Editorial de la Universidad de Santiago de Compostela.

Masip, P. et al. (2010). International research on online journalism: hypertext, interactivity, multimedia and convergence. *El profesional de la Información, 19*(6).

Salaverría, R., García-Avilés, J. A., & Masip, P. (2010). Concepto de convergencia periodística. In X. López García & X. Pereira Fariña (Coords.), *Convergencia digital. Reconfiguración de los medios de comunicación en España* (pp. 41–63). Santiago de Compostela: Servicio editorial de la Universidad de Santiago de Compostela.

Part V
Corporate and Institutional Communication

Chapter 27
The Situation of Digital Strategic Communication in Ecuador and Other Countries in Latin America: The Management of the Community Manager

Fanny Paladines, Carlos Granda Tandazo
and Valentín Alejandro Martínez Fernández

Abstract At present, companies face the challenge of addressing the new forms of communication by positioning their brands. This in turn results in the emergence of new professional profiles, such as the Community Manager. The present research analyses quantitative information from experts in the digital area of Ecuadorian, Mexican, Argentinian, Colombian, Venezuelan, and Uruguayan organizations. The study is focused on the study of the role of Community Managers in strategic communication. Inputs of particular interest for this research include: (a) the importance attached to the training and tools handling to address strategic issues; (b) the role of the manager of virtual communities related to companies, in order to know and control communication flows and consolidate a brand or business; (c) the participation, interaction, creation of engaging content, positioning and online reputation. This paper is aimed at contributing to research on the subject. In the region, figures providing evidence in this field are scarce and of little relevance.

Keywords Community manager · Strategic communication · Digital

27.1 Theoretical Framework

During the last years one of the most emblematic communication projects in Ecuador was the first "Strategic Communication Observatory, with the purpose of knowing the status and trends in Ecuador and other countries, including areas such

F. Paladines (✉) · C. Granda Tandazo
Technical Particular University of Loja (UTPL), Loja, Ecuador
e-mail: fypaladines@utpl.edu.ec

C. Granda Tandazo
e-mail: cwgranda@utpl.edu.ec

V.A. Martínez Fernández
University of A Coruña, A Coruña, Spain

© Springer International Publishing Switzerland 2017   213
F.C. Freire et al. (eds.), *Media and Metamedia Management*,
Advances in Intelligent Systems and Computing 503,
DOI 10.1007/978-3-319-46068-0_27

as internal, corporate, institutional, commercial, and digital communication, as well as CSR strategies. The research group *Gestión de la Comunicación Estratégica* from the Department of Communication Science of the Technical Particular University of Loja—Ecuador—and scholars from the National University of Cordoba—Argentina—, carried out this initiative. The study follows the lines already sketched by the same group of lecturers, taking as a reference the works carried out in Europa and the United States. Also, it opens the way to new questions and challenges of new trends in strategic communication.

27.1.1 Digital Strategic Communication

Digital communication means any way of communication through digital techno-logical devices. It has emerged thanks to the new information and communication technologies—the Internet, mobile and satellite communication, and digital tele-vision, among others (Islas and Baird 2006). Digital technology has worked mainly with electronics and telecommunications, and information coding and transmission, beyond the limited scope of analogue technologies (Gómez Aguilar and Martínez García 2010). The possibility to send data, voice and pictures via new digital devices closes even more the gap between virtual and on-site services (Osorio 2002). Digital communication is an inescapable truth, offering more and more options of interacting.

As in the marketing mix—price, product, place, and promotion—, Gálvez Clavijo (2010) claims that "*digital marketing is based on the 4F—flow, function-ality, feedback and loyalty (fidelidad in Spanish)—, since these are the variables that make up an effective strategy*" (p. 20).

27.1.2 Enterprise 2.0

Until only recently no one would have imagine how the use of the Web 2.0 could modify the entrepreneurial management. Its emergence led to the evolution towards new forms of communication, not only for assisting customers via online, but also to make them an active part of companies. Customers now have the possibility of creating content and recommending products and services to other users (Carballar Falcón 2013).

Some authors such as Celaya Barturen (2009) refer to a new age in commercial communication, in which former marketing and advertising practices are now ineffective in the Social Web. Ramos Serrano et al. (2009) consider that conven-tional advertising is becoming extinct. Companies are losing huge amounts of money because they do not reach their targets and stakeholders. As the authors above-mentioned note, "*Google, AOL, Yahoo, MySpace and Microsoft are likely to become the new kings of advertising, without disregarding the significant market power of general-interests TV*" (p. 25).

27.1.3   Work Team 2.0

The problem companies are facing today is the coherent implementation of a customer-oriented strategy (Cuesta Fernández and Alonso 2010). The complexity of the media and social networks, and their management, planning and community managing require a specialized team. Some of the new profiles of the Social Media team are: Social Media Manager/Strategist, Content Manager; Content Curator; Professional Blogger; Web Analyst; Online/Digital Marketing Manager.

27.1.4   Community Managers

The explosion of social media aroused growing interest among companies. The figure of the Community Manager is of particular relevance among social media strategies. If we look forward a definition, we may see that institutions disagree. In fact, AERCO—the Spanish Association of Responsible for Online Communities—states:

> Community Managers are responsible for sustaining, increasing and somehow protecting digital relations between companies and their customers. They have the knowledge about the needs and strategies of the organizations and the interests of customers. Also, they know the objectives of their companies and act accordingly. Speaking very generically, community manager are those who preserve thee digital identity of a company (Rojas 2011: 45).

> They are responsible of the community and thinkers of medium and long-term strategies to address communication and interaction services. Their main role is to bring together members around a common interest (Paladines Galarza 2012: 128).

Community Managers should play a social and agitator role, since until publics foster conversations naturally, they have to propose topics that will make the community a dynamic space (Martínez Priego 2009). Organizations should aim at having community managers, although Burgos García and Cortés Ricart (2009) note that their presence in companies is compatible with external collaborations to implement an action plan.

27.2   Methodology

This study is descriptive, quantitative, and aims at knowing the current situation and the management of community managers in various Latin American countries. Starting from the database of the "Strategic Communication Observatory in Ecuador", 100 Ecuadorian companies were selected for undertaking a telephone survey, but only 22 answered. The goal was to know whether there are specific decision-makers in digital communication. After an online search, 7 more companies and agencies were selected. These ones were sent a 12-item question-naire by email, obtaining 15 valid responses. We also conducted a search for international experts in related digital forums and social networks—*Twitter*—and *Facebook*. It has been identified 18 community managers from countries such as Mexico, Argentina, Colombia, Venezuela and Uruguay, but replied only 10. The research tool was a questionnaire divided into five parts: (a) Community Manager profile; (b) Management of digital communication; (c) Digital communication tools; (d) Contents; and (e) Assessment and results.

27.3 Results

Digital activities in Ecuador are under the responsibility of graduates in Marketing and Advertising and public relations responsible. In other countries, that respon-sibility lies with graduates in community management or with people with a Master in Social Media Planning and Marketing and Digital Business. Digital communi-cation managers in Ecuadorian companies consider that a community manager is supposed to be a strategist and creative, and to have comprehensive knowledge about digital tools. 100 % of those polled at an international level accept that a community manager should be a good writer. At the national level, the results show that the national management of digital communication contributes 53 % to the achievements of companies´ achievements. At the international level, there is a high percentage is much higher.

Figure 27.1 shows that the majority of national companies have passed the learning stage in the construction of a social media strategy. 33 % of them are planning their strategies and have presence on social networks. Also, 20 % have been able to gain the loyalty of their publics through digital actions.

Nonetheless, according to previous research by Paladines Galarza et al. (2014), the investment in digital communication in Ecuador not as good as it could be,

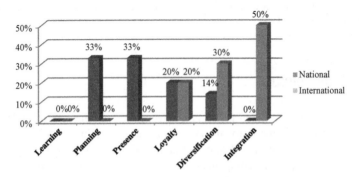

Fig. 27.1 Development stages of a digital strategy. *Source* Research by the Strategic Communication Observatory in Ecuador

especially if it is compared with companies in developed countries. On the other hand, international companies are in the stage of maturity of the development of their digital strategy. Also, 50 % are in the advanced stage of integration (Fig. 27.2).

In Ecuador, the most used social networks for external communication activities are *Facebook* and *Twitter*. The main objective is to promote their brands and achieve reputation, generating interaction with their publics. The use of Twitter is increasing among other Latin American companies, and the level of activity is similar to Facebook. There is greater use of YouTube and Skype.

According to Fig. 27.3, the most used social network in internal communication is Facebook. In Ecuador, the use of the Internet is increasing, but it could be better exploited. In Twitter for example, there can be identified opinion leaders who generate content and, in YouTube and other social networks, it can be reached and measured virality, and thus increased the sense of belonging to the company. At the international level, 70 % of companies are using Facebook and Twitter to improve their identity. Unlike national companies, there is greater use of social networks, such as blogs.

In Ecuador, unlike some other countries, there is little use of apps. 87 % of companies use apps to specific promotions. 67 % of companies use Whatsapp and SMS. The analysis shows that the most employed tools for measuring activity in social networks are: *Hootsuite, Google Analytics, Facebook Insights and SumAll.*

When consulting respondents on the evolution of networks in the nearest future, 100 % notes that they should unify and complement digital strategies. They consider that social networks would enable the public to interact with companies. Also, 93 % consider that viral contents will be in some way sponsored. 87 % say that

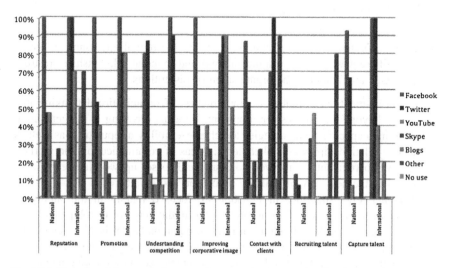

Fig. 27.2 Use of social networks in external communication—national and international. *Source* Research by the Strategic Communication Observatory in Ecuador

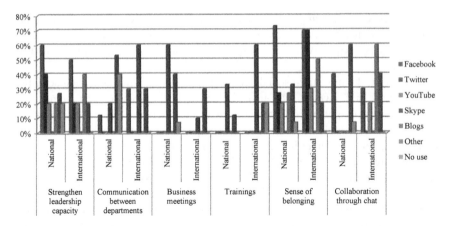

Fig. 27.3 Use of social networks in internal communication at the national and international level. *Source* Research by the Strategic Communication Observatory in Ecuador

companies will have to adapt their ways of communicating contents. Pictures and video are more attractive, and social networks such as Instagram and Snapchat are increasing their number of users. 90 % of experts in other countries consider that investment companies make in social media, together with a proper planning of the community manager, will increased the return on investment in shorter periods. Likewise, 90 % affirm that companies will change the way to attract talents, decrease one-on-one interviews and offer work from home.

27.4 Discussion and Conclusions

- Ecuadorian organizations are required to act in the digital environment, which demands different ways of managing internal and external relationships, restructuring budgets and adapting the organization of institutions.

- The above-mentioned implies a new structure of organizational thinking that goes far beyond the business environment. It means investing in the generation of internal capacities and strategies to achieve visibility a reputation, as well as effective relationships with stakeholders. Areas of communication and marketing have to reinvent themselves, to learn, and to set up working teams specialised in the new way of building reputation and loyalty.
- Social networks strategies in Ecuador are focused mainly in the supply of products and services and are located in the initial stage of development. The implementation of digital culture in internal communication is still a challenge for companies in Ecuador. On the other hand, international organizations make greater use of digital tools and engage their employees in a creative manner.

- The crisis communication is the most difficult part of managing companies. National and international experts use Facebook and Twitter to monitor clients, but few implement a prevention plan.

In short, online strategic communication will open up new possibilities for creating close engagement with clients, who had hitherto felt unconnected to them. At present, customers are part of changes, structures, and designs of products and services.

References

Burgos García, E., & Cortés Ricart, M. (2009). *Iníciate en el marketing 2.0. Los social media como herramientas de fidelización de clientes.* Spain: Gesbiblo, S. L.

Carballar Falcón, J. (2013). *Social Media. Marketing personal y profesional.* México: Alfaomega Grupo Editor, S.A.

Celaya Barturen, J. (2009). *La empresa en la Web 2.0. El impacto de las redes sociales y las nuevas formas de comunicación online en la estrategia empresarial.* Barcelone: Gestión 2000, Grupo Planeta.

Cuesta Fernández, F., & Alonso, M. (2010). *Marketing directo 2.0.* Spain, Barcelone: Centro Libros PAPF.

Gálvez Clavijo, I. (2010). *Introducción al Marketing en Internet: Marketing 2.0.* España, Madrid: Editorial INNOVA.

Gómez Aguilar, A., & Martínez García, M. (2010). Gabinete de Comunicación y Educación de la Universidad Autónoma de Barcelona. Retrieved from: http://www.gabinetecomunicaciony educacion.com/files/adjuntos/Redes%20sociales%20y%20dispositivos%20m%C3%B3viles% 20oportunidades%20y%20amenazas%20de%20la%20conexi%C3%B3n%20permanente.pdf. Accessed June 5, 2015.

Islas, O., & Baird, J. L. (2006). *Mediamorfosis de la televisión e Internet desde la óptica mcluhaniana.* In s3.amazonaws.com.

Martínez Priego, Ch. (2009). Escuchar a los social media en E. Sanagustín (Coord.), *De 1.0 al 2.0: Claves para entender al nuevo marketing.* Spain: eBook.

Osorio, C. (2002). Enfoques sobre la tecnología. *Revista Iberoamericana de Ciencia, Tecnología, Sociedad e Innovación.* Retrieved from: http://www.campus-oei.org/revistactsi/numero2/ osorio.htm. Accessed June 3, 2015.

Paladines Galarza, F. (2012). Tesis. *Gestión de la comunicación de la marca en las redes sociales: Estudio de tres casos de campañas con Facebook en Ecuador.* University of Santiago de Compostela, Faculty Communication Science. Santiago de Compostela, Spain.

Paladines Galarza, F., Granda Tandazo, C., & Velásquez Benavides, A. (2014). La marca ecuatoriana y su gestión en redes sociales. *Razón y Palabra, 86.*

Ramos Serrano, M., Garrido Lora, M., & Rodríguez Centeno, J. C. (2009). *Publicidad y comunicación corporativa en la era digital.* Madrid: Grupo Anaya, S.A.

Rojas, P. (2011). *Community Management en una semana.* Barcelone, Spain: Editorial Gestión 2000.

Chapter 28
Patronage and Sponsorship in the Online Communication Management of Ecuadorian Companies for Improving Visibility in Digital Social Media

Mónica-Patricia Costa-Ruiz, Verónica-Alexandra Armijos-Buitrón, Jhoana-Elizabeth Paladines-Benítez and Raquel Tinoco-Egas

Abstract The advertising channel through the social media shows a lack of efficiency due to the so called "saturation effect", which stimulates audiences attention and protects them from advertising impacts. Therefore, advertisers have turned to new ways of communication, such as the action communication because of the importance and application it has. Action communication comprises two complementary tools such as patronage and sponsorship. Action communication has emerged as a major strategic tool in corporate and commercial communication for the management of Corporate Social Responsibility and the relationship with stakeholders through accountability shown in their websites. While in the United States and Europe this strategic and operational approach of communication has been an important development in the past twenty years, in Latin America it is at an early stage, and in the case of Ecuador, it is still in a hatching stage. To determine the current state in Ecuador, an exploratory research was carried out by analyzing the contents of the websites of the twenty five first companies which have mostly turned over as well as their presence in social media and the perception from their audiences.

Keywords Action communication · Sponsorship · Patronage · Corporate social responsibility

M.-P. Costa-Ruiz (✉) · V.-A. Armijos-Buitrón · J.-E. Paladines-Benítez
Private Technical University of Loja, Loja, Ecuador
e-mail: mpcosta@utpl.edu.ec

V.-A. Armijos-Buitrón
e-mail: vaarmijos@utpl.edu.ec

J.-E. Paladines-Benítez
e-mail: jepaladines@utpl.edu.ec

R. Tinoco-Egas
Technical University of Machalaa (UTMACH), Machalaa, Ecuador
e-mail: raqueltinocoegas@gmail.com

© Springer International Publishing Switzerland 2017
F.C. Freire et al. (eds.), *Media and Metamedia Management*,
Advances in Intelligent Systems and Computing 503,
DOI 10.1007/978-3-319-46068-0_28

221

28.1 Literature Review

28.1.1 Communication by Action

The development of information technology has led significant changes in the model of social communication and its process. From a linear approach, in which media plays an essential role in the dissemination and extension of the message, it has evolved into a network communication or net-communication supported by "nodal expansion", immediacy of the response and viral feedback; which gives media an important role and to a large extent the control of the message to the recipients, which in turn, form a new type of audience made up by active individualities, thus creating a thick "neuronal wave" that gives rise to the origin of "synaptic connections communication " where the reformulation of the messages takes place.

The net communication supposes a new behavior on the part of the consumer of information products, both, in the access to information and its relationship with the social media. Media loses the predominance in the communication process in favor of the empowerment of the audience. Now, it is not the media, it is the audiences who create statements of opinion. Obviously this model of communication also affects the relationship between the social agents with their public and, consequently, in the strategies which they carry out to achieve its objectives basically, in regards to create a positive image and influence on the decisions of the receivers of their messages.

The main problem of conventional advertising is based on the loss of effectiveness of "analog media" and its progressive decreases of their audiences, and the questioning of its credibility enhanced by the saturation of advertising messages on websites and the subsequent stimulation to the drop of attention to the receivers of messages.

Along with this change in social communication, companies have met another change on the requirements of their target audiences in terms of the role they have to fulfill within the society in which they operate their business. A requirement based on the practice of corporate responsibility which transcends to the social sphere and to the indispensable display of itself in order to create or strengthen a positive public image and position it in the minds of the receivers of the messages.

Consequently, it is the boost, which from the beginning of this century experiences unconventional advertising tools and within them the important role of the Action Communication which includes two currently essential tools for communication policies of big companies, which are the sponsorship and patronage. Tools, which on the other hand, have already a significant use in the communication strategies of medium-sized companies too.

Action Communication should be considered as a means that allows the association and promotion the image and reputation of a brand and/or a company, through an event or a leader opinion that have a degree of appeal to a specific target audience (Bello Acebrón et al. 1996) and contribute to create or strengthen a favorable positioning. In this regard, Rodríguez del Bosque et al. (1997) state that action communication must have the following characteristics: (a) the contribution which given in cash or material to a company conducts it to promote or support an activity of general interest, it could be leisure, sports, culture, humanitarian, environmental, and so on; (b) the activity does not have to be part of the main task of the company, otherwise it would be considered as a sales promotion; (c) the company always gets a profit in terms of reputation and public image.

Action Communication supposes certain relationship between the company, its brands and the market which it operates in, through a new connecting element, specified in an event that, in turn, can take different forms in which the public identify a specific image associated with the brand and/or sponsoring company and all of it a particular environment (Martínez Fernández 2004).

In general terms, Patronage is considered as financial or material support given to a work or person in order to execute activities that are matter of public interest and without direct compensation for the beneficiary. On the other hand, Sponsorship is the given contribution to a person, to a product or to an organization, as an exchange for getting the benefit of including its brand relating it to the sponsoring work in such a way that it highlights from any other type of advertising (Piñuel Raigada 1997). Nevertheless, in today's Information and knowledge Society there is certain convergence between patronage and sponsorship, which allows them to be considered as an investment from which the company expects to get certain intangible and tangible returns (Rabanal 2006).

28.1.2 The Communication by Action in Latin America

Action Communication in Latin America still has an incipient character, although in recent years, it shows some growth, as consequence of the boost that governments are giving, especially in relation to the conservation of cultural heritage, the development of cultural programs and high impact of sporting events. In this regard Brazil is the most outstanding country, it promotes rules that encourage patronage and sponsorship, whose positive impact has been observed in most organized events, as it is the case of the World Cup and Olympic games, where it has been, and still is, essential the participation of the private sector.

However, in Latin America, it is important to emphasize that the model of donation subsists. In a legal system different from the one that Action Communication supports and in which by tax exemptions different benefits to companies that contribute to the development of general interest activities are stipulated. Thus, the possibility of getting tax credit as compensations for donations for legal persons exists in Brazil, Chile, Colombia, Ecuador, Paraguay, Peru and Uruguay (Antoine 2009). For example, in Brazil, with the Rouanet law of 1991, it is possible that corporate sponsors may make a deduction of up to 4 % in income tax, which, with its application increased the total amount of contributions to culture from 14 to 270,000,000 of Reales between 1994 and 1998. Another country in which the implementation of public policies and enactment of laws driving patronage and sponsorship have led to positive results is Chile.

An interesting reflection on Patronage in Latin America is stated by Antoine (2009) which concludes that it has not been possible to adopt a more cultural patronage and flexible policies, partly because of the inability to produce statistical data to plan and evaluate the effects of reforms tax or perhaps due to the dominant orientation of attending the most urgent or essential need with the resources gotten from the taxes as well as because of the lack of dissemination of laws and regulations which make possible donations through tax benefits.

28.2 Methodology

The research is exploratory, descriptive and correlational whose objectives are: To define action communication status in the Ecuadorian companies, to set the current state of sponsorship and patronage in Ecuador and to determine the relationship between the level of sales and the actions of patronage and sponsorship. Also, the inductive method was used, it allowed to get specific information on the current status of the action communication in Ecuadorian companies. For this, the universe was large companies according to the ranking giving by Superintendencia de Compañías del Ecuador, 2014, it helped to determine the sample. The sample selected was based on three factors: sales, number of employees and market.

Identified and selected the companies, the situation of action communication was diagnosed, through the analysis of web pages, taking into account as variables the actions of patronage and sponsorship of these companies. The diagnosis was made through the use of a sheet, which allows data collection. The data obtained allowed a quantitative analysis to determine the application of action communication (patronage and sponsorship) in the companies under study. Also it was determined the relationship between the level of sales in the most representative segment and the action communication.

28.3 Results

28.3.1 Ecuadorian Case

According to secondary sources it is has been determined the Ecuadorian legislation does not promote polices of patronage and sponsorship actions, which limits the private sector to implement such activities by not having tax incentives.

Thus by Executive Decree 580, the President of the Republic of Ecuador, amended the Regulations for the implementation of the Incentives Law on Production and Tax Fraud Prevention, which took effect on January 1, 2015; this law determines that the limit of 4 % for costs and advertising expenses, set by the Incentives Law on Production will not apply in the case of expenses incurred by patronage and organization of sporting, artistic and cultural activities; which shows the lack of support from the government for the development of these actions.

The Ministry of Culture, faced with this reality, has empowered of this issue and seeks to promote patronage and sponsorship activities, by the creation of the appropriate legislation. In addition to this, the Ministry of Culture, encourages different projects such as Development and Diversity Program, and Development and Diversity Cultural for the poverty reduction and social inclusion Project.

28.3.2   Characteristics of the Companies Analyzed

Based on the above said, the current state of best positioned twenty five companies in the market were analyzed in relation to its sales, number of employees, geographic location and coverage. As shown in Figs. 28.1 and 28.2, in 2014, 64 % of companies reported sale levels between 401 and 800 million of dollars, and the 38 % of these companies employed between 1 and 1500 people. Also Ecuadorian companies are located mostly in the province of Pichincha, 52 % of the companies are located there; 44 % are located in Guayas and 4 % in Manabí, The 56 % of these companies have nationwide coverage and 44 % have international coverage.

Fig. 28.1 Sale level millions USD. *Source* Authors

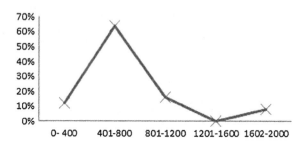

Fig. 28.2 Numbers of employees. *Source* Authors

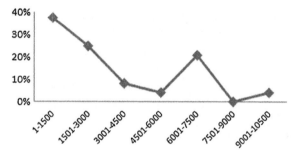

28.3.3   Action Communication in Ecuador

The action communication through web pages is an activity that has gained importance in recent years, so much so that according to the review done in October 2015, 68 % of the companies show information, through various communication strategies, about company linking with their stakeholders. Companies show higher patronage than sponsorship since most 56 % of them benefit different activities and only 28 % of them record sponsorship activities, which show that sponsorship is still an uncommon practice in Ecuadorian companies (Fig. 28.3).

Patronage strategies are mainly shown in actions aimed at: education, sports, community action, health care, entrepreneurship support, food donating, and environment. As shown in Fig. 28.5 62 % of researched companies focus their efforts on patronage of educational activities; 54 % of them aim actions towards community through foundations and programs that allow linking with this stakeholder (Fig. 28.4).

Regarding sponsorship activities, the companies develop activities to support the national talent in the craft and musical fields; and the rescue and promotion of the cultural heritage of Ecuador.

Fig. 28.3 Action communication in Ecuador. *Source* Authors

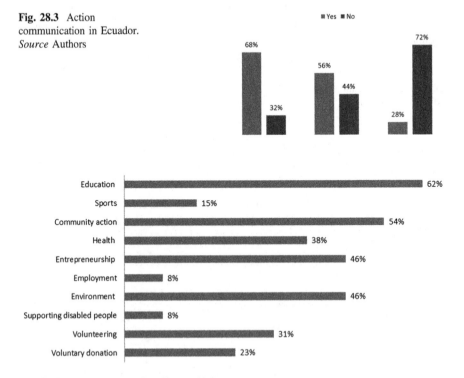

Fig. 28.4 Patronage strategies. *Source* Authors

Fig. 28.5 Action
communication by sale range.
Source Authors

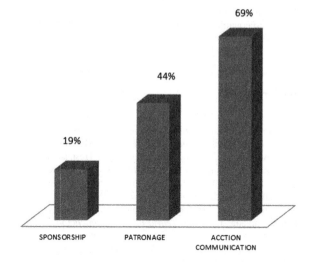

As show in Fig. 28.2 the most representative company segment is placed in the sales range of 401 and 800 million dollars, so, it is evident that in this group, that 69 % does develop action communication activities, 44 % of them different sponsor activities and 19 % does sponsorship activities.

28.4 Discussion and Conclusion

From the study done in October 2015, it about action communication through websites of the twenty-five first companies placed within sale range in 2014 in Ecuador, it can be concluded that 68 % of them show their linkage with their stakeholders, which can be seen through the information displayed on their websites.

According to the analysis in the theoretical framework about the state of the art of patronage and sponsorship in Ecuador, it's implementation is considered as an strategy of social responsibility, in order to meet the expectations of stakeholders in relation to the improvement of the economic, social and environmental areas. The study shows that the patronage has a greater representation in companies, 56 % of them develop actions in this area, while sponsorship is less applied, because only 28 % is aimed to the cultural sponsorship actions. This could be due to the Ecuadorian government's policy inexistence that encourages this type of strategies.

Based on the turnover level of the Ecuadorian companies, it was found that they are mostly located in a sale range between 400 and 801 million dollars during 2014. These companies are more focused on patronage actions than sponsorship.

Acknowledgments This research had the support of Dr. PhD. Valentin-Alejandro Martinez-Fernandez in the Prometheus—SENESCYT linking program with number CEB-008-2015.

References

Antoine, C. (2009). *América Latina y las políticas públicas de mecenazgo privado a la cultura: ¿Una nueva oportunidad perdida?*. Chile: Boletín Gestión Cultural.

Bello Acebrón, L., Vásquez Casielles, R., & Trespalacios Gutierrez, J. (1996). *Investigación de Mercados y Estrategia de Marketing*. Madrid: Cívitas.

Martínez Fernández, V. A. (2004). Comunicación por Acción: patrocinio y mecenazgo. En I. Bel Mallén (Coord.), *Comunicar para crear valor: la dirección de comunicación en las organizaciones* (pp. 245–258). Pamplona: Eunsa.

Piñuel Raigada, J. (1997). *Teoría de la comunicación y gestión de las organizaciones*. Madrid: Síntesis.

Rabanal, M. (2006). *El Patrocinio y Mecenazgo Empresarial en España. Ejercicio 2005*. Madrid: Círculo de la Responsabilidad Social.

Rodríguez del Bosque, I., De la Ballina Ballina, J., & Santo Vijande, L. (1997). *Comunicación comercial. Conceptos y aplicaciones*. Madrid: Cívitas.

Superintendencia de Compañías. (2015). Ranking empresarial al 2014. Recuperado el 9 de agosto de 2015 de http://appscvs.supercias.gob.ec/rankingCias/

Chapter 29
Online Organizational Communication: The Communications Department *Version* 2.0 in Tech Companies

Berta García Orosa

Abstract In the last decade of the 20th Century, communications departments opted to create an active online presence in order to deliver information to their audiences. After a phase of relatively homogeneous strategies and parallel paths, those companies with online communications departments have chosen innovation and unique tactics adapted to their specific situation as a strategic element. As this real-world evolution was taking place, academic literature tracked its progress and analyzed this transformation of online communications departments featuring, until the present day, a panorama of relative creativity tending towards multimedia content and hypertextuality but with smaller steps being taken towards interactivity with the end user (Dader et al. 2014; Sánchez Duarte and Rodríguez Esperanza 2013). This paper aims to describe this progress and the current status of online organization communication, in addition to painting a picture of the most recent tendencies and innovation proposed by the most innovative tech companies in Spain.

Keywords Organizational communication · Internet · Participation · Social networks · Public relations

29.1 Theoretical Framework and Methodology

The current structure of organizational communication in Spain has been consolidated over the last 50 years through processes of institutionalization, diversification and specialization of such activity. The recent history of communications departments and consultancies shows boundaries that are still vague and imprecise but with some milestones indicating change. The first public relations agency in Spain came to being in 1960 (Orosa 2005:60) and was followed by other practices that the arrival of multinational companies introduced in the country (Martín 1998: 37).

B. García Orosa (✉)
University of Santiago de Compostela (USC), Santiago de Compostela, Spain
e-mail: bertago@gmail.com

© Springer International Publishing Switzerland 2017
F.C. Freire et al. (eds.), *Media and Metamedia Management*,
Advances in Intelligent Systems and Computing 503,
DOI 10.1007/978-3-319-46068-0_29

The transition to democracy represents an important turning point in the history of communications departments, which truly begin to take off in the 1980s. With more functions and larger departments increasingly linked to company leadership, communication directors begin to take on various roles and challenges in the communication between company and audience.

In this evolution of organizational communication the creation of digital work and communication environments became strategic points among communication directors.

The turn of the century and the arrival of social networks as communication are two important points in this story, though they are not the only ones. The creation and above all the expert management of companies' online presence has been a constant demand made in both the professional (VVAA 2013)and academic worlds in recent years (Andreu 2006; Esparcia and Martínez 2005; Del Hoyo 2006; Sardà et al. 2013; Arroyo et al. 2013; Gómez Nieto et al. 2012).

This demand was in response to communications departments' slow adaptation to the online environment and a general failure to take advantage of the Internet's potential. The inclusion of technological advances in the daily work of communications departments began to blossom as the Internet became immensely popular and made a place for itself in the vast majority of communications departments. Nonetheless, the physical presence of the latest technological innovation on companies' websites did not always entail a true and efficient use of online tools—social networks, hypertextuality, multimedia, etc.—in the company's communication strategy. Rather, it was simply part of the company's image of modernity that may very well not have featured any actual activity. In this first phase some 90 % of communications departments created websites fundamentally dedicated to dumping content created offline in an online environment without a true modification of the communicative process. Thus, these were basic, version 1.0 communications departments that essentially went hand in hand with a unidirectional, asymmetric information model in which the departments would use the Internet as just another element of their plan which only aimed to transmit information to those who would receive it (Orosa 2009) Some changes began to bring this version 1.0 closer to the so-called transparent department, or department 2.0, which entailed the audience's participation in a symmetrical, bi-directional manner.

This department 2.0 would lead to, in its most advanced phase, the creation of a collaborative community that would participate in the development of messages. It would mean the beginning of a new communicative model in which participative structures would be created in order to obtain mutual advantages through which the communications department would achieve greater online presence and earn a good return stemming from its positioning in social networks. This virtual community would allow communications departments to take advantage of all of the Internet's resources, including feedback received through the receptors' various channels. In this way, the communications department would become just another and cease being the sole transmitter of information, despite a privileged position, as it had been in previous phases; instead it would participate (as a particularly legitimized actor in some cases) in the process.

Over the last decades, the evolution of communications departments has varied depending on the company. Whereas many companies operate practically within communication model 2.0, with the creation of new professions such as community manager, social media strategist, social media analyst or social media manager, others have failed to adapt to the internet or even lack an online presence altogether (Olvera Lobo and López Pérez 2013).

In this context, researchers in the field tend to describe the model as a continuation of the propaganda model—persuasive, providing unidirectional and asymmetric information—rather than an attempt to establish open communication (Dader et al. 2014: 116) with an enduring presence (Sánchez Duarte and Rodríguez Esperanza 2013).

This article aims to carry out an in-depth analysis of the situation in tech companies, one of the sectors most essentially linked to change, when five years had passed since the game-changing turn of the century. The methodology, in addition to a review of the relevant literature, consists of content analysis of a sample of data. The list of companies was obtained from the "List of Innovative Companies" from the credit line of the Official Credit Institute Technological Fund 2013–2015 (Instituto de Crédito Oficial Fondo Tecnológico in Spanish).

For this paper, we will use the generic term "communications department", though in reality communications departments can have many structures, functions and different names. We take digital or online communications department to mean the communications department responsible for the planning, implementation and evaluation of communication policy in the digital realm. Therefore, it includes the development of an autonomous online communication space which, among other things, allows for the different audiences to interact with the company. In its most advanced phase, we would be dealing with a CCARU communicative model with four major activities or functions: (a) Creation of content; (b) Conversation; (c) Awareness and research; (d) Updating content.

Information and communication technology are just another instrument with specific functions within each communication plan, but with the potential to achieve changes in the central elements in the field of communication.

Within the online communications department or newsroom, we analyzed the following variables: (1) the newsroom's presence on the company's website; (2) the importance of the newsroom; (3) the evaluation of the CCARU communicative model; (4) information and communication tools; (5) the on-going adaptation to the internet; (6) the specific online function of each work tool; (7) social networks.

29.2   The Online Communications Departments
in the ICO Innovative Companies

A first look at the online newsrooms of these "innovative" companies shows a wide range of businessmen and women, communication professionals and new professions that are experimenting with new channels of communication that seek to bring the company closer to a changing and complex audience. The break with the tendency towards homogenization in terms of selecting and using technological tools in online newsrooms and the search for specific communication strategies tailored to the needs and circumstances of each company make it difficult to determine a common thread among every organization.

With these observations, if we make a generalization regarding the use of online newsrooms, what the companies have in common indicates the continuation of the previous communicative model. The heterogeneity of the various names (newsroom, media, news, etc.) and the absence of these pages on the homepage (normally it takes two or three clicks to reach them) are still apparent. In the same manner, the news, press releases and company information still tend to be the primary communicative instruments. Thus, we see preeminence of the unidirectional and asymmetric transmission of information from the source, in this case the company, and a heterogeneous audience. Nor do we see any significant efforts to adapt these information tools to the online environment; hypertextuality is rare and multimedia content, practically nonexistent.

Advances in technology are not leading to a significant transformation of the online newsrooms of the companies analyzed. In terms of the newsroom alone, we see a consolidation of the traditional relationships with the audiences. The traditional instruments of communication still dominate, and the CCARU communication model, featuring the creation of content, the ability to converse with the user, the possibility of gaining awareness and researching topics relevant to the company and the users' habits, and updating of content is nowhere to be found on these websites (Fig. 29.1).

The most advanced aspect is the first: the creation of messages with little adaptation to the Internet, primarily focusing on hypertextuality and multimedia content and, to a lesser extent, on updating content and fomenting interaction with the user. There are various possibilities from linking with external issues to using tags that build a new discourse for the company. We have also observed only limited use of GPS instruments and applications for mobile devices.

Updates are rarely addressed within the newsrooms, which still create stories and, accordingly, disseminate them at a rhythm much like that of the traditional media.

The webpages are still quite static and the changes mainly revolve around press releases and calendar updates.

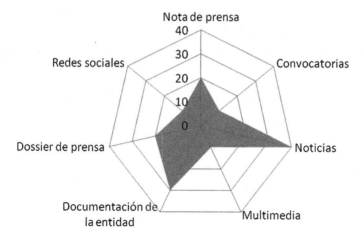

Fig. 29.1 The main instruments of online communications departments. *Source* Created by author (translation clockwise: press release, employment opportunities, news, multimedia, about us, press kit, social networks)

The communications departments' greatest interest revolves around the creation of content for a passive receptor, and department 2.0 is still far off. As we have pointed out in previous research, the newsrooms' receptor can participate in a limited fashion in the messages designed by the communications department, but he can rarely influence those same messages.

In recent years, the rapid *mise-en-scène* of social networks may have changed this panorama. In fact, although there is no single defined strategy in the utilization of social networks (Segado-Boj et al. 2015), some research has revealed concrete tendencies towards such a strategy in particular fields (Dasilva et al. 2013). Again, in the companies analyzed such tendencies are rare. Furthermore, social networks and other new technology, instead of bringing communications departments and their audiences closer to the so-called version 2.0, are used mostly as propaganda tools that adhere to a, once again, unidirectional, asymmetric communication model.

At the same time, novel strategies for communicating with different audiences outside of the newsroom reveal a tendency for newsrooms to innovate in other online arenas besides the news page. Take social networks for example: although they are nonexistent when it comes to newsrooms, they are frequently employed in innovative ways outside of this area. Thus, in some companies social networks encourage the user to interact from the time he or she lands on the page. See the two examples in Fig. 29.2 in which an associate seems to offer the user help or information as soon as the he or she reaches the website.

Fig. 29.2 Initial contact with the client. *Source* http://www.wirelessmundi.com/es/index.html and http://adacomputer.es/

29.3 Conclusions

This analysis does not reveal any elements particularly different from those ana-lyzed in our previous research of other sectors. Rather, it confirms the tendency towards the diversification of online communication strategies, the adaptation to the needs and opportunities of each organization, and the notable formation of a gap between a reduced number of companies that see innovation as a strategy for using new technologies in the communications department and the majority, which have no active online presence. Furthermore, in no way can we point to any fulfillment of the CCARU communication model.

We see an uncertain and ambiguous panorama, exacerbated by the economic crisis that has created challenges and tendencies that transcend various fields of communication. Regarding the first point of the CCARU model (content), an adaptation to the internet is needed, given that, as we observed in this study, both hypertextuality and multimedia content are generally lacking. On the same token and in keeping with the search for true communication with the receptor, technology has made it possible for different actors and institutions to collaborate in the creation of content.

The model's second point, regarding awareness and research, is one of the most addressed, yet it barely figures in most organizations' online presence. The ability of organizations—and their audiences—to conduct research holds a great deal of potential for the company's proprietary knowledge and as a strategic element in the creation of new messages.

Conversation or interaction with the user is one of the major demands made by academia and the profession of communications; yet these aspects are among the least advanced. The options offered through the creation of a truly bidirectional, asymmetrical channel with the receptor are significant. Nonetheless, the fact that fostering an open stance towards the public and creating communities represents one of organizations' most valuable instruments can also jeopardize the positions of those with insufficient resources or poorly devised strategies. Currently, the supposed interaction does not always aim to connect or communicate with the user; rather it may simply reflect a marketing strategy, or more concretely, a means of earning the loyalty of the target audience.

Lastly, the possibilities for the final axis of the communication model, updates, would allow for great strides in individualizing messages.

The four axes of this new model of communication are linked with technological innovations that are only slowly implemented by the so-called innovating companies in Spain. Furthermore, in no case do they aim to bring about fundamental changes to the central elements of the field of communication implemented on the Internet.

Acknowledgments This paper is part of the research project "Innovation and the development of online media in Spain. The make-up of journalistic interactivity on multiple platforms: news, conversation and services" (reference number CSO2012-38467-C03) funded by the Ministry of the Economy and Competitiveness.

References

Andreu, A. (2006). Medida de la reputación corporativa. Internet, intranet e información. *Telos: Cuadernos de Comunicación, Tecnología y Sociedad, 66*, 95–98.
Arroyo, I., Baños, M., Van-Wyck, C. (2013). Análisis de los mensajes audiovisuales del Tercer Sector en Youtube. *Revista Latina de Comunicación Social, 68*, 328–354.

Dader, J. L., Cheng, L., Campos, E., Quintana, N., & Vizcaíno-Laorga, R. (2014). Las webs de los partidos españoles en campaña electoral. Continuismo entre 2008 y 2011. *Trípodos, 1*(34), 115–152.

Dasilva, J. Á. P., et al. (2013). Companies on facebook and twitter. Current situation and communication strategies. *Revista Latina de Comuniación Social, 68*, 676–695. doi:10.41185/RLCS-2013-996.

Del Hoyo, J. (2006). Nuevas redes y empresa: tecnologías web y su aplicación a la comunicación corporativa. *Telos: Cuadernos de comunicación e innovación, 66*, 79–82.

Esparcia, A. C., & Martínez, A. A. (2005). Relaciones públicas y tecnología de la comunicación. Analisys de los sitios de prensa virtuales. *Organicom. Revista brasileira de comunicacao organizacional e relaçoes públicas, 3*, 133-149.

Gómez Nieto, B., Tapia Frade, A., Díaz Chica, o. (2012). La comunicación corporativa a través de las páginas web: el caso de las ONGs españolas. *Revista de Comunicación Vivat Academia, 120*, 1–20.

Martín, F. (1998). *Comunicación empresarial e institucional*. Madrid: Universitas

Olvera Lobo, M., & López Pérez, L. (2013). La divulgación de la ciencia española en la web 2.0: El caso del consejo superior de investigaciones científicas en andalucía y cataluña. *Revista Mediterránea de Comunicación: Mediterranean Journal of Communication, 4*(1), 169–191.

Orosa, B. G. (2005). *Los altavoces de la actualidad. Radiografía de los gabinetes de comunicación*. A Coruña: Netbiblo.

Orosa, B. G. (2009). *Gabinetes de comunicación online. Claves para generar información corporativa en la red*. Zamora: Comunicacion social ediciones y publicaciones.

Sánchez Duarte, J. M., & Rodríguez Esperanza, S. (2013). La extrema-derecha en Facebook. España 2000 y Democracia Nacional durante la campaña electoral de 2011/The extreme right in Facebook. España 2000 and Democracia Nacional during the election campaigns of 2011. *Revista Mediterránea de Comunicación/Mediterranean Journal of Communication, 4*(1).

Sardà, A. M., Rodríguez-Navas, P. M., & Rius, M. C (2013). La información de las administraciones públicas locales. Las webs de los ayuntamientos de Cataluña. *Revista Latina de comunicación social, 68*, 21-27.

Segado-Boj, F., Díaz-Capo, J., & Lloves-Sobrado, B. (2015). Líderes latinoamericanos en Twitter. Viejas costumbres para nuevos medios en tiempos de crisis políticas. *Revista Latina de Comunicación Social, 70*, 156–173. http://www.revistalatinacs.org/070/paper/1040uni/10es.html doi:10.4185/RLCS-2015-1040

VVAA. (2013). *El Dircom del futuro y el futuro del Dircom*. Madrid: Burson-Marsteller.

Chapter 30
Personalisation of Galician Politics in YouTube

Pablo Vázquez-Sande and Andrea Valencia-Bermúdez

Abstract The growing trend toward Americanisation of election campaigns has led to the adoption of characteristic features of personalisation in politics with a double focus of attention: from the political party towards the politician and, within it, to their most personal lines. This work will reflect on the causes and consequences of personalisation in politics, presenting the results of our research, in which 400 YouTube videos from 23 Galician politicians have been analysed. This study will study the features of customization in formal aspects, such as video titles, recordings descriptions, and the accounts from which videos are uploaded. Then, we will contrast the results obtained from the aggregate sample of videos with that videos including personal stories, with the aim of seeing if a higher degree of contents personalisation also means greater customisation in these components.

Keywords Political communication · Personalization · Youtube · Storytelling

30.1 Theoretical Framework

One of the main trends of post-modern election campaigns is the adoption of distinctive of American election processes—known as Americanisation (Salgado 2002; Canel 2006). Among its characteristics are the professionalization of campaign teams, the deideologisation; the launch campaign as a horse race; the permanent campaign; the tendency to sensationalism; the strength of infotainment; the hypermediation and the emergence of catch-all parties; and the personalisation (Adam and Maier 2010).

The personalisation process operates in two ways: firstly, it moves the focus of attention from the political party to the candidate in particular; secondly, it focuses

P. Vázquez-Sande (✉) · A. Valencia-Bermúdez
University of Santiago de Compostela (USC), Santiago de Compostela, Spain
e-mail: vazquezsande@gmail.com

A. Valencia-Bermúdez
e-mail: andrea.v.bermudez@gmail.com

© Springer International Publishing Switzerland 2017
F.C. Freire et al. (eds.), *Media and Metamedia Management*,
Advances in Intelligent Systems and Computing 503,
DOI 10.1007/978-3-319-46068-0_30

attention on the personal and private aspects of the candidate itself, dismissing political and programming questions. García (2009) notes, "the candidate has become a central axis around which all contemporary politics revolve" (García 2009:27). On the other hand, Van Aelst et al. (2012) emphasise the complementarity of both displacements, calling them individualisation and privatisation, respectively.

This phenomenon, which can be approached from a three-way perspective—the political system, the media, and the electorate's behaviour—(Rodríguez et al. 2014), is one of the interpretative keys in which current political narratives are presented, as it will be shown below. That is what Laguna (2011) calls "cognitive shortcut" for electors, who will vote a candidate instead of a programme or proposals.

Among the causes of personalisation in politics are the increased ease with which personal leadership of politicians generates ties with citizens (Cheresky 2003); the attraction of voters for knowing details and anecdotes of leaders (Berrocal 2003); the mediatisation or adaptation of politics to the need of the media (Giglioli 2005); and the greater users' understanding of human categories than policies (Del Rey 1996).

The effects on political communication and politics are, among others: the modification of political messages, which consequently changes the *framing*; the eclipse of parties, becoming "more vertical, personality-centred, and adjusted to the personal aspects of leaders" (Laguna 2011:46); the deideologisation of campaigns because "the electorate tend to decide their vote on the basis of standards related to the image and personality of candidates, instead of political and ideological aspects" (García et al. 2005:133); the increasing strength of *catch all parties*, which avoid defined ideologies; the trivialisation of contents; and the tendency to sensationalism, using emotions as a strategic resource: "the sacrifice of the rational for the rise of the emotional, the opacity of arguments for the rise of emotions, the deepness, giving space to the superficial" (Cárdenas 2013).

30.2 Methodology

The employed research technique is the content analysis, using a sample of 398 videos, corresponding to the 23 Galician local politicians with representation and at least one video on YouTube. Criteria for the selection of documents were objective, since the selection was composed by the 20 most watched videos from each candidate.

This sample and technique will allow us to answer two main questions: (1) is there a trend toward personalisation in Galician politics, taking as a reference the candidate´s strategy on YouTube? (2) Is there a correlation between content personalisation and a greater personalisation of formal aspects—video title, description, and accounts-?

To this end, it has been chosen a compared methodology between the total of analysed units (M1) and the videos with personal elements (M2), which will produce a particular view of personal-content videos and their weight with regard to the comprehensive overview.

30.3   Results

The results obtained have been divided in order to examine separately each of the established objectives. Nevertheless, it has to be clarified that the specific weight of personal-content videos is low compared to the total sample, since it accounts 5, 3 % of it (in absolute terms, M2 consists of 21 out of 398 videos of M1).

30.3.1   Video Titles

The first element of analysis is the title, since it allows us to know its identity and whether it has personal components of the politician and the party; so we are aware that both options (party and candidate) may be or not present in the title. Therefore, we distinguish four possibilities for video titles: (1) the party representing the politician; (2) the candidate; (3) both options; and (4) none of them.

Results obtained showed that most often titles only contain the personal identity of the candidate (67.8 %), while the name of the party appears in 5.5 % of videos. However, both elements also share space in titles—11.8 %—and, conversely, none of them appear in the remaining 14.9 %.

Focusing on the titles applied to videos using personal storytelling, it has been found that the name of the politician is the most common, as 67.8 % that this group implied in the large sample now represents 80.9 %. Secondly, 9.5 % of documents contain both elements, 4.8 only the party and 4.8 none of the components.

In short, this reveals that despite the fact that the trend to personalisation (understood as video titles containing only the personal identity of politicians) is a constant in the aggregate sample (2 out of 3 from the M1), as 398 recordings are indexed this way. This becomes even clearer with personal storytelling, since the figure is 4 out of 5 videos. By contrast, the other three options are infrequent in personal videos, when compared to the aggregate sample.

30.3.2   Video Description

Secondly, using the same analytical protocol, it is analysed the video description, defined as the space below videos reserved for text.

In contrast with the convergence of titles, the four options are more balanced in descriptions. Descriptions including the personal identity of candidates represent 32.4 %; those including the name and the political party reach 32.1 %; and 30.7 do not incorporate any of the elements. Far behind (4.8 %) are those where it only appears the name of the party.

Phase 2 has analysed what happens when the sample is reduced to videos including personal content. The contrast of M2 and M1 shows that the proportion of those videos with the personal identity of candidates increases (38.1 % in the second sample compared with 32.4 in the first one), as well as those that omit both the name of the party and the politician (38.1 in contrast to 30.7 of all videos). On the other hand, videos identified with the party (4.8) maintain the percentage, and those that include personal and political identities decrease—from 32.1 to 19 %.

In short, personalisation of political videos remains evident and is emphasised in recordings with personal content.

30.3.3 Accounts from Which Videos Were Uploaded

The third part, based on the same analysis explained above, is focused on the study of the YouTube accounts from which videos were uploaded. The trend is different from that shown in titles and descriptions. First, the identity of the candidate does not prevail anymore over the party, not both or any of these elements. 57 % of videos use neither of the two, 24.1 % include the name of the party, and 18.9 % only have references to the candidate.

When analysing channels including personal storytelling, the majority (57.1 %) of M2 have been uploaded on YouTube from channels without references to political parties and candidates. Thus, statistics are in line with the aggregate sample, in which this possibility was reflected on 57 % of recordings. On the other hand, there are variations when the name of channels only includes the political party, as the major trend of all analysed videos is reversed. Thus, personal videos in M2 are uploaded through channels with personal identities (28.6 % of documents). Political parties are present in 14.3 % of videos, unlike what happened in M1, where the candidate was identified in 18.9 % of documents, and political parties in 24.1. Comparatively, this means that in videos with personal contents of M2, channels with the identity of candidates increased by 9.7 % compared to M1. It can this be concluded that the trend towards personalisation is still evident and has increased in personal-content documents, although this trend is less significant than the case of descriptions and titles.

30.4 Discussion and Conclusions

The main purpose of this research was to check whether the trend to personalisation was reflected on online strategies of Galician candidates for local elections in 2011. The units of analysis were their videos on YouTube, studying in a second phase whether those recordings followed personal storytelling patterns. The purpose was to know if further contents personalisation implies greater personalisation of formal aspects such as titles, description, and accounts from which those videos were uploaded. In this regard, the following are the four key findings.

Personalisation in titles and descriptions is evident, as the most common option is to include only the identity of the candidate (compared to alternatives such as the name of the party, the combination of both elements, and the exclusion of the two). This is reflected both in M1 (overall sample), in 67.8 and 32.4 % of cases for titles and descriptions, respectively; and in M2 (sample including videos with personal contents), in 32.4 and 38.1 %.

The scenario differs when analysing YouTube accounts from which videos were uploaded. The identity of channels rarely allows us to associate them with the full name of politicians: 18.9 % in M1 and 28.6 % in M2.

Thirdly, a greater personalisation of contents always involves a greater personalisation in formal aspects, as the trend of including only the identity of the politician is increased in the three parameters: titles, from 67.8 % of M1 to 80.9 % in M2; descriptions, from 32.4 to 38.1 %; and accounts, from 18.9 to 28.6 %.

Finally, personalisation in politics implies the annulment of parties, since the percentage of videos responding to the pattern "only the name of the party" even reaches 25 % in six categories (title, descriptions and accounts in M1 and M2). In fact, titles remain at 5.5 % in M1 and 9.5 % in M2; as regards descriptions, only 4.8 in both samples are included in this classification; and when it comes to YouTube channels, only 24.1 % of M1 and 14.3 % of M2 may be included in the pattern.

References

Adam, S., & Maier, M. (2010). Personalisation of politics: A critical review and agenda for research. In *Communication yearbook* (Vol. 34). Londres: Routledge.

Berrocal, S. (coord.) (2003). *Comunicación política en televisión y nuevos medios*. Barcelona: Ariel Comunicación.

Canel, M. J. (2006). *Comunicación política. Una guía para su estudio y práctica*. Madrid: Tecnos.

Cárdenas, J. D. (2013). Storytelling y márquetin político: Humanidad y emociones en la búsqueda de la visibilidad legitimada. *Poliantea, IX*(16), 33–50. Retrieved from http://www.academia.edu/4317051/Storytelling_y_m%C3%A1rquetin_pol%C3%ADtico_humanidad_y_emociones_en_la_b%C3%BAsqueda_de_la_visibilidad_legitimada. Accessed November 15, 2013.

Cheresky, I. (2003). En nombre del pueblo y de las convicciones. Posibilidades y límites del gobierno sustentado en la opinión pública. *PostData, 8*.

Del Rey, J. (1996). *Democracia y Posmodernidad. Teoría General de la Información*. Madrid: Editorial Complutense.

García, L. J. (2009). Y el protagonista es el candidato: la personalización como enfoque en comunicación política. In R. Zamora (Ed.), *El candidato marca*. Madrid: Fragua.

García, V., D'adamo, O., & Slavinsky, G. (2005). *Comunicación política y campañas electorales*. Barcelona: Gedisa.

Giglioli, P. P. (2005). *Invito allo studio della società*. Bolonia: Il Mulino.

Laguna, A. (2011). Liderazgo y Comunicación: La Personalización de la Política. *Anàlisi: Quaderns de Comunicació i Cultura, 43*, 45–57.

Rodríguez, J., Jandura, O., & Rebolledo, M. (2014). La personalización en la cobertura mediática: una comparación de las campañas electorales en España y Alemania. *Trípodos, 34*, 61–80. Retrieved from http://www.tripodos.com/index.php/Facultat_Comunicacio_Blanquerna/article/view/166/71. Accessed January 14, 2015.

Salgado, L. M. (2002). *Marketing politico. Arte y ciencia de la persuasión en democracia*. Barcelona: Paidós.

Van Aelst, P., Sheafer, T., & Stanyer, J. (2012). The personalization of mediated political communications: A review of concepts, operationalizations and key findings. *Journalism, 13* (2), 1–18. Retrieved from http://www.academia.edu/1513659/The_personalization_of_mediated_political_communication_A_review_of_concepts_operationalizations_and_key_findings. Accessed September 26, 2013.

Chapter 31
Social Media in Crisis Communication: Germanwings Flight 4U9525

Diego Rodriguez-Toubes and Yolanda Dominguez-Lopez

Abstract Negative events test the resilience and leadership of a company in order to amend mistakes and adapt to changes required by crisis and disasters. Effective crisis management and communication help reduce damages and uncertainty and contribute to a quick recovery. This paper analyses the early reactions regarding the Germanwings air crash in March 2015. We motorised and analysed online information to check whether the communication management is performed effectively.

Keywords Airline crash · Crisis communication · Disaster · Germanwings · Social media

31.1 Theoretical Framework

The evolution of means of transportation has prompted the tourism industry's growth and expansion over the past decades. Airplanes are the most commonly used mean of transportation, representing more than half of all international trips all over the world (United Nations World Tourism Organizations (UNWTO) 2014). There is a mutual connection between both industries in that any problem for the tourism industry can have severe consequences for the transportation sector and vice versa (Kozak et al. 2007).

Every crisis has its own causes, impacts, and recovery patterns inherent to all living situations. Tailor-made answers must be developed in order to adapt to the different situations' needs (Blackman and Ritchie 2010). Despite the fact that crises do not always follow the same defined patterns, there are authors who have proposed general crisis-management models (Faulkner 2001). In any case,

D. Rodriguez-Toubes (✉) · Y. Dominguez-Lopez
University of Vigo (UVigo), Vigo, Spain
e-mail: drtoubes@uvigo.es

Y. Dominguez-Lopez
e-mail: ydominguez@uvigo.es

© Springer International Publishing Switzerland 2017
F.C. Freire et al. (eds.), *Media and Metamedia Management*,
Advances in Intelligent Systems and Computing 503,
DOI 10.1007/978-3-319-46068-0_31

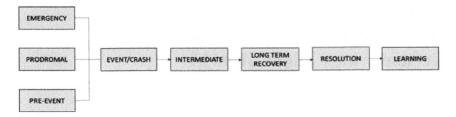

Fig. 31.1 Stages in airline crash crisis management. *Source* Henderson (2003: 281)

crisis-management models require providing specific actions to confront the different situations a tourism destination or enterprise might face.

In the event of an airplane crash, Henderson (2003) noted that Faulkner's model's first three steps (pre-event, prodromal, and emergency) were likely to be brief and overlap into one stage (Fig. 31.1). In this case, the first stages hasten and immediately lead to the intermediate step.

31.1.1 Crisis-Management Communication

Terrorist attacks and airplane crashes are events that acquire more media coverage, as they usually entail drama, uncertainty, and suffering (Henderson 2003). In order to properly overcome a crisis situation, developing a thoughtful communication and recovery plan becomes crucial. A lack of planning will imply a slow recovery by the tourism enterprise or destination in jeopardy (Scott et al. 2010; Zerman 1995). Crises test the resilience and leadership of a company in terms of amending mistakes and adapting to the changes required by different events. Thus, the tourism industry should focus their efforts on developing risk- and crisis-management models to overcome delicate situations and encourage the development of prevention, action, and communication programs.

The use of traditional media in communicating institutional crises has been increasingly supplemented by social-media use. Over the last five years, several studies have analysed the use of social media in crisis-communication protocols, and international organisations specialised in tourism disasters have submitted some recommendations (International Air Transport Association (IATA) 2014; UNWTO 2012). The choice to use social media versus traditional media to spread information has a direct impact on the user's perception and reactions regarding sharing information. Companies must properly integrate the environment in which they issue the message to the source and, in times of crisis, integrate the phase they find themselves in (Liu et al. 2011). In terms of Henderson's (2003) study, social media begins to play a leading role in crisis communication from the intermediate period onwards, whereas in the early stages—pre-event, prodromal, and emergency—traditional media and word-of-mouth are the most appropriate ways to build a credible image (Jin and Liu 2010).

The generation of tweets after a major disaster depends on various factors, such as technology use, culture, the population's behaviour, the severity of the disaster, and politics. Socioeconomic factors, such as access to technology and the social groups' different motivations, are more important variables for the prognosis of disaster-related tweets than the size of the population or the level of damage (Xiao et al. 2015). By understanding the risk perception of different tourist segments, companies can design tailor-made messages addressed to counteract a user's mis-perceptions about the different markets (Schroeder and Pennington-Gray 2015). It is key that meta-media managers promote the user's engagement (Huertas et al. 2015) and, even in crisis situations in which it is recommended to amplify the message through photos and retweets, to be present, listen, and engage stakeholders everywhere to talk about the company (Gruber et al. 2015).

Some authors have noted that social networks are not being strategically used, remarking on an obvious lack of uniformity regarding the airlines' responses (Hvass and Munar 2012). In a changing media landscape, Jin and Liu (2010) proposed a social-media-based crisis-communication model as a framework to succeed in crisis-communication management. Meanwhile Kietzmann et al. (2011) developed a model that brought together seven pillars—*identity, conversations, sharing, presence, relationships, reputation, and groups*—that companies should monitor to understand how social-media activities change depending on their role and impact. Strong and effective crisis management and communication help to reduce the negative effects of a crisis and contribute to a quick recovery. Hence, it is important that organisations and tourism destinations develop effective communication programs in an attempt to overcome negative events as soon as possible. This paper analyses to what extent airlines follow the recent crisis-communication protocols established by the major international organisations in the field of social media.

31.2 Methodology

On 24 March 2015, Germanwings, Lufthansa Group's low-cost carrier, experienced a fatal crash while operating a flight between Barcelona, Spain and Dusseldorf, Germany. There were 144 passengers and six cabin crew travelling on board; all of them lost their lives. Research sources have hinted at the possibility of a premeditated mass murder performed by the airplane pilot, who locked himself in the cabin and voluntarily crashed the plane. This information is based on the flight voice recordings extracted from the black boxes.

The current research performs a content analysis of the management of communication in social media concerning Germanwings flight 4U9525's crash. The content analysis was developed following the next sub-stages: information reading, publishing-time monitoring, and news-update check. As a result, the gathered and classified information led to the timeline in which the most important publications were highlighted. Following Henderson's model (2003) and other recommendations (Jin and Liu 2010), the research background was based on tweets issued during the intermediate stage. In this stage, other answer-related procedures were launched that complemented the communication, such as victims' relatives' care and research, corresponding to Wilks and Moore's model's response stage (2003).

From 24 March to 27 April 2015, over 200 tweets and retweets were read and analysed from both Germanwings's and Lufthansa's Twitter profiles. The content was monitored by the authors in order to check the information updates regarding the airline crash. Among the wide range of social networks that are used in the field of crisis communication, particularly by airlines, Twitter was chosen to explore and analyse the incident. There are compelling reasons to choose Twitter, as it has proved to be the post-crisis communication leader, especially taking into account the information volume and the content aspects (Sung and Hwang 2014). Nowadays, Twitter has become the most consulted and monitored source concerning breaking news (IATA 2014; Gruber et al. 2015).

Whereas in advertising and sales departments, the use of other platforms is widespread, information dissemination is currently linked to digital media and public relations, being strongly represented by Twitter (Hvass and Munar 2012). 'In times of crisis, people are turning to Twitter and other social media to share information, react to the situation, and make affective displays' (Lachlan et al. 2016: 652).

The advice of UNWTO (2012) and IATA (2014) are the bedrock on which we developed an assessment tool—the check-lists—that will provide different items that will allow us to rate the communication-management effectiveness applied to social media. The rating is based on press releases and actions carried out by Germanwings. In this last stage, the final results are drawn from a dichotomous scale. The reason why we chose a dichotomous scale versus a Likert is to objectify responses and avoid the likely possibility of having to perform a subjective rating for a subject as delicate as a plane crash.

31.3   Results

As soon as Germanwings and Lufthansa were aware of the crisis, they performed a non-stop research enquiry to ascertain the causes of the fatal air crash. Table 31.1 summarises the most important statement posted by the airlines on their Twitter profiles.

The main statements posted by both companies make clear they were willing to assume responsibility, clarify the causes of the event, provide updated information, and offer support to the victims' relatives. The main actions they undertook regarding communication recovery were as follows:

– Remaining accessible to the media
– Offering cooperation to research sources

Table 31.1 Germanwings's timeline communication in intermediate phase

24/03/15	Air crash acknowledged
	Toll-free line to support and assist the victims' relatives
	Sympathy and condolences statement
25/03/15	Caster Spohr, Lufthansa's CEO, gives a statement about the accident
	Specific website to pay tribute to victims and talk about the event
	First use of the hashtag #indeepsorrow
26/03/15	Victims' relatives are offered free flights to travel to the accident area
	A press conference provides new research data
	Important announcement: co-pilot Andreas Lubitz is believed to have crashed the aircraft deliberately
27/03/15	Victims' family-care centre installed in Marseille
	The Lufthansa Group adopt new security measures requiring two pilots to remain in the cockpit at all times (*Two person rule*)

Note Compiled from Lufthansa's and Germanwings's Twitter profiles
Source Prepared by the authors

– Priority communication to the victims' families before providing any information to the mass media
– Up-to-date information translated into English, German, and Spanish
– Press conference announcements through Twitter.

Table 31.2 displays the checklist that has been reviewed by the authors according to UNWTO (2012) and IATA (2014) best practices. The data collected

Table 31.2 Check-list: required items for effective crisis communication

Recommended actions	Actions to avoid
– Issue a provisional press release as soon as possible	– Early judgements or hypotheses with no background source
– Designate a crisis-management and communication team	– Discussing unknown aspects of the research
– Frequently post updated information	– Communicating improperly
– Accuracy and unity in both companies' messages	– Lack of solidarity
– Openly communicate condolences and accept responsibilities	– Lack of coordination between the companies and the media
– Quickly feedback and answer the users' questions	
– Link the different social-media networks' content	
– Create a specific webpage based on the event	

Note Compiled from UNWTO (2012) and IATA (2014) directions
Source Prepared by the authors

reflect both must-do actions and aspects to avoid in times of crisis. The ticked statements are the ones Germanwings and Lufthansa performed.

According to IATA (2014) a provisional statement should be broadcasted within the first 15 min to one hour after the disaster, in this case we did not find evidence that this has been made. However, the results showed a good performance by both companies, reflecting their commitment and professionalism regarding airplane-crash communication.

31.4   Conclusions

This research tracked the communication process in the event of an airline crash. In this timeline, both Lufthansa and their low-cost subsidiary, Germanwings, performed the management of the crisis according to the most recent recommendations in this field. The intense communication and efforts undertaken by both companies helped reduce negative effects and allowed a quick recovery. Lufthansa's positive reputation also helped restrain the negative image that could have been caused by this tragedy in the long run. In 2014, Lufthansa and Germanwings were given 7-star safety rating, the highest score possible, being included in the global top 10 for security among aviation companies (AirlineRatings 2016). However, Lufthansa Group has undertaken actions to avoid allowing their subsidiary company's image to decay. Indeed, from October 2015 onwards, Germanwings has gradually merged into Eurowings, becoming a part of their group and adopting their name on the flight operations.

Most academic research has focussed on risk-management strategies after the crisis has already occurred. There is a lack of proactive crisis-management models which focus the companies' efforts on anticipating future problems. Anticipating a crisis is a challenge that the tourism industry must face today. However, there are ways to improve and implement new protocols in risk-prevention matters. In light of the Germanwings crash, this task has once again come to debate, implying the need to adopt new security measures throughout the airline industry.

References

AirlineRatings. (2016). *Airline & safety ratings*. Retrieved January 27, 2016 from http://www.airlineratings.com/news/402/worlds-top-ten-airlines-

Blackman, D., & Ritchie, B. W. (2010). Tourism crisis management and organizational learning: The role of reflection in developing effective DMO crisis strategies. In N. Scott, E. Laws, & B. Prideaux (Eds.), *Safety and security in tourism: Recovery marketing after crises* (pp. 45–57). London: Routledge, Taylor & Francis Group.

Faulkner, B. (2001). Towards a framework for tourism disaster management. *Tourism Management, 22*, 135–147.

Gruber, D. A., Smerek, R. E., Thomas-Hunt, M. C., & James, E. H. (2015). The real-time power of Twitter: Crisis management and leadership in an age of social media. *Business Horizons, 58,* 163–172.

Henderson, J. C. (2003). Communicating in a crisis: Flight SQ 006. *Tourism Management, 24,* 279–287.

Huertas, A., Setó-Pàmies, D., & Míguez-González, M. I. (2015). Comunicación de destinos turísticos a través de los medios sociales. *El profesional de la información, enero-febrero, 24* (1), 15–21.

Hvass, K. A., & Munar, A. M. (2012). The takeoff of social media in tourism. *Journal of Vacation Marketing, 18*(2), 93–103.

IATA. (2014, December). *Crisis communications and social media. A best practice guide to communicating in an emergency.* Montreal: International Air Transport Association.

Jin, Y., & Liu, B. F. (2010). The blog-mediated crisis communication model: Recommendations for responding to influential external blogs. *Journal of Public Relations Research, 22,* 429–455.

Kietzmann, J. H., Hermkens, K., McCarthy, I. P., & Silvestre, B. S. (2011). Social media? Get serious! Understanding the functional building blocks of social media. *Business Horizons, 54,* 241–251.

Kozak, M., Crotts, J. C., & Law, R. (2007). The impact of the perception of risk on international travellers. *International Journal of Tourism Research, 9,* 233–242.

Lachlan, K. A., Spence, P. R., Lin, X., Najarian, K., & Greco, M. D. (2016). Social media and crisis management: CERC, search strategies, and Twitter content. *Computers in Human Behavior, 54,* 647–652.

Liu, B. F., Austin, L., & Jin, Y. (2011). How publics respond to crisis communication strategies: The interplay of information form and source. *Public Relations Review, 37,* 345–353.

Schroeder, A., & Pennington-Gray, L. (2015). The role of social media in international tourist's decision making. *Journal of Travel Research, 54*(5), 584–595.

Scott, N., Laws, E., & Prideaux, B. (2010). Tourism crises and marketing recovery strategies. In N. Scott, E. Laws, & B. Prideaux (Eds.), *Safety and security in tourism: Recovery marketing after crises* (pp. 1–13). London: Routledge, Taylor & Francis Group.

Sung, M., & Hwang, J.-S. (2014). Who drives a crisis? The diffusion of an issue through social networks. *Computers in Human Behavior, 36,* 246–257.

UNWTO. (2012). *Toolbox for crisis communications in tourism. Checklist and best practices.* Madrid: World Tourism Organization.

UNWTO. (2014). *Tourism highlights* (2014th ed.). Madrid: World Tourism Organization.

Wilks, J., & Moore, S. (2003). *Tourism risk management for the Asia Pacific Region: An authoritative guide for managing crises and disasters.* Commonwealth of Australia, APEC. International Centre for Sustainable Tourism (AICST).

Xiao, Y., Huang, Q., & Wu, K. (2015). Understanding social media data for disaster management. *Natural Hazards, 79,* 1663–1679.

Zerman, D. (1995). Crisis communication: Managing the mass media. *Information Management & Computer Security, 3*(5), 25–28.

Chapter 32
University-Society. Proposal for New Forms of Communication in the University of the Basque Country Through Service-Learning

Andoni Iturbe Tolosa and Monike Gezuraga Amundarain

Abstract The University as an educational institution and substantial element of society requires a constant revision in its ways of interaction and communication with the surrounding environment. This work aims to influence the new forms of communication that are being implemented in Higher Education due to overcoming traditional models, hence striving to strengthen the communicative ties and participatory involvement of public citizens. To do this, we will insist on various communication strategies that the university incorporates; and we will pay special attention to the case of the University of the Basque Country (UPV/EHU). So, we will reflect on its field and possibilities for improvement. In this scenario we consider it important to highlight the contribution of the Service-Learning proposal. We are referring to a pedagogical approach that integrates the development of skills in college students with community service. We will emphasize the opportunity that this proposal gives us in order to integrate Teaching, Research and Extension, assignments of the University in relation to the Society; and new challenges and opportunities for communication that can be derived from this practice.

Keywords Communication · University · Service-learning · Missions

32.1 Introduction

Our work presented here takes a diachronic approach that puts its focus on the near future and on the significant and increasingly present contribution of Service-Learning, both in our close environment and internationally. Through content analysis, we examined the policy of University communication, and

A. Iturbe Tolosa (✉) · M. Gezuraga Amundarain
University of the Basque Country (UPV/EHU), Leioa, Spain
e-mail: aiturbetolosa@gmail.com

M. Gezuraga Amundarain
e-mail: monike.gezuraga@ehu.eus

© Springer International Publishing Switzerland 2017
F.C. Freire et al. (eds.), *Media and Metamedia Management*,
Advances in Intelligent Systems and Computing 503,
DOI 10.1007/978-3-319-46068-0_32

specifically the Plan of Communication held by the University of the Basque Country.

32.2 Communication Strategies University-Society

32.2.1 University's Third Mission. Forms of Communication University-Society

Decades ago we started debating on what the substantial missions of university ought to be. It seems that nowadays we reached the consensus for incorporating, next to the two historical university missions of Teaching and Research, the mission of Extension. We should remark here that the mission of Extension adopts different meanings according to the context approached: University's *social function*, *third mission* of the University, etc. (Torres 2013). As we shall see, it is impossible to understand social function without communicative approach.

University as an educational institution centralizes its communicative actions, both internal and external, in a department of communication that is traditionally in charge of informative relations (management of interviews, press conferences, press releases, etc.). Aguilera et al. (2010: 105) are of the opinion that the audiences of the University communication demand the institutions for a reconversion of the channels of communication.

Some recent forms of communication are linked to communication based on action (Costa 1999); community micro-communication (Zallo 2011), ethics (Bilbeny 2012), accentuated by a change of paradigm that is driven by Internet and new technologies and affecting society as a whole, as well as the traditional model established between University and Society.

32.2.2 The Communication Strategy of UPV/EHU

The *Office of Communication* is in charge of all the communicative actions of the University of the Basque Country UPV/EHU. This Office counts with the collaboration of those responsible for communication in the different centers and departments; all of them together represented in an Advisory Council.

When we review the Plan of Communication of UPV/EHU, we see that the main message (UPV/EHU 2011: 17) transferred to society is that the University of the Basque Country is a public, accessible and genuine university. However, there doesn't seem to exist a more defined strategy in the actions directed at society: "To organize more events in spaces close to society and to take activities of the UPV/EHU out onto the streets. This concerns taking to the streets some of the activities already being realized by the University, as well as creating 'ad hoc' ones to gain presence outside the institution. For example, a fair could be organized in stands on the squares of the main cities of the Autonomous Community of The Basque Country" (UPV/EHU 2011: 17, Translation).

Another aspect of this communication strategy susceptible to improvement would be the broadening of its message in the projection towards society, since the main message here is the following: "UPV/EHU is a research entity" (2011: 12, Translation). Is it merely this what it is all about?

32.3 Service-Learning in Higher Education

32.3.1 Service-Learning: Conceptual Approach, Protagonists and Impact

Service-Learning (S-L) can appear as a confusing concept for people who come to it for the first time. We shall state that the application of S-L is channelled through educational projects that contain clear learning objectives, clear objectives of Service to the Community, an unequivocal protagonist role of the students immersed in the process, and a continuous reflection by all parties involved (Rubio 2008).

Service-Learning is not about practicing charity or volunteering. It is not a new way of labeling the already existing activities of field practice either. It concerns a proposal that integrates elements of Service and of Learning, situating these elements on the same level. The following example can help us to contextualize this proposal more clearly (Table 32.1):

Above we highlighted the protagonist role of the students, but we must not forget the required presence of two other agents in these projects: the teachers; and the collectives and social groupings that settle the collaboration, to whom "the service is rendered".

Authors such as Puig et al. (2007), Simonete (2008) and Tapia (2004), among others, argue that this proposal causes undoubted benefits in: the students; the teaching staff; the community with which it collaborates; and the educational institution itself and its relationship with the community, where the possibility of generating networks and its high democratic value stand out.

Table 32.1 Project example

Project: "¿ASÍ TE CUADRA?" (drug use prevention)	
Collaboration-service	Learning
Design students plan and develop an awareness campaign with the aim of preventing drug use associated with nightlife	Knowledge about the impact of the consumption of drugs among adolescents; familiarization with various marketing strategies; etc.

Source Adaptation from Batlle (2013: 63)

32.3.2 S-L in Higher Education: Integrating Teaching, Research and Extension

Let's take a look at two examples of this integration: The Universitat Politècnica de València (UPV), which incorporates S-L as a valuable tool for the promotion of a participatory technology transfer (Schlierf et al. 2010).

Within a line of work that seeks to incorporate community-based research into the university, we found the *Taller de Barris (Barris' workshop)*. This is where a collaboration group is established in which teaching and research staff, UPV students, residents of the neighborhood *El Palleter*, and other community agents participate. This alliance allows to generate an in-depth diagnostic on the situation of the neighborhood, and for the contribution of different proposals that intend to provide solutions to emergent needs.

In 2003 the National University of La Plata (Buenos Aires) initiated a project that aims to respond to nutritional deficiencies detected in the girls and boys who attend the community kitchens in several deprived communities. Teachers and researchers who were already working on alternative, low-economic cost alimentation and the impact on the inhibition of pathogens for diarrhea, proposed to realize a project with kefir.

The students involved participated in all the research activities related to microbiology; anthropology of food; distribution of kefir; and evaluation of the nutritional status, making comparisons among the children who took kefir and the ones who didn't (Herrero 2010).

In such a way, S-L provides an utterly rich space for facilitating the opportunity to integrate the missions of Teaching, Research and Extension.

32.3.3 S-L in the UPV/EHU. A Communication Challenge and Opportunity

As we mentioned previously, S-L presents itself as a valuable communication tool in the University-Society relationship, recognizing agents outside the academic sphere as an indispensable part of the continuous formation of students and teachers.

With this knowledge, a group of teachers from UPV/EHU decided to incorporate S-L in their teaching and research activities. Evidence of this were the academic courses of 2011–2012 and 2012–2013, which motivated a "boom" of projects with more than a dozen of developed initiatives being identified in the degrees of: Nursing, Sociology, Social Education, Pedagogy, Psychology, Computer Science Engineering, Nautical Engineering and others (Gezuraga 2014).

We must emphasize that, to a large extent, the agents who participated in this study all put in value, in one way or another, the capacity for relation and communication between university and society through Service-Learning. Let's focus for a moment our attention onto these voices:

Focus Group with Community Partners-SC3: "We cannot have universities, educational centers that exist in a bubble, in an outer space, can we? Or we start educating for the society, or we are finished" (Gezuraga 2014: 329, Translation).

Interview with the Director of University Social Responsibility: "Yes, there is a clear contribution, of course, in terms of extension of the projection unto society, but I think that it is the feedback from society to university that is especially enriching" (Gezuraga 2014: 304, Translation).

Focus Group with Students-A3: "(...) the level of inaccessibility that universities usually have, is very high, and as long as you don't make a breach, and I think this is a good way of doing it, a good pickaxe to break down that wall" (Gezuraga 2014: 286, Translation).

In the light of the dominance of a value-based communication policy, Service-Learning contains two essential components for understanding the scope of the pedagogy of communication in a public institution such as the university.

32.4 Final Considerations

The review carried out in this work has allowed us to reach the following conclusions:

- In its message addressed to society, the University of the Basque Country evokes the idea that it is a nearby, easily accessible university. However, it seems that the institution has encountered difficulties at the moment of undertaking concrete actions that genuinely establish a direct communication with society.
- Furthermore, much of the content communicated by the university is related to Research, which may lead to a certain undervaluation of the other two missions: Teaching, and Extension, domains where also very important efforts are being made by this university.
- The external communication strategy of UPV/EHU should establish new strategies that lead to strengthen the links of the university with its environment. Since the University of the Basque Country is a public university, we would suggest that they could make use of the communicative potential of this very proposal.
- It concerns a big communication challenge since it involves a new form of presenting itself to society: new ways of doing, both for the academic world as for "the social world". The very development of Service-Learning itself in our university still has a long way to go, but the horizon presented to us by other universities has convinced us that it is a path with great educational and communicative potential.

When we take into account that the three pillars of Departments of Communication are relations with the media; marketing-publicity; and society, it appears obvious that the latter should have further development. It is precisely in that terrain where we have tried to make our contribution.

References

Aguilera, M., Farias, P., & Baraybar, A. (2010). La comunicación universitaria: modelos, tendencias y herramientas para una nueva relación con sus públicos. *Revista Icono14*, 8(2), 90–124. Retrieved from http://www.icono14.net/ojs/index.php/icono14/article/view/248/125. Accessed November 10, 2015.

Batlle, R. (2013). *60 buenas prácticas de Aprendizaje-Servicio. Inventario de experiencias educativas con finalidad social.* Bilbao: Zerbikas.

Bilbeny, N. (2012). *Ética del periodismo. La defensa del interés público por medio de una información libre, veraz y justa.* Barcelona: Universitat de Barcelona.

Costa, J. (1999). *La comunicación en acción. Informe sobre la nueva cultura de la gestión.* Barcelona: Paidós.

EHU-UPV. (2011). *Plan de Comunicación 2012–2015.* Retrieved from https://www.ehu.eus/es/komunikazio-bulegoa/komunikazio-plana. Accessed October 15, 2015.

Gezuraga, M. (2014). *El Aprendizaje-Servicio (A-S) en la Universidad del País Vasco (UPV/EHU): En el camino hacia su institucionalización.* (Unpublished doctoral dissertation). Universidad Nacional de Educación a Distancia, Facultad de Educación, España. Retrieved from http://e-spacio.uned.es/fez/eserv/tesisuned:Educacion-Mgezuraga/Gezuraga_Amundarain_Monike_Tesis.pdf. Accessed November 9, 2015.

Herrero, M. A. (2010). Una nueva forma de producción de conocimientos: El Aprendizaje-Servicio en Educación Superior. *Tzhoecoen, 5*, 63–80.

Puig, J. M., Batlle, R., Bosch, C., & Palos, J. (2007). *Aprendizaje Servicio. Educar para la ciudadanía.* Barcelona: Octaedro.

Rubio, L. (2008). *Guías Zerbikas 0: Aprendizaje y Servicio Solidario, guía de bolsillo.* Bilbao: Fundación Zerbikas.

Schlierf, K., Boni, A., & Lozano, J. F. (2010). La transferencia de tecnología participativa desde la universidad: hacia un cambio tecnológico. In M. Martínez (Ed.), *Aprendizaje Servicio y Responsabilidad Social de las universidades* (pp. 193–217). Barcelona: Octaedro.

Simonete, D. (2008). *Service—Learning and academic success: The links to retention research.* Minnesota: Campus Compact. Retrieved from http://www.compact.org/wp-content/uploads/resources/downloads/MN-SL_and_academic_success.pdf. Accessed October 30, 2015.

Tapia, M. N. (2004). *Aprendizaje y Servicio Solidario.* Madrid: CCS.

Torres, A. (2013). Misión Social de la Universidad: un Estudio de Caso en el Sector Salud en México. Paper presented at the *Conferencia Internacional LALICS 2013 "Sistemas Nacionais de Inovação e Políticas de CTI para um Desenvolvimento Inclusivo e Sustentável"*: 2013, Río de Janeiro. Retrieved from http://www.redesist.ie.ufrj.br/lalics/papers/117_Mision_Social_de_la_Universidad_un_Estudio_de_Caso_en_el_Sector_Salud_en_Mexico.pdf. Accessed November 1, 2015.

Zallo, R. (2011). *Estructuras de la comunicación y de la cultura. Políticas para la era digital.* Barcelona: Gedisa.

Chapter 33
Management Strategies and Online Communication Tools for Value Creation in Media Companies

María Victoria-Mas and Iván Lacasa-Mas

Abstract This chapter seeks to clarify the way in which media companies can create value in a digital context and maintain the competitive advantage that the strength of their brand gives them. Based on a literature review, we have identified some of the possible strategies and tools to manage the brand value of media companies. These strategies and tools are those that have to do with the management of corporate communication through the Internet. Furthermore, the *La Vanguardia* case study shows us how one of the main news brands in Spain manages its value. The chapter suggests that future research analyze the effectiveness that the implementation of a communication system based on corporate social responsibility (CSR) could have to generate credibility and brand value. Specifically, it proposes an analysis of the utility that digital platforms in which news brands are already present could have in that communication.

Keywords Intangibles · News brands · CSR · Corporate communication · Online platforms

33.1 Theoretical Framework

In the last two decades, the value that news companies create for their external and internal stakeholders—advertisers, audiences, investors, journalists, and society—has diminished (Picard 2010: 56–59). This chapter seeks to analyze some ways in which news corporations can once again generate value for their stakeholders through the management of intangibles.

M. Victoria-Mas (✉) · I. Lacasa-Mas
Universitat Internacional de Catalunya (UIC), Barcelona, Spain
e-mail: mvictoria@uic.es

I. Lacasa-Mas
e-mail: lacasa@uic.es

© Springer International Publishing Switzerland 2017
F.C. Freire et al. (eds.), *Media and Metamedia Management*,
Advances in Intelligent Systems and Computing 503,
DOI 10.1007/978-3-319-46068-0_33

33.1.1 Value Creation Through the Management of Intangibles in Media Companies

The management of intangibles can be defined as a comprehensive corporative strategy focused on creating value for stakeholders and producing a competitive advantage based primarily on reputation and brand value (Gutiérrez 2008: 30; Van Riel and Fombrun 2007: 7). These two intangibles are, according to experts, what should be prioritized; corporate communication is one of the best ways to manage them (Carreras et al. 2013; Van Riel and Fombrun 2007). Since the end of the 90s, many communication departments have evolved towards models based on corporate social responsibility (CSR), which seek to guarantee the company's good behavior and to ensure that it complies with the commitments it has made to its different strategic audiences. The objective is to improve the corporate reputation and to generate brand value (Gutiérrez 2013: 21–24; Casado et al. 2013: 54; Van Riel and Fombrun 2007).

33.1.2 Management and Communication of CSR Through Digital Platforms

Corporate social responsibility should be considered a corporate strategy that is composed of a set of voluntary actions and seeks to highlight the commitments made with every interest group (Campos et al. 2011: 35; Manfredi 2009: 138). CSR is based on the identification of such groups and on the hierarchy of obligations and responsibilities that each company has towards them, depending on its specific mission (Manfredi 2009: 139). In the case of news corporations, the mission is to create a well-informed public opinion. Consequently, ahead of stockholders, the primary interest groups for news corporations are: (a) readers or citizens; (b) their employees, who must have the adequate tools and conditions so that their practices and codes of conduct contribute to high-quality journalism; and (c) advertisers, who must make the political and economic independence of journalism viable (Fernández 2012: 112; Lavine and Wackman 1992: 76). When a news company adopts this order of priorities, it integrates CSR in the decision-making process about operations and value creation (Manfredi 2009: 142).

The different studies that have been conducted regarding CSR management in media companies show that this corporate philosophy has not yet been established in this sector. Although the mission statement and references to different social actions appear on the corporate websites of newspapers (Campos et al. 2011: 36–38; Fernández 2012; Manfredi 2009), they are not explicitly expressed, and they must be deduced from the foundational history or from the way the company represents itself on said websites. Almost all news companies focus on the part of their mission that has to do with its social and cultural goals (Campos et al. 2011; Manfredi 2009: 147), but they do not provide concrete information about the real application of each one of their initiatives (Manfredi 2009: 147). Furthermore, they orient their communication primarily towards stockholders, whose interests are economic (Fernández 2012). The research cited above indicates that there is no unanimous consensus about a catalogue of CSR measures specific to media companies.

There are authors who propose applying some of the good governance principles of public institutions to media companies and explicitly mentioning the mechanisms that ensure the preservation of the mission of such companies (Sánchez De La Nieta et al. 2012: 24; Manfredi 2009: 147; Arrese 2006: 67). These mechanisms must guarantee support for high-quality journalist practices. It is necessary for media companies first to strengthen their ethical commitments internally and then to meet the expectations of their external audiences and to generate credibility. One of the main concrete strategies in this area is the drafting and publication of style guides, which include the company's editorial principles and statutes. By publishing this guide, the company makes a moral contract with their primary interest groups: professionals and the public (Aznar 1999: 77–79). The principles outline the brand's mission (Nieto and Iglesias 2000; Campos et al. 2011: 37). These style guides also specify the professional norms, codes of conduct, and style guidelines that regulate the activities of the members of news companies (Aznar 1999: 86–100; Echeverri 1995: 204). The statutes consider various ways to monitor the application of the principles and the company's compliance with its social responsibility—for example, the establishment of the Professional Committee or the Ombudsman.

Some authors believe that for this philosophy to really contribute to value creation and to generate credibility, it is necessary not just to define the mission and to create operative policies that make the editorial statutes effective but also to include the implementation of concrete actions that make such policies visible (Cobo 2008; González-Esteban et al. 2011; Sánchez De La Nieta et al. 2012: 24). Beyond the internal management of the company's CSR, it is necessary to communicate the channels through which it will be managed, whether it be through the publication of the editorial statutes or, for example, informing about the actions of the Professional Committee. In this way, it will be possible to gain credibility with audiences because a greater commitment on the part of the company will be evident. According to the cited studies, the online platforms where news brands are already present currently can serve as one of the main tools to communicate CSR. This chapter, through an analysis of a media company case study, explores (a) to what extent it has defined its mission and corporate social responsibility and (b) what type of CSR management and communication it performs on its digital platforms.

33.2 Methodology

For this case study, we chose one of the Spanish news outlets with the most brand value: *La Vanguardia* (Victoria-Mas and Lacasa-Mas 2015). We analyzed materials generated by the company that speak about its mission and CSR: the corporate website, the online edition, social media profiles, and the style guide (Camps et al. 2004). Furthermore, we conducted thirteen in-depth interviews with the directors and middle managers of the digital edition[1] and of the different departments concerned with brand management and strategy.[2] In this way, we analyzed how this brand manages and communicates its CSR.

33.3 Results

33.3.1 *Definition of Mission and CSR of* La Vanguardia

Our study indicates that *La Vanguardia* has defined its mission—forming public opinion—in the editorial principles included in its style guide (Camps et al. 2004: 475). This commitment made by the brand originates in what it affirms is its raison d'être: "the exercise of the right to information on the basis of pluralism, liberty, responsibility, and rigor" (Camps et al. 2004: 475). This is, according to *La Vanguardia*, "the backbone that inspires its actions as a newspaper committed to high quality journalism and fully aware of the relevant social function of the press" (Camps et al. 2004: 476).

The editorial principles define the social responsibility that, in keeping with its mission, the newspaper has towards each of its main stakeholders. *La Vanguardia* identifies itself with "public service journalism" and asserts that subscribers, readers, and advertisers are "its main social asset," in addition to the rationale of its

---

[1]One-hour interviews, conducted in May 2014, about the editorial organization, editorial principles, management of quality and communication, and understanding that the members of the newsroom have about the identity and mission of the brand. Those interviewed were Enric Sierra, Deputy Director of *LaVanguardia.com*; Txema Alegre, Chief Editor of *LaVanguardia.com*; Josep Maria Calvet, Information Chief of *LaVanguardia.com*; Joel Albarran, Cover Chief for *LaVanguardia.com*; Sílvia Colomé, Head of Multimedia Production for *LaVanguardia.com*; Toni Rubies, Head of participation for *LaVanguardia.com*; and Patricia Ventura, Social Media Manager for *LaVanguardia.com*.

[2]One-hour interviews, carried out in May 2014, about the strategic principles that affect the editorial organization, the editorial principles, the management of quality and communication, and the understanding that the members of *La Vanguardia* have about the identity and mission of the brand. Those interviewed were David Cerqueda, General Director of Digital Business for Grupo Godó (ADM and IDM); José Luís Rodríguez, Managing Director of *LaVanguardia.com*; Ismael Nafría, Digital Innovations Director for ADM; Ferriol Egea, Head of Web Analytics for *LaVanguardia.com*; María Boria, Head of Subscriptions for *La Vanguardia Ediciones* S. L.; and Óscar Ferrer, Director of Marketing for *La Vanguardia Ediciones* S. L.

independence and own existence (Camps et al. 2004: 476). The brand also specifies its responsibilities towards the Catalonian society and public and private institutions; "without damaging its influence", it commits itself to being independent, to exercising the right to criticize within the legal framework of the Spanish democratic system, and to protecting this democratic system (Camps et al. 2004: 476). Lastly, the responsibility that the company undertakes with regards to its employees is "the respect to individual liberty and everyone's opinion" and "the mutual commitment from the company and the journalistic staff to a critical loyalty to the editorial line of the journal" (Camps et al. 2004: 477).

33.3.1.1 Definition of Mission on Digital Platforms

On *La Vanguardia*'s corporate website (www.grupogodo.com), there is no explicit mention of the brand's mission. The mission must be deduced from the institutional history that is published there. Neither the "Who we are" section of the online edition nor the different social media profiles reference the mission or the editorial principles.

33.3.2 Management and Communication of La Vanguardia's CSR

La Vanguardia has several tools that can help guarantee the implementation of the principles and compliance with its mission and CSR. These tools are part of the Statute, which regulates professional relationships within the newsroom. First, it establishes several rights—such as professional secrecy and the conscience clause —and responsibilities of the members of the newsroom (Camps et al. 2004: 477–483). Second, it envisions the formation of a Professional Committee that would be tasked with ensuring compliance with the principles and norms of the Statute and with representing the newsroom staff before the director and the publishing company in any professional matters. The statutes define that the Editorial Board's responsibility is to communicate with the Professional Committee regarding changes in the editorial line, changes in the relevant corporate strategies, and to hear its members. In addition, the statutes determine the Editorial Board's role in the coordination of editorial work according to the rights and responsibilities aforementioned. Third, the Statute creates the Ombudsman position, which is tasked with "protecting and guaranteeing the rights of readers, attending to their questions, grievances, and suggestions about content" and "to ensure that the treatment of the texts, headlines, and graphics comply with the ethical and professional rules of journalism" (Camps et al. 2004: 489). It is only in describing the role of the Ombudsman that the Statute makes explicit under what conditions and through which platforms information about the way these tools work must be made public.

33.3.2.1 Management and Communication of CSR on Digital Platforms

The responses we obtained during the interviews with different professionals from *La Vanguardia* allowed us to analyze the everyday management of the CSR that the brand carries out online. It should be noted, first, that the journalistic staff who work on the digital edition are separate from the staff who work on the printed edition, and the two sets of staff normally work independently. However, beyond the style guide, *La Vanguardia* has not established a way to communicate either the mission or the editorial principles to the audience and to the digital editorial staff or the values or norms that spring from these principles. Their definition and implementation in the digital context has not been a strategic priority, according to those in charge of the management of the digital enterprise whom we interviewed. This explains in part the different understandings that they have about the mission and the differences that exist between these understandings and what the brand says is its mission. Not one of the thirteen professionals defined it in clear terms or in the same way.

All who were interviewed did agree on the challenges to journalistic quality that arise from media convergence. However, only two of them mentioned lacking a communication channel or mechanism that would orient them in their task. The head of information, for example, said that an update to the style guide would be useful to define the professional norms that must be implemented when faced with the new situations that arise from working online but also the incorporation of a person who would be tasked exclusively with advising journalistic staff on how to apply editorial norms to the new needs that constantly emerge.

The responses of the interviewees highlight that the tools that the Editorial Statute outlines to ensure compliance with the editorial principles do not affect the work of those working on the digital edition. This staff also does not have access to the Professional Committee; it has no representation on this body, and it is left on the sidelines of assemblies, meetings with management, and decision making. The Ombudsman, according to the person in charge of the participation area on the digital edition team, often forwards the concerns he receives about the online edition to the online staff so that they may resolve them autonomously.

An analysis of the digital platforms where the brand is present (corporate website, online edition and social media profiles) has allowed us to study the online communication that the brand carries out regarding its CSR. These platforms are not one of the resources used to make visible the mechanisms established to ensure the implementation of social principles and responsibility. The Professional Committee and the Ombudsman lack its own space on the web, and contact with them is not facilitated. The statute is not published in the online edition nor on the corporate website, nor is it usually named in the information that is published in the different editions.

33.4 Discussion and Conclusions

Despite the fact that the value of the *La Vanguardia* brand is one of its main competitive advantages, the company has not implemented a CSR strategy that would help maintain it. It has defined its mission and responsibilities towards its stakeholders in its style guide. Furthermore, its editorial statute outlines mechanisms such as the Professional Committee and the Ombudsman to ensure compliance with its principles. In this way, it has manifested its will to preserve the quality of journalism and its commitment to the social responsibility that characterizes its mission. However, these tools are focused on the professionals' commitment to the quality of journalism, not on the company's responsibility towards workers. Furthermore, the lack of a comprehensive CSR strategy is more evident in the mechanisms for quality regulation that are available to staff who work on the digital edition. The members of the digital edition newsroom do not have access to any of these mechanisms, except for the style guide, which does not make reference to an online context. The brand also does not take advantage of digital platforms to make these tools visible.

In light of the literature concerning the management of intangibles, the case study undertaken here makes us think that future studies should analyze the case of other media organizations with the goal of (a) developing an understanding of the status of this issue with more empirical foundations and (b) proposing possible routes for the management of media companies' value. Specifically, we propose that future research assess the effectiveness that communication systems based on CSR and digital platforms have or could have in terms of generating credibility, a good reputation, and brand value for all strategic stakeholders.

References

Arrese, A. (2006). Reflexiones sobre el buen gobierno corporativo en las empresas periodísticas. *Doxa Comunicación, 4*, 59–81.

Aznar, H. (1999). *Comunicación responsable. Deontología y autorregulación de los medios*. Barcelona: Ariel Comunicación.

Camps, M., Agustí, M., Baladoch, J., Cadilla, E., Castro, M., & Izquierdo, R., et al. (Coords.). (2004). *La Vanguardia. Libro de Redacción*. Barcelona: Ariel.

Campos, F., Yaguache, J., & Rivera, D. (2011). Credibilidad de la prensa: Misión y responsabilidad social corporativa socio-económico-mediático. *Chasqui. Revista Latinoamericana de Comunicación, 113*, 34–39.

Carreras, E., Alloza, Á., & Carreras, A. (2013). *Reputación corporativa. Biblioteca corporate excellence*. Madrid: LID Editorial Empresarial.

Casado Molina, A. M., Méndiz Noguero, A., & Peláez Sánchez, I. (2013). The evolution of dircom. *Comunicación y Sociedad, 26*(1), 47–66.

Cobo, C. F. (2008). La ordenación de las relaciones profesionales en los medios informativos españoles. Del Estatuto de Redacción de El País al Estatuto de Informativos de la CRTVE. *Revista de Comunicación, 7*, 29.

Echeverri, A. L. (1995). *Recursos Humanos en la empresa informativa*. Salamanca: Universidad Pontificia de Salamanca.

Fernández Vázquez, J. (2012). La responsabilidad social corporativa en los principales grupos de comunicación españoles: incorporación, gestión y análisis de la información a través de sus páginas web. *Correspondencias & Análisis, 2*, 112–128.

González-Esteban, J. L., García-Avilés, J. A., Karmasin, M., & Kaltenbrunner, A. (2011). La autorregulación profesional ante los nuevos retos periodísticos: Estudio comparativo europeo. *Revista Latina de Comunicación Social, 66*, 426–453.

Gutiérrez, E. (2008). *Corporate communication in corporate governance: Why should it be managed strategically? The Spanish Case*. Trabajo presentado en EUPRERA 2008 Congress Institutionalizing Public Relations and Corporate Communications, Milan.

Gutiérrez, E. (2013). Una cartografía conceptual para la comunicación de instituciones. In E. Gutiérrez & M. T. La Porte (Eds.), *Tendencias emergentes en la comunicación institucional* (pp. 53–75). Barcelona: UOC Press.

Lavine, J., & Wackman, D. (1992). *Gestión de empresas informativas*. Madrid: Ediciones Rialp.

Manfredi, J. M. (2009). Indicadores de RSC en la empresa periodística. *Ámbitos, 18*, 137–148.

Nieto, A., & Iglesias, F. (2000). *Empresa informativa*. Barcelona: Ariel.

Picard, R. G. (2010). *Value creation and the future of news organizations: Why and how journalism must change to remain relevant in the twenty-first century*. Lisbon: Media XXI.

Sánchez De La Nieta, M. A., Monfort, A., & Fuente, C. (2012). El estatuto de redacción en la empresa periodística: Criterio básico de Responsabilidad Social Corporativa para la recuperación de la credibilidad del periodismo. In J. L. Fernández & S. Paz (Eds.), *Construir confianza: Intuiciones y propuestas desde la Ética para la empresa sostenible*. Madrid: Cátedra de Ética Económica y Empresarial de la Universidad Pontificia.

Van Riel, C. B. M., & Fombrun, C. J. (2007). *Essentials of corporate communication. Implementing practices for effective reputation management*. Great Britain: Routledge.

Victoria-Mas, M. & Lacasa-Mas, I. (2015). Gestión del valor de marca en las empresas de prensa. El caso de La Vanguardia. *El profesional de la información, 24*(4), 405–412.

Chapter 34
Online Communication Management in Sporting Events. Case: EUC Handball 2013

Carla López Rodríguez, Lorena Arévalo Iglesias and Jessica Fernández Vázquez

Abstract The benefits and the increasing importance of sports events in our society, have been converted into a powerful communication tool for institutions and companies. Concerning to this the desired audiences should be ensured and reached through it. Moreover, sports events are going beyond the experience, and represent a profitable element in economic and social terms. In addition, we are living in a digital world nowadays. Concerning to this it has to be highlighted that the Internet rules most of our lives, and therefore it is used by almost all institutions and organizations to establish contact with their audiences. In this context the present current study should analyse the communication of the organizers of the European universities championship (EUC) of handball in Poland, 2013. The goals of this study are the presentation of how the online communication is used within sport events, which are the common messages, how can "online" contacts between the organizers of events and their followers of sports events be established and how are their reactions. All this information should help us to understand and know how the new media affect this sector and show the common practises on it.

Keywords Sporting events · Social media · Online communication management · Sports communication

C. López Rodríguez (✉) · L. Arévalo Iglesias · J. Fernández Vázquez
University of Vigo (Uvigo), Pontevedra, Spain
e-mail: carla.lopez@uvigo.es

L. Arévalo Iglesias
e-mail: larevalo@alumnos.uvigo.es

J. Fernández Vázquez
e-mail: jessicafdez@uvigo.es

© Springer International Publishing Switzerland 2017
F.C. Freire et al. (eds.), *Media and Metamedia Management*,
Advances in Intelligent Systems and Computing 503,
DOI 10.1007/978-3-319-46068-0_34

34.1 Theory

Internet has radically changed our habits and our ways of communication. Its evolution and development lead us to talk about Web 2.0, transparency, immediacy and inevitability, concerning social media. Also this social and technological phenomenon stands for complexity and a rapid change of nature. Further this technological process of democratization plays a decisive role facilitating the transition from unidirectional media to a bidirectional communication.

Nowadays we are confronted with a new social and media framework. In it the participatory culture and collective intelligence (Jenkins 2008) imply the base of new online communication where users play a decisive role. Therefore its important role is to reflect the own definition of social media as "digital communication platforms that give users the power to generate content and share information through private or public profiles" (IAB 2012).

The relation of new technologies and interactivity of new public roles and new models of communication requires radical measures. Changes in the structure of organizations, which have to modify their communication management and their relation to the public (Benavides 2001: 20–29).

This topic is the origin of our study. We question sports organizations concerning their events and how to adapt its online communication management into the new social reality. To answer this question we study the online communication management done by AZS (University Sport Association from Poland), in the European Universities Handball Championship (EUC), in Katowice in 2013. By analysing the performance of social media on this sporting event we seek to know how such communication works, which content is the most important to share in the social media and what the most interesting issues to the user are.

34.2 Methodology

The case of study is the most appropriate method to deal with this study because it investigates empirically a contemporary phenomenon within its context, using different sources and converging multiple variables of interest. A method, which, through the triangulation of the material compiled, allows to generate theories about a social phenomenon (Yin 2005: 2).

It is true that the decision for this case is an uncommon example. Thus, our area of work, the university, determines the reasons why we chose the EUC Handball 2013 for our study. An institution which, with the rise of Internet and Web 2.0, is immersed in a change having to update their social function (Losada 1998). In this context, university sporting events are conceived as an action that brings popularity to the university and it influences its image and reputation.

Consequently we decided to analyse the communication in social media for sporting events with international impact. Therefore, we selected European competitions of European University Sports Association (EUSA) that applied contemporarily and methodologically of the study of 2013. As an example 17 EUC were organized between May and November. The main factors to choose the handball championship were the active use of social platforms to promote the event and its synchronous development with the investigation (June 2013).

To collect the information, we monitoring daily the website and social platforms from the previous month until the month after the competition.

34.3   Results

The Event The 7° EUC Handball was held in Katowice, Poland, between 23 and 30 of June 2013. More than 300 people from 20 universities and 10 countries had participated.

34.3.1   How was the Online Communication?

The promotion of the event started at their website, Facebook page and Twitter a year before the championship.

34.3.1.1   The Activity

At Twitter and the website the activity did not show considerable differences (concentrated before and during the event). Only on Facebook the activity was concentrated on the period before the event (Table 34.1).

The Website We observed a quantity of publications before and during the event. The activity reached 3.43 posts per day during the competition. Before and over the entire event, the Web was frequently used as an information and promotion portal. However, the activity on the Website after the event was hardly remarkable (only two posts).

Table 34.1 Activity in the social platforms of EUC Handball 2013

Activity phases	Website: 47 posts (%)	Facebook: 183 posts (%)	Twitter: 198 tweets (%)
Pre-event	45	70	42
Event	51	23	48
Pos-event	4	7	10

Source Self made

In Facebook They made 183 posts in total, most of them before the event (70 %), which illustrates a phase where they published weekly. However, during the event the number of posts reached the top with six publications per day. Concerning to this the users have the possibility to follow the championship. Fewer posts were observed after the event (only 12).

Twitter Has illustrated with 198 tweets a balance between the number of tweets before and during the event. However, it is during when, with 13 tweets a day, helps to monitoring the event. As we have seen on Facebook, the users' activity has decreased significant after the event (only 19 tweets).

34.3.1.2   The Content and the Answer

To classificate the content we have based on contribution of Fernández (2005). The author sets the following issues to organize an event: PR, technical-sports, administration, logistics and protocol. After the analysis, we consider to incorporate the tourist variable as well.

The website The main content belongs to PR category. During the three moments of the event this category has shown a lot of posts to promote and document the event through newsletters. Followed by the technical-sports category, the posts are focused on teams, schedules and results. This category shows, like the previous one, activity during the three phases of the event, increasing during. Secondarily we find issues related to logistics, for housing and transport, tourism to promote the city and the country and the protocol with the official speeches. Finally it should be highlighted that the administrative content is scarce and shows a similar activity before and during the event with posts about the registration and official information.

The answer There is no option to participate.

Facebook The same trend happened on Facebook. Here too, PR is the main category with activities in the three moments of the event. The activity it is concentrated before the championship with posts to promote the event, the sponsors or future championships. In progress and after the event its activity is scarce with posts about news of the competition. The second category is technical-sports which increases its activity during competition showing its development. In the period before the event there were only a few posts concerning volunteering and participating teams. With less prominence, but also with activity in the three moments of the event, we find the protocol category whose issues were the opening ceremonies, the closing ceremonies and the complementary activities. In the last positions are the administrative, logistic and tourism categories. With scarce activity before the event, the first focus is spotted on reporting and registration items, the second shows the installations and the third promotes the city.

The answer The favorite user activity is the Like: there were 1260 likes compared to 144 comments and 109 shares. The most popular content that has reached more interactions belongs to PR category, which has the highest number of comments, shares and likes; and technical-sports. The protocol category, thanks to awards, dinners and ceremonies posts, was able to increase its presence and became more and more interesting for the users. Protocol and technical-sports are categories more susceptible to be shared and to generate likes. On the other side, logistics, administrative and tourism are unattractive categories.

Twitter Shows similar activity and content like Facebook. Most of the tweets have the same content because they provide links that redirect to Facebook or the website in order to complete the information. The main difference is detected during the event. In this period of time, Twitter provides direct links to other platforms to watch the games, to check the results and the daily newsletters.

The answer The interaction in this network did not really exist. The actions by the users are reduced to three RT´s and to two favourites. Of these tweets (one tweet has one RT and favourite), two are made before the event (about the transmission of another championship and a player), and two during (about the final of the championship).

34.4   Conclusions

Based on the objectives and the results obtained, it is confirmed that the website is the best option for a daily summarize of a sport event. However, the website makes it impossible for the users to participate on it. This lack was supplemented with its presence in Facebook, giving a good combination of the benefits of both of them.

Also, the choice for this network promoting the championship is suitable because of its popularity in Poland and the specific characteristics of the user profile. By contrast, Twitter has gone unnoticed, whose election to communicate the event has been a failure because it is not popular enough in this country and its use has been limited to direct the traffic to the other platforms and copy its content.

It stresses that online content in sporting events goes beyond the championship, because the PR category has emerged as the favourite content to communicate in social media. Also we recognized a similar content in the website and in the social networks, where PR and technical-sports have the most presence, including the protocol category, these three categories showed the most interesting topics for the users. On the other hand, administrative, logistical and tourism issues lost presence on social platforms and are not interesting for the users.

References

Benavides, J. (2001). Nuevas reflexiones sobre Internet. In Juan Benavides, David Alameda, & Nuria Villagra (Eds.), *Los espacios para la comunicación* (pp. 153–165). Madrid: Fundación General de la Universidad Complutense.

Fernández, J. (2005). *Vademécum de protocolo y ceremonial deportivo: La organización de los distintos eventos deportivos*. Barcelona: Paidotribo.

Jenkins, H. (2008). *Convergence culture. La cultura de la convergencia de los medios de comunicación*. Barcelona: Paidós.

Libro blanco de IAB. (2012). *La comunicación en medios sociales*. Madrid: Edipo. Retrieved from http://www.iabspain.net/libros-blancos/

Losada, A. (1998). *La comunicación institucional en la gestión del cambio: El modelo universitario*. Universidad Pontifica de Salamanca.

Yin, R. (2005). *Investigación sobre estudio de casos: Diseño y Métodos*. Londres: SAGE Publications.

Chapter 35
The Community Manager: Responsibilities Assigned by Companies

Carmen Silva Robles

Abstract The advent of the social web has changed corporate communication through new tools, strategies and, above all, a new sensitivity to connect with audiences and users. One of the major consequences is the real need that has arisen in companies to hire professionals suitable and dedicated to the management of their online reputation, the adaptation of their brand and messages to the digital environment and thinking. However, employers are often incapable of defining what they need, and there is no consensus about the terms that should be used to designate the various tasks associated with online communication. Nonetheless, without a doubt, there is a term that has become the king of the digital jobs: the community manager. The question is: what is a community manager? This article presents the results of a content analysis of advertisements of community manager job vacancies, which aims to find out how employers/companies conceptualise this profession.

Keywords Community manager · Corporate communications · Social web · Professional profiles · Digital communication

35.1 Theoretical Framework

35.1.1 Community Managers

For José Antonio Gallego, President of the Spanish Association of Online Community Managers (AERCO, according to its initials in Spanish), a community manager is "the person in charge of managing and maintaining the community of a brand's loyal followers, and the link between the needs of followers and the pos-

C. Silva Robles (✉)
University of Cádiz (UCA), Cádiz, Spain
e-mail: carmensilva.robles@uca.es; csilvaro@uoc.edu

C. Silva Robles
Open University of Catalonia (UOC), Barcelona, Spain

© Springer International Publishing Switzerland 2017
F.C. Freire et al. (eds.), *Media and Metamedia Management*,
Advances in Intelligent Systems and Computing 503,
DOI 10.1007/978-3-319-46068-0_35

sibilities of the company" (in AERCO and Territorio Creativo 2010: 4). Silva argues that the community manager is a figure that is completely linked to public relations, which is a conception shared by this work. In addition, Silva points out that the community manager is responsible for generating constant flows of communication between companies and their publics through online tools, and "seeks to adapt the objectives of companies to the Web 2.0, by identifying points of common interest between companies and users' preferences" (Silva-Robles 2012: 197).

Moreover, Mejías (2013: 50) adds that the community manager "is the person in charge of preserving the digital identity of the company". Aced (2010) and Elorriaga (2014) share this view and go one step further when they affirm that the community manager is the person responsible for the online reputation of a brand and for positing organisations in the digital environment.

35.1.1.1 Responsibilities

Barra (2008), Ríos (in Cortes 2009), Benito (2010) and Aced (2010: 40–46) point out that the community manager has the following responsibilities:

- Monitor and listen to online conversations.
- Initiate and participate in online conversations and respond to users' questions and comments quickly.
- Create content related to the organisation (without using a corporate discourse) and share it across the most suitable platforms.
- Establish transparent relationships with bloggers and users who deal with issues related to the sector the organisation belongs to, in order to create direct and honest relations with opinion leaders.
- Moderate conversations and put new energy into them to sustain them.
- Become an ambassador for the company without acting as a corporate spokesperson.

35.2 Method

This study is based on the content analysis of advertisements of community manager job vacancies, published between 1 January and 30 June 2013 in: *Infojobs,* the most used employment website among Spanish people; LinkedIn, a social networking website for people in professional jobs, with the largest number of users; and AERCO, the only Spanish Association of Online Community Managers and Social Media Professionals. The analysis only took into account the advertisements that included the terms "community manager" or "community management", either as a job position, a speciality, or previous experience.

A total of 309 advertisements of job vacancies were analysed: 261 found in Infojobs, 22 in LinkedIn, and 26 in AERCO.

The aspects taken into account in the analysis of advertisements were:

1. Community manager responsibilities.
2. Public relations responsibilities.
3. Management responsibilities.

The collected data were coded in Microsoft Excel and subsequently treated with SPSS to find correlations and make inferences between them. Finally, graphs were created in Excel to represent the results.

35.3   Analysis of Results

35.3.1   Responsibilities

35.3.1.1   Community Manager Responsibilities

The most mentioned responsibility is "management of social media and their communities", which is present in more than half (58.9 %) of all the analysed advertisements. This finding highlights the essence of the professional profile and, above all, its conceptualisation and identification by employers.

In second place, with a presence 30 % points below, are: website management (26.2 %) and search engine marketing/search engine optimisation (25.6 %). The third place is occupied by monitoring and active listening (14.6 %). These results clearly show that the responsibilities/functions that are mentioned the most in the sample of advertisements of community manager job vacancies are those that are strictly digital, i.e. those that emerged from the web environment and were not adapted from the off-line world.

The other responsibilities that were mentioned in the advertisements are: implementation of the digital communication strategy (12.6 %), management of advertising campaigns in social media (10 %), development of social profiles and communities (7.4 %), online reputation management (6.8 %), care and attention to user through platforms 2.0 (6.5 %), customer relationship management (6.1 %), online branding (5.2 %), public relations with users (4.2 %), organisation of events 2.0 (3.6 %), routing of customer needs to appropriate departments (3.2 %), identification of problems and opportunities (1.6 %), blog-based public relations (1.3 %), and participation in conversations about products (1 %) (Chart 35.1).

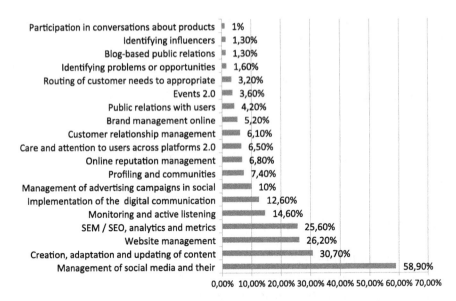

Chart 35.1 Community manager responsibilities. *Source* Author's own creation

35.3.1.2 Public Relations Responsibilities

There are very few advertisements of job vacancies that mention public relations among the responsibilities of the community manager.

In fact, only 11.70 % of the ads mention the development and review of communication material; 9.1 % mention communication management in general; and 7.8 % mention the production of reports and public relations with strategic targets.

Other less-mentioned responsibilities are: event organisation (5.8 %), public relations with the media (3.6 %), achievement of communication goals (2.9 %), branding (2.6 %), news monitoring and analysis (2.3 %), development, management and control of budgets (1.9 %), institutional relations (1.6 %), development, design and implementation of communication strategy (1.3 %), reputation management (0.3 %), management and inspection of communication campaigns (0.3 %), management of corporate social responsibility and management of visibility (0 %) (Chart 35.2).

35.3.1.3 Management Responsibilities

As in the previous case, not many advertisements of community manager job vacancies include this type of responsibilities. The management-related responsibilities that were mentioned in the ads include: the design of communication

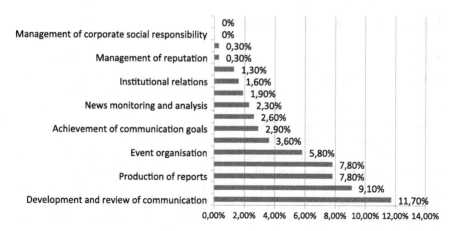

Chart 35.2 Public relations responsibilities. *Source* Author's own creation

strategies and policies, with 13.3 %, which indicates that the need to cover this task is greater than the need to cover the most-mentioned public relations responsibility: the development and review of communication material. Other managerial tasks were mentioned in lower percentages of the sample of ads: management of work groups, with 5.2 %; monitoring of the competition, with 2.6 %; analysis of the environment/diagnosis and prevention of problems, with 2.3 %; decision making, with 1.3 %; and development of proposal of objectives, with 1 %.

The following variables are hardly mentioned in sample of ads: development of control systems, with 0.6 %; and allocation of tasks and resources, attendance to events and crisis communication management, all with 0.3 % (Chart 35.3).

Chart 35.3 Management responsibilities. *Source* Author's own creation

35.4 Discussion and Conclusions

When a company seeks a community manager, the function that it mostly aims to cover is the management of social channels and their communities. And this is the role that identifies this job profile, although other responsibilities are also assigned to it, such as: development, adaptation and updating of online content; website management; and search engine marketing/search engine optimisation.

If we compare these results with those obtained by *Les community managers en France* (2014), we can notice that the French conception of the community manager is closer to marketing, since its responsibilities focus on the concept of customers rather than on the concept of communities. As mentioned, from our perspective of public relations, the idea of community is much closer to the community manager role given that the ultimate goal of the latter is not sell something and thus the concept of customer result inaccurate. According to the results, the community manager has the following responsibilities:

1. Improve brand visibility, 81 %.
2. Attract new customers, 62 %.
3. Increase loyalty among existing customers, 57 %.

In short, employers consider that the main responsibility of the community manager is the management of social media and their communities, and the creation, updating and adaptation of content to social media.

The absence of other responsibilities in the ads of community manager job vacancies suggest a certain lack of knowledge in the people who created the ads. However, it is also possible that advertising companies do not wish to mention certain functions to the community manager job vacancies.

In the light of data obtained from the study, employers' conception of community manager is far from the one offered by theorists and researchers from the world of digital marketing and communication. We therefore believe that, although the conception of the community manager is becoming increasingly more consolidated and defined, is still has a long way to reach a meeting point between the business and scientific worlds, and especially between professionals and employers.

References

Aced, C. (2010) *Perfiles profesionales 2.0*. Barcelona: Editorial UOC.
ADECEC. (2008). *La comunicación y las relaciones públicas en España: radiografía de un sector*. Madrid: ADECEC.
AERCO., & Territorio Creativo. (2010). *Las funciones del community manager*. February 2009. Retrieved May 22, 2010 from http://etc.territoriocreativo.es/etc/2009/11/community-manager-whitepaper.html
Barra, A. (2008). *Community managers, Animadores y Dinamizadores de Comunidades*. Retrieved August 10, 2011 from http://www.albertbarra.com/marketing-hotelero/marketing-hotelero-20/community-managers-animadores-dinamizadores-de-comunidades/

Benito, M. (2010). Retrieved November 3, 2011 from http://www.youtube.com/watch?v=CbWJ62WJ5eQ

Cortes, M. (2009). *¿Qué hace un community manager?* Retrieved August 16, 2011 from http://marccortes.blogspot.com/2009/02/que-hace-un-community-manager-nos.html

Elorriaga, A. (2014). *El marketing y las redes sociales: la figura del community manager en las empresas de la Comunidad Autónoma Vasca y al Comunidad Foral de Navarra.* Retrieved December 11, 2014 from http://www.slideshare.net/angetxu/presentacin-tesis-doctoral-sobre-el-community-manager

Etude: les community managers en France (2014). Retrieved January 16, 2015 from http://www.slideshare.net/captainjob/enquete-community-managers-2014?ref=http://www.ludosln.net/etude-cm-community-manager-france-2014/

Mejía, J. C. (2013). *La Guía del community manager.* Madrid: Ediciones Anaya Multimedia.

Silva-Robles, C. (2012). Community managers: la dirección de RR.PP. en la red. *Revista Internacional de Relaciones Públicas, II*(3), 193–216. Retrieved July 7, 2012 from http://revistarelacionespublicas.uma.es/index.php/revrrpp/article/view/88/68   http://dx.doi.org/10.5783/RIRP-3-2012-10-193-216

Chapter 36
Creating a New Tool for Corporate Communication: The Open Communication Room (OCR)

María Ruiz Aranguren, Leire Iturregui Mardaras and Rosa Martín Sabarís

Abstract This paper forms part of a research project which aims to frame and plan a corporate communication tool to systematise the information flows of different businesses, in a communication environment characterised by hypermedia convergence. The traditional channel of communication between corporate and institutional sources of information and journalists has come to be identified with Press Offices which, with the appearance of the internet, have evolved into Virtual Press Offices. However, on many occasions, the information they offer is incoherent, unconnected and excessive. The project proposes the design of what we have denominated an Open Communication Room (OCR). It is intended to organise the work of communication departments to disseminate information, catering to the productive routine and social function of the media.

Keywords Corporate communication · Hypermedia convergence · Internet · Press rooms · Media

36.1 Theoretical Framework

The interaction between communication departments (belonging to businesses, institutions and associations) and journalists has undergone significant changes in recent decades due to the arrival of the Internet. This new landscape for commu-

M. Ruiz Aranguren (✉) · L. Iturregui Mardaras · R. Martín Sabarís
University of the Basque Country (UPV/EHU), Leioa, Spain
e-mail: maria.ruiza@ehu.eus

L. Iturregui Mardaras
e-mail: leire.iturregui@ehu.eus

R. Martín Sabarís
e-mail: martin.sabaris@ehu.eus

© Springer International Publishing Switzerland 2017
F.C. Freire et al. (eds.), *Media and Metamedia Management*,
Advances in Intelligent Systems and Computing 503,
DOI 10.1007/978-3-319-46068-0_36

nication has its own distinctive characteristics, including standardisation in content creation and a need to be concrete and specific in the face of abundant information and diverse uses. As such, communication professionals have experienced changes to their practices and routines.

In that regard, the studies that have analysed the relationships between journalists and communication departments agree that the most widely-used tool is the Virtual Press Room (VPR), which is, "a series of pages within a company website which offer the traditional communication department services but now via the internet" (Sánchez Pita and Rodríguez Gordo 2010: 5–6).

Three fundamental conclusions can be drawn about the use and nature of VPRs. First, they display uneven development, with private corporations' VPRs showing a level of development unmatched by those of public institutions or other types of social organisation (Vázquez-Burgos 2004; Cantalapiedra et al. 2012a, b). Second, the informative architecture and the content documentation systems lack uniformity, which corresponds to a lack of general criteria in relation to VPRs. In the case of the different bodies that make up a public institution, this lack of unity is particularly noticeable (Xifra 2011). Third, journalists have criticised the failure of VPRs to provide credible, frequently updated online content (Masip 2008; Fortunati et al. 2009).

36.1.1 The Uneven and Limited Development of Virtual Press Rooms (VPRs)

Among the guidelines set out in the Europe 2020 strategy, one is for public administrations to achieve an effective online presence. If we focus on Spain, reports by ONTSI (2009–1914) regarding the digitalisation of local administration show that municipal administrations have adapted very little to the new uses and experiences enabled by the internet.

Furthermore, there are few recent studies of VPR management and those that exist focus on private companies, suggesting that the use of technologies in relation to the media is poor in this area (Castillo Esparcia 2006).

In the case of public institutions, one striking example is the VPRs of the Spanish ministries, which exhibit differences in terms of architecture and usability (Xifra 2011). The same is true of Spanish universities: although half of all public and private education centres maintain and update a VPR, they are a long way from taking full advantage of the resources the internet offers (Busto Salinas 2013).

If we consider the case of the 18 NGOs with the highest gross annual budget, we find that a large proportion of them (41 %) call their VPR "News", meaning that communication is based on very few components (Soria Ibáñez 2011).

36.1.2 Background: MeneXtra, Pilot Project

This paper is part of a research project financed by the Ministry of Economy and Competitiveness, but its seed was planted several years before. The Virtual Press Room (VPR) project, which ran from 2009 to 2010 with finance from the Regional Government of Biscay, marks the beginning of this process. The project's work continued for a further two years with the project "Hermes: A new institutional platform for journalistic communication," which was financed by the SPRI, a public agency dependent on the Basque government's Department of Industry, by means of the Saiotek scheme. As part of the VPR project, the local area was analysed, and the municipal institutions of Biscay in particular. This study proved the absence of an efficient common model for transferring and organising information destined for journalists.

The second project created and developed a tool to organise, structure and standardise one of the most common—and, paradoxically, least consistent—types of content in relations between communication departments and journalists: media alerts. The multimedia platform developed in this project (*MeneXtra*) is an extremely useful communication and dissemination tool for both journalists and for local councils, public administrations, businesses and institutions. As well as simplifying the creation, publication and dissemination of media alerts, the platform aims to act as a shared diary and thus facilitate the coordination of information events. Taking the expertise and knowledge gained from running both projects, we now intend to develop this further.

36.2 Methodology

Our objective, moving forward from the work carried out so far, is to develop the design and implementation of a model that responds to the needs of corporate communication professionals and journalists, via the internet, under the name Open Communication Room (OCR).

The fundamental hypothesis of this works is that current VPRs are not facilitating the work of communication professionals. The following sub-hypotheses are derived from that:

- H1: The lack of consistency in VPR content makes the journalist's work more difficult and impedes the transmission of information to the public.
- H2: VPRs' inefficiency originates from the fact that the information resources they offer are not organised according to journalistic routines and criteria.

Based on that, our objectives are the following:

1: Develop an integrated vision of corporate communication underpinned by the digital context.
2: Improve the efficiency of interaction between the media and businesses.

3: Strengthen the usual functions of communications departments by exploiting the inherent advantages of the digital medium.
4: Establish the information resources needed for corporate communication to be efficient in different contexts.

The proposed research project follows a methodology based on both quantitative and qualitative techniques, which will be implemented in five phases. So far, the first two phases have been put into effect.

First, we have developed a developed an integrated vision of corporate communication underpinned by the digital context.

In the second phase, a study of the practical dimension of corporate communication has been planned, including the creation of digital questionnaires aimed at media professionals at national, regional and local levels, and communication officers of institutions, businesses and associations. The purpose of this task is to identify both parties' routines in their daily interaction and to test whether the design and performance of VPRs are effective in these professionals' work.

36.3 Results

The results are listed below:

- VPR performance is very limited and they do not meet the needs of communication professionals.
- Current VPRs create confusion. Therefore, the OCR must be a corporate communication tool that standardises the different sections through which organisations will share their information.
- The needs of businesses and journalists are such that the new tool (OCR) must coordinate different functionalities using the possibilities offered by social networks, a newsletter system, e-mail alerts, SMS or WhatsApp alerts and the need for text, photos, images etc.

36.4 Discussion and Conclusions

This project aims to contribute to the Information Society by putting forward a virtual tool to improve the management of information flows. The development of OCR lets us ask new questions; that is to say, the OCR can function as both a professional platform and at the same time as a tool for academic research.

Based on the results of our investigation, we are able to outline the need to respond to the following issues:

(1) It is essential to create a more accurate profile of the information needs of the media, on the one hand, and the public, on the other. This will limit the cacophony of information and clarify the informative roles each party should play to avoid overlapping and confusion.

(2) We have detected a need to strengthen the role of communication departments with highly-qualified professionals, in order to manage efficiently the different information flows they create and administrate.

(3) The University has the opportunity to become a social agent capable of rising to the challenges of corporate communication. Institutional communication is one of the principal sources of employment for professionals graduating from the Faculty of Communication Science.

References

ADECEC. (2002). *La comunicación y Relaciones Públicas en España. Radiografía de un sector.* Barcelona: Pirámide.

Almansa Martinez, A. (2011). *Del gabinete de prensa al gabinete de comunicación.* Madrid: Comunicación social.

Alvarez, J., & Timoteo, P. (2005). *Gestión del poder diluido.* Madrid: Pearson-Prentice Hall.

Aragonés, P. (1998). *Empresas y medios de comunicación.* Barcelona: Gestión 2000.

Barquero, J. M. (2000). *Manual de las Relaciones Públicas, comunicación y publicidad.* Madrid: Gestión 2000.

Benavides, J., Costa, J., & et al. (2001). *Dirección de comunicación empresarial e institucional.* Barcelona: Gestión 2000.

Brujó, G. (2008). *La nueva generación de valor.* Madrid: Lid.

Busto Salinas, L. (2013). Trascendencia de los gabinetes de comunicación en la práctica periodística. El caso concreto de la nota de prensa. *Historia y Comunicación Social, 18,* 601–612.

Cantalapiedra, M. J., Iturregui, L., & García, D. (2012a). La comunicación entre gabinetes y periodistas a través de la web 2.0: el caso de meneXtra.com. *Estudios sobre el Mensaje Periodístico, 18,* 213–222.

Cantalapiedra, M. J., Iturregui, L., & García, D. (2012). Una aproximación a la comunicación entre gabinetes y periodistas. In D. CALDEVILLA (Coord.), *Aplicaciones del EEEs a partir de la web 2.0 y 3.0* (pp. 17–39). Madrid: Visión Libros.

Capriotti Peri, P. (2009). *Branding Corporativo. Fundamentos para la Gestión Estratégica de la Identidad Corporativa.* Santiago de Chile. Retrieved from http://www.dircomsocial.com/profile/PaulCapriotti

Capriotti, P. (1999). *Planificación estratégica de la imagen corporativa.* Barcelona: Ariel.

Cárdenas Rica, M. L. (1999). Profesionalización de los gabinetes de prensa municipales. *Revista Latina de Comunicación Social, 15.* Retrieved from http://www.ull.es/publicaciones/latina/a1999c/117luisa.htm

Carrillo Durán, M. V., & Parejo, M. (2009). *Las salas de prensa virtuales de los gabinetes de comunicación de las universidades públicas españolas.* In IV Congreso de la CiberSociedad. Crisis analógica, futuro digital. Retrieved from http://www.cibersociedad.net/congres2009/

Castells, J. M. (1992). La participación ciudadana en la Administración Pública. *Revista vasca de administración pública, 34,* 39–85.

Castillo Esparcia, A. (2006). Public relations and press room. Análisis de las salas de prensa virtuales de las grandes empresas de España. *Razón y Palabra, 49*. Retrieved from http://www.razonypalabra.org.mx/anteriores/n49/mesa7.html

Castillo Esparcia, A., & Almansa Martínez, A. (2005). Relaciones públicas y Tecnología de la Comunicación. Análisis de los sitios de prensa virtuales. *Organicom, 3*, 133–149.

Cebrián Herreros, M. (1996). Concepto y exigencias de la calidad de información. Comunicación de empresas e instituciones. *Telos, 45*.

Celaya, J. (2008). *La empresa en la web 2.0. El impacto de las redes sociales y las nuevas formas de comunicación on line en la estrategia empresarial.* Madrid: Gestión 2000.

Comisión Europea. (2013). Agenda Europea: calidad de la administración pública. Available on http://ec.europa.eu/europe2020/pdf/themes/34_public_administration.pdf

Callison, C. (2003). Media relations and the internet: How fortune 500 company web sites assist journalists in news gathering. *Public Relations Review, 29*(1), 29–41.

Cutlip, S. M. (2001). *Relaciones públicas eficaces.* Barcelona: Gestión 2000.

DIRCOM, Asociación de Directivos de Comunicación. (2012). *Anuario de la Comunicación 2012.* Madrid: Enlaze3 Print Management.

Elías, C. (2003). Adaptación de la metodología de 'observación participante' al estudio de los gabinetes de prensa como fuentes periodísticas. *Empiria. Revista de metodología de ciencias sociales, 6*, 45–159.

Estevez, L. (2014). *Cómo conseguir presencia en los medios sin un gabinete de prensa.* Barcelona: Editorial UOC.

Fernández Asenjo, G., & Torre Alfaro, N. (2010). *Gabinetes de comunicación en mínimos.* Madrid: Fragua.

Flores Vivar, J., & Guadalupe Aguado, G. (2005). *Modelos de negocio en el ciberperiodismo.* Madrid: Fragua.

García Jiménez, J. (1998). *La comunicación interna.* Madrid: Díaz de Santos.

García Orosa, B. (2009). *Gabinetes de comunicación on line. Claves para generar información corporativa en la red.* Sevilla: Comunicación Social.

García Orosa, B. (2013). Los gabinetes de comunicación on line de las empresas del Ibex 35. *Historia y Comunicación Social, 18*, 295–306.

García Orosa, B., & Capón García, J. L. (2005). Gabinetes on line y redes sociales virtuales. In G. López García (Coord.), *El ecosistema digital: modelos de comunicación, nuevos medios y público en Internet* (pp. 197–228). Valencia: Universitat de Valencia.

Garrido, F. (2004). *Comunicación estratégica.* Barcelona: Gestión 2000.

Genaut, A., García, D., & Maurari, I. (2012). Análisis de los contenidos audiovisuales en los agregadores de prensa. In *Crisis y Políticas. La Radiotelevisión pública en el punto de mira* (pp. 239–252). País Vasco: Servicio Editorial de la Universidad del País Vasco.

González-Herrero, A., & Ruiz De Valbuena, M. (2012). Trends in online media relations: Web-based corporate press rooms in leading international companies. *Public Relations Review, 32*, 267–275.

González Molina, S. (2012). Procesos de convergencia comunicativa en las fuentes de información: una visión desde los gabinetes de prensa. *Anàlisi, 47*, 75–89.

Hoyo, J. (2008). Nuevas redes y empresa. Tecnologías web y su aplicación a la comunicación corporativa. *Telos, 66*, 79–82.

Losada Díaz, J. C. (2004). *Gestión de la comunicación en las organizaciones: Comunicación interna, corporativa y de marketing.* Madrid: Ariel.

Losada Díaz, J. C. (2010). *Comunicación en la gestión de crisis. Lecciones prácticas.* UOC: Barcelona.

Maarek, P. (1997). *Marketing político y comunicación: claves para una buena comunicación política.* Barcelona: Paidós.

Maintz, R. (1985). *Sociología de la Administración Pública.* Madrid: Alianza.

Martín Martín, F. (1998). *Comunicación empresarial e institucional.* Madrid: Universitas.

Matilla, K. (2009). *Conceptos fundamentales en la planificación estratégica de las Relaciones Públicas.* Barcelona: UOC.

Molina Cañabate, J. P. (2011). *Introducción a la comunicación institucional a través de Internet.* Madrid: Grupo 5.

Ministerio de Industria, Energía y Turismo ONTSI. (2013a). *Informe anual de la Sociedad en Red.* Retrieved from http://www.ontsi.red.es/ontsi/es/estudios-informes/informe-anual-la-sociedad-en-red-2012-edición-2013

Ministerio de Industria, Energía y Turismo ONTSI. (2013b). Contenidos digitales en España. Retrieved from http://www.ontsi.red.es/ontsi/sites/default/files/presentacion_informe_contenidos_digitales_edicion2012.pdf

Pérez Beruete, C., & Sánchez Galindo, M. (2010). *Nuevos modelos de gestión y función de los responsables de comunicación.* Madrid: Fundación EOI.

Piñuel, J. L. (1997). *Teoría de la comunicación y gestión de las organizaciones.* Madrid: Síntesis.

Pozo Lite, M. (2007). *De las Organizaciones como fuente de Información Periodística Especializada.* Madrid: InSpain Publications.

Ramírez, T. (1995). La influencia de los gabinetes de prensa. Las rutinas periodísticas al servicio del poder. *Telos, 40,* 47–57.

Robledo, M. Á., & Macías Castillo, A. (2009). *Comunicación corporativa. Las relaciones con los medios de comunicación.* Salamanca: Universidad Pontificia de Salamanca.

Rodríguez, M. M., & Marauri, I. (2013). Políticas de comunicación proactivas. El control de la reputación online para gestionar y prevenir una crisis. *Telos, 95,* 98–107.

Rodríguez, M. M., Marauri I., & Cantalapiedra, M. J. (2013). Proactive crisis communications in public institutions. *Revista Latina de Comunicación Social, 68,* 676–695. Retrieved from http://www.revistalatinacs.org/068/paper/985_Bilbao/19_Cantalapiedra.html

Rodríguez, M. M., Marauri, I., & Pérez Dasilva, J. (2006). La comunicación institucional y de servicios. Las páginas web municipales de las capitales de provincia españolas. *Estudios sobre el Mensaje Periodístico, 12,* 431–442.

Rojas, P. (2011). *Community Management.* Barcelona: Gestión 2000.

Rojo Villada, P. A. (2005). Convergencia tecnológica y privatizaciones en el hipersector de la información europeo. *Zer, 12.*

Rojo Villada, P. A. (2008). *Modelos de negocio y consumo de prensa en el contexto digital.* Murcia: Universidad de Murcia.

Sánchez Pita, F., & Rodríguez Gordo, C. (2010). *Tendencias en la construcción de salas de prensa virtuales de las principales empresas del índice bursátil IBEX 35.* In II Congreso Internacional de Comunicación 3.0. Universidad de Salamanca. Retrieved October 4–5, 2010 from http://campus.usal.es/~comunicacion3punto0/comunicaciones/050.pdf

Soria Ibáñez, M. A. (2011). La interacción de los públicos en las ONG 2.0: El estado actual de la comunicación social. *Revista Internacional de Relaciones Públicas, 2,* 175–195.

Túnez Lopez, M. (2012). *La gestión de la comunicación en las organizaciones.* Madrid: Comunicación social.

Túñez López, M., & Martínez Solana, M. Y. (2014). Análisis del impacto de la función, las actitudes y las condiciones laborales del periodista en la producción de noticias: Hacia un periodismo de empresa. *ZER, 19*(36), 37–54.

Urkiza, A. (2007). *Enpresa Komunikazioa profesionalen ikuspegitik.* Bilbao: UEU.

Wilcox, D. L. (2006). *Relaciones Públicas Estrategias y Tácticas.* Addison Wesley.

Chapter 37
Menextra: Designing a Professional Service for Media Alert Management and Distribution

Aingeru Genaut Arratibel, Iñigo Marauri Castillo, María José Cantalapiedra and María del Mar Rodríguez

Abstract The aim of this investigation is to design and to implement a media alert management and distribution service, called *Menextra*. Its goal is to establish a more effective communication between the media and corporate communication departments, organising information around three main axes: time, interests and content. As part of this platform's design process we have consulted previous studies of the use of the internet as a distribution tool for corporate information and an analysis of the services currently on offer. The interface and platform have been designed and implemented, and their performance observed, with the information gathered in mind. *Menextra* is not just a theoretical study. It is a real and operative service that is intended to serve as a test to investigate the extent to which it satisfies the information needs of the media and communication departments. This first version of *Menextra* will be tested with real users (journalists and communication officers) who can identify its benefits and drawbacks in their daily work, and a second version can then be developed better adjusted to their needs.

Keywords Journalists · Mass media · Communication departments · Media alert management · Virtual press rooms

A. Genaut Arratibel (✉) · I. Marauri Castillo · M.J. Cantalapiedra · M. del Mar Rodríguez
University of the Basque Country (UPV/EHU), Leioa, Spain
e-mail: aingeru.genaut@ehu.eus

I. Marauri Castillo
e-mail: inigo.marauri@ehu.eus

M.J. Cantalapiedra
e-mail: mariajose.cantalapiedra@ehu.eus

M. del Mar Rodríguez
e-mail: mirenr@gmail.com

© Springer International Publishing Switzerland 2017
F.C. Freire et al. (eds.), *Media and Metamedia Management*,
Advances in Intelligent Systems and Computing 503,
DOI 10.1007/978-3-319-46068-0_37

37.1 Theoretical Framework

Since its beginnings, Web 2.0 has offered the world of corporate communication a practically infinite array of ways to communicate. A quick look over the results of the investigations carried out in recent years shows, without a shadow of doubt, that these multiple stages of development have not advanced in a uniform way, and that the convergence process is extraordinarily diffuse. Sonia González Molina explores this complexity in her doctoral thesis (González Molina 2011: 85 and foll.) and applies it to the whole area of corporate communication when she refers to the communication convergence process as "a complex process that impacts on the organisation of companies, the technology that they use, the content they produce and the professionals who work in them" (González Molina 2012: 47).

37.1.1 Convergences in Web 2.0

As early as 2005, Castillo and Almansa suggested that the communicative possibilities offered by Web 2.0 allowed "some improvement on the journalistic channel in supplying information to stakeholders" (Castillo Esparcia and Almansa Martínez 2005: 136), and these tools had already been included in corporate communication as an integral part of future communication strategies, both external and internal (Celaya and Herrera 2007: 58). Nevertheless, the difficulty such a transformation entails remains apparent in even the most optimistic texts, which quite rightly recognise that "the communication proposals from VPRs [Virtual Press Rooms] are increasingly numerous and, to a great extent, unmanageable" (Muelas Navarrete 2015: 203).

What we need, then, is a point of contact enabled by technology "in favour of an improvement in relational flows between organisations and media" (Castillo Esparcia 2004: 205), but which is organised around specific information criteria. Professional communication between two defined agents: communication departments and journalists (Fig. 37.1).

37.2 Methodology

After an initial analysis of the existing bibliography, a study was carried out of the VPRs of institutions and companies in the Independent Community of the Basque Country, as well as of the digital tools available online which facilitate relationships between journalists and communication departments through the distribution of corporate information.

The second phase involved several sessions of work to identify the needs of journalists and communication officers in their daily interaction. To do so, a

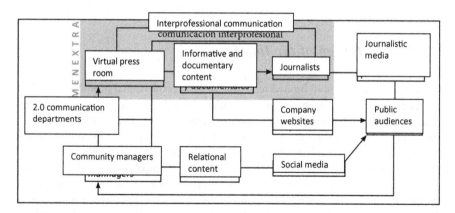

Fig. 37.1 Menextra in 2.0 communication department models of communication. *Source* Created by the authors

methodology was designed that centred on simulation (Izquierdo et al. 2008), in which each team member played different roles, performing communication actions as both a communication officer and as a journalist. In this way, the information needs of each participant in the process have been identified.

The third phase of the work converts these needs into different proposals for the design of the platform's interface.

In the fourth phase, the technical problems that the platform presents are identified and solved by introducing fictitious events into each section, and including in them as many resources as possible.

The fifth phase revolves around another simulation in which each member of the work group analyses the extent to which the information needs assigned to them have been met.

The fifth phase corresponds to a wider dissemination and use of *Menextra* beyond the web. To achieve this, the interface is being redesigned with a view to greater multi-platform compatibility, mainly with mobile devices. In parallel to this, professional workshops are being planned to make contact with working journalists and communication officers. These workshops are dedicated to gaining first-hand knowledge of their needs, and to establishing whether the new version meets those requirements.

Finally, a control group is to be created within the professional environment. To do so, we first intend to call upon the professionals who have participated in the workshops, giving them access to the platform, and seeing the level of acceptance that it may have in their daily work.

37.3   Results

37.3.1   The Design of the Interface

Its central axis is the more limited resource for both professional profiles: time. The interface has been created based on a calendar and a configurable diary that can be adjusted to the individual's daily routine. This service makes the process of preparing a work diary more agile and flexible for journalists and communication officers in their daily routine. The second axis is made up of a double criterion for event selection—by location and by topic. Selecting by topic allows a choice of one to three subjects, taken from the divisions into which information is normally organised in journalism, with which the search filters are established. Once these filters have been set, the events stored in the platform will appear in the application's main panel. In this section, different colours are used to differentiate between future events, to organize the journalist's diary, past events, for documentation, and present events—those that are taking place at that moment. In addition, it has a vertical bar in which a miniature view of the information box shows the events that the user has selected (Fig. 37.2).

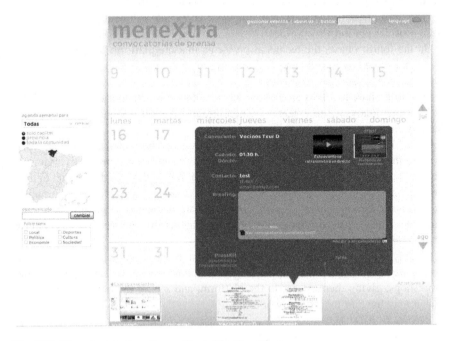

Fig. 37.2 Menextra interface. *Source* Created by the authors

37.3.2   Access and Control

Menextra has been designed to appear as a closed platform for the emitters, but not for the receivers. In this way, anyone wishing to use it to place journalists at an event, must always register first themselves. On this registration file, they will have to give their name, a contact telephone number and e-mail address, as well as indicate the institution, company or association with whom they work. Once this is done, *Menextra* administrators verify the person's identity, the validity of the contact details and their profession. Only after carrying out this verification process will the emitter be able to use *Menextra*. Although it seems a somewhat restrictive process, previous studies of generic, free-access press release aggregators have demonstrated that the stored information has a sharp tendency to turn into an annoying and useless kind of spam (Genaut et al. 2012). We believe that this restriction greatly benefits the emitter, because if receivers can be sure of finding identifiable, accurate and useful information, they will naturally be much more predisposed to taking it into consideration.

With regard to the receiver, although we are clearly aiming for the professional market, the possibility of the platform being accessible to all types of user has not been ignored. This free access would avoid content being limited to just the traditional journalistic market and, furthermore, would serve as a promotional showcase to attract future communication officers.

37.3.3   Content and Media

Once the event has been selected, the journalist will have access to a brief event summary, which the event organiser must provide. In addition to this briefing or summary, the organiser may also include further files for the journalist to use in preparation for the event.

At the bottom left of the information box, there is a section entitled "Press kit". This section is dedicated to the tasks of preparation and extension of information that journalists will need to conduct prior to an event.

Finally, as this is a media alert service, the possibility of streaming these alerts directly has always been present throughout the design of *Menextra*. If an external service is used for streaming, it is simply a case of providing a link and the video will be visible in the event's information box. To avoid depending on third-party services, and to make sure this direct transmission is available even to communication departments with limited resources, we are considering implementing the streaming service within the platform of the second version of *Menextra*.

37.3.4 Interactivity

The platform's main interactive tool is the search function. It acts another way of accessing events, as an alternative to the main way is gradual access through selection by location and topic.

The calendar and diary, on the other hand, offer a more specific and personalisable channel for interaction. Any event the user selects can be transferred to their personal diary with one click, allowing them to organise a list of events that interest them and integrate it into their daily routine.

37.4 Discussion and Conclusions

Menextra began thanks to a Saiotek project funded by the Basque government's Department of Industry, Innovation, Business and Tourism in 2011, and the platform was in operation by 2013. Now, the new version of *Menextra* is integrated into a much wider and more ambitious research project: a project titled "Application of hypermedia convergence in corporate communication: Open Communication Room (OCR)" which began in 2015 and is financed by the Ministry of Economy and Competitiveness. *Menextra* is integrated into this new project and has the following fundamental objectives:

(a) to reinforce its function as a professional tool via a new design and greater compatibility with mobile devices;
(b) to extend the platform to include more of the information functions that are characteristic of 2.0 communication departments, and to start creating Open Communication Rooms (OCRs);
(c) to act as a methodological tool: a lab experiment to find out the needs of journalists and communication officers first-hand, via the workshops, and put them into practice in closed control groups, monitoring the use of the platform at all times.

References

Álvarez Rodríguez, M. L., Martí Pellón, D., & Domínguez Quintas, S. (2010). Reputación y Responsabilidad desde webs corporativas. *Área abierta, 26*. Accessed December 13, 2013, at http://revustas.ucm.es/indez.php/ARAB/article/view/ARAB1010230001A

Capriotti, P. (2014). De los media/canal a los media/público. La relación de las organizaciones con los medios de comunicación desde la perspectiva de las relaciones públicas. *Hologramática, 6, 2*, 19–32. Accessed September 16, 2015, at http://www.bidireccional.net/Blog/2007Hologramatica6.pdf

Castillo Esparcia, A. (2004). Comunicación empresarial e institucional. Estrategias de comunicación. *ZER, 17*, 189–207. Accessed January 22, 2013, at http://www.ehu.eus/zer/hemeroteca/pdfs/zer17-10-castillo.pdf

Castillo Esparcia, A., & Almansa Martínez, A. (2005). Relaciones Públicas y Tecnología de la Comunicación. Analysis de los sitios de prensa virtuales. *Organicom, 3*, 132–149. Accessed September 16, 2015, at http://www.eca.usp.br/departam/crp/cursos/posgrad/gestcorp/organicom/re_vista3/132.pdf

Catalina García, B. (2015). Gabinetes de prensa como principal fuente documental de los medios de comunicación. Fuenlabrada como caso de estudio. *Index Comunicación, 5*, 121–143. Accessed September 16, 2015, at http://journals.sfu.ca/indexcomunicacion/index.php/indexcomunicacion/article/view/184/169

Celaya, J., & Herrera, P. (2007). *Comunicación empresarial 2.0: La función de las nuevas tecnologías sociales a la estrategia de comunicación empresarial.* Barcelona: BPMO. Accessed December 13, 2013, at http://libros.metabiblioteca.org/handle/001/162

Domínguez Quintas, S., Álvarez Rodríguez, M. L., & Martí Pellón, D. (2012). Dirección de Comunicación en internet. Estudio y recomendaciones para los espacios de prensa en webs corporativas desde el análisis de portales en internet de grupos empresariales en Galicia. *Revista Internacional de Relaciones Públicas, 3, 2*, 45–70. Accessed December 13, 2013, at http://dialnet.unirioja.es/descarga/articulo/3966572.pdf

García Orosa, B. (2013). Los gabinetes de comunicación on line de las empresas del Ibex 35. *Historia y Comunicación Social, 8*, 295–306. Accessed December 13, 2013, at http://revistas.ucm.es/index.php/HICS/article/viewFile/43967/41573

Genaut Arratibel, A., Marauri Castillo, I., & García, D. (2012). Análisis de los contenidos audiovisuales en los agregadores de notas de prensa. In *II Congreso Internacional de Comunicación Audiovisual y Publicidad: Crisis y Políticas* (pp. 239–252). Bilbao, 4 and 5 October 2012. Accessed April 8, 2013, at https://web-argitalpena.adm.ehu.es/listaproductos.asp?IdProducts=UWLGSO8649

González Molina, S. (2011). *La convergencia en els gabinets de prensa i comunicació: Les oficines especialitzades en seguretat viària.* Barcelona: Facultad de cièncs de la comunicación Blanquerna, Universidad Ramón Llull. Accessed March 19, 2014, at http://hdl.handle.net/10803/21791

González Molina, S. (2012). Procesos de convergencia comunicativa en las fuentes de información: Una visión desde los gabinetes de *prensa. Análisi, 47*, 75–89. Accessed on December 13, 2013, at http://dialnet.unirioja.es/servlet/articulo?codigo=4387168&orden=411119&info=link

Izquierdo, L., Galán, J. M., Santos, J. I., & Del Olmo, R. (2008). Modelado de sistemas complejos mediante simulación basada en agentes y mediante dinámica de sistemas. *Empiria, Revista de Metodología de Ciencias Sociales, 6*, 85–112.

Marauri Castillo, I., Rodríguez González, M., Genaut Arratibel, A., & Iturregui Mardaras, L. (2013). El muro de las críticas. El uso de las redes sociales por los sectores más denunciados por los consumidores. *Estudios Sobre el Mensaje Periodístico, 20*(1), 159–175. Accessed October 21, 2014, at doi:10.5209/rev_ESMP.2014.v20.n1.45225

Marca Francés, G., Matilla, K., & Mateos Rusillo, S. (2014). Museos y periodistas. Análisis de la sala de prensa virtual como espacio de relaciones públicas. *Historia y Comunicación Social, 19*, 105–115.

Muelas Navarrete, P. (2015). De fuente corporativa a diario digital: La adaptación de la comunicación corporativa a Internet en la Comunidad Valenciana. *Dígitos, 1*, 99–210. Accessed September 16, 2015, at http://revistadigitos.com/index.php/digitos/article/view/10

Pérez Dasilva, J. Á., Genaut Arratibel, A., Meso Aierdi, K., et al. (2013). Las empresas en Facebook y Twitter: Situación actual y estrategias comunicativas. *Revista Latina de Comunicación Social, 68*, 676–695. Accessed December 4, 2013, at http://www.revistalatinacs.org/068/paper/996_Bilbao/RLCS_paper996.pdf

Ruiz Mora, I., Solar Olmedo, S., & Álvarez Nobell, A. (2010). Salas de prensa virtual, redes sociales y blogs: Posibilidades de la comunicación 2.0. Estudio de las diez empresas españolas líderes en el IBEX 35. In *V Congreso Internacional de Investigación en Relaciones Públicas*. Barcelona, 13 and 14 May 2010. Accessed March 18, 2013, at https://aalvareznobell.files. wordpress.com/2010/03/salas-de-prensa-redes-sociales-y-blogs-corporativos-2010.pdf

Sánchez González, M., & Paniagua Rojano, F. J. (2013). Estrategias de comunicación 2.0 en asociaciones profesionales: Estudio del caso de los Colegios de Médicos en España. *Revista Mediterránea de Comunicación, 4,* 21–51. Accessed March 19, 2014, at http://www. mediterranea-comunicacion.org/Mediterranea/article/view/40

Sánchez Pita, F., & Rodríguez Gordo, C. (2010). Tendencias en la construcción de salas de prensa virtuales de las principales empresas del índice bursátil IBEX 35. In *II Congreso Internacional Comunicación 3.0*. Salamanca, 4 and 5 October 2010. Accessed March 18, 2013, at http:// campus.usal.es/ ∼ comunicacion3punto0/comunicaciones/050.pdf

Part VI
Marketing, Advertising and Tourism

Chapter 38
Graphic Design and Social Networks: Methodological Proposal Supported by the Open Innovation and Co-creation

Blas José Subiela Hernández

Abstract This paper proposes a methodology for the process of graphic design in which public participation in decision-making through social networks is contemplated. And it takes advantage of a particular case study to show the proposal in detail. The proposal is novel in two ways: a creative technique and public participation are used in the selection of the final design. Graphic design is, above all, a communication tool that works with visual signs. And these signs must convey meanings evenly decodable by receivers. Hence, the theory of communication is a fundamental discipline for graphics. In addition, the design is also a semiotic process. In this sense, brand's graphic design is a symbolic exercise of identification and differentiation, as brands are symbolic translations of the organizations they represent. And within the universe of semiotics, rhetoric is of particular relevance, since the creative technique proposed is based on the development of visual metaphors. The public participation in the selection of the final design is carried out through social media, showing followers some sketches and asking them to votes, according to "open innovation" and co-creation theories. Through the case study allows doing a meta-communication work, while the proposed methodology is applied for the design of the head of an academic journal of communication.

Keywords Graphic design · Visual metaphor · Methodology · Open innovation · Co-creation

38.1 Theoretical Framework

Graphic design is, above all, a communication process that works with visual signals, which have to transmit meanings decodable by receivers. The theory of communication is not unrelated to the discipline of design, since when the receiver interprets the message under similar parameters that was encoded by the issuer, "an

B.J. Subiela Hernández (✉)
Catholic University of Saint Anthony of Murcia (UCAM), Murcia, Spain
e-mail: bsubiela@ucam.edu

© Springer International Publishing Switzerland 2017   297
F.C. Freire et al. (eds.), *Media and Metamedia Management*,
Advances in Intelligent Systems and Computing 503,
DOI 10.1007/978-3-319-46068-0_38

sufficient communication is established, and therefore the process has been more or less successful" (García Jiménez 2007: 52).

Besides, the design is a fundamentally semiotic discipline. As mentioned above, the visual signs with which it works should have meaning. In this sense, the design of the head of a publication is a very specific task of graphics, but to which can be applied all the theoretical principles of brands' graphic design. And, as any brand, it is a symbolic exercise of identifying and then differentiating. As Buttle and Westoby (2006: 1181) note:

> While the use of logos may primarily be aimed at creating a distinctive way to associate a brand name with a graphic representation, logos should also convey the ethos of the brand they represent. Hence, if a brand is meant to represent quality and exclusiveness then the logo should do the same.

In the same vein, Villafañe (1999: 67) states that a brand should be a symbolic translation of the organization to which identifies. However, for Chaves (2003: 25), "it is a serious mistake to confuse the signs of institutional identification with the means to communicate corporate image attributes and positioning". In any case, there is no doubt of the need to seek maximum consistency between the organization and the selected set of graphical symbols to represent it.

It could be helpful to get a graphic that meets these requirements involving the online community in the decision-making process, in line with proposed theories of open innovation (Chesbrough and Crowther 2006) and co-creation (Piller et al. 2012) and, more generally, the principles of web 2.0 (Levine et al. 2000). Thus, users are becoming more active in building their brands and taking part in decisions, which until recently were taken on a purely internal way. A good example of this influence is the case of GAP and its attempt to redesign its brand in 2010. One week after the introduction of its new graphics, the brand was forced to come back to its original design due to strong campaign to reject the new proposal on social networks (Alandete 2010).

38.2   Methodology

Any design process begins with a comprehensive gathering of information. To this end, it is used a specific model of briefing for graphic design, in which it is collected, together with the general description of the organization, information of visual aspects, such as previous graphics, graphic styles of the sector, and mandatory and forbidden visuals.

From the information obtained during the research, the creative process begins. We believe that the visual metaphor is one of the main creative design resources (Rivera Díaz 2008; Jardí 2012; Batey 2014; Llorente-Barroso and García-García 2015). In this case it is proposed the development of an array of visual metaphors. It is a creative technique focused on visual creativity, in which a two-dimensional matrix is generated: first, the attributes with which the organization seeks to be identified (previously extracted from the briefing); and then, concepts visually representative of the sector in which the organization operates are introduced. By crossing the two variables we have generated images that are visually representative of the sector and the values of the organization. From these images, a series of sketches are published in the Facebook page for his followers to vote for the more appropriate option. Once a decision is made, the sketch becomes the ultimate graphics, adapting it to a proportional system and providing it with a particular graphic style.

38.3 Results

From the developed briefing is extracted the most relevant information on the origin and evolution of the publication, as well as its personality and features. The first aspect in which we focus is the name "Sphera Publica". Its first part, Sphera, comes from the Latin concept Sphaera, but takes the form of old Castilian.[1] From the multiple meanings assigned by the Spanish Academy to this concept, we are interested in the following: "Scope, space that extends or reaches an agent under the powers and role of a person, etc." And then he adds that the sphere of action or activity is "a space that extends or reaches under any agent." Moreover, the adjective published (note that the accent is eliminated, since in Classical Latin there are no accents) defines the space of the one who speaks about the Academy. And that results in the first definition provided by Habermas (1982) to refer to those places, physical and virtual concepts, which are taking shape of public opinion, which has a relevant value in the media, the major players in the public sphere.

Therefore, the name of the publication is placed at the centre of contemporary communication research, which, in turn, has an obvious graphic representation by the figure of the sphere.

The original design of the head, conducted by Fernando Contreras, only uses a typographic resource and it flees from the source of the sphere, although it is suggested by the larger size of the letter "e" (Fig. 38.1).

According to information gathered in the briefing, this header, which has been used continuously since the beginning of the publication to the present, must change significantly. The new stage of the magazine[2] wants to pair with new

---

[1]As we can see in some scientific treatises of the sixteenth century cataloged in the Virtual Library Miguel de Cervantes; for example, the "Treatise of Sphera" composed by Dr. Ionannes of Scrobusto.

[2]The magazine was founded in 2000 and has risen to tenth place in the classification of scientific journals of communication of IN-RECS, with an impact factor of 0.075. However, it suffers from the lack of assessment by ANECA of the work of management and direction of scientific publications, so over a period of time it becomes inactive and stops publishing. This situation causes a significant decline in their quality indices. During 2014/2015 it was decided to give a new impetus and within the new strategy defined for the publication, redesigning its head is on the plan.

SphEra Publica

Fig. 38.1 Original design of the head of "Sphera Publica". *Source* http://sphera.ucam.edu/index.php/sphera-01/about

graphics that do not keep the connection to the old style, beyond maintaining the name.

The concepts with which the magazine wants to be identified are: scientific and analytical approaches, interest in the media and the public sphere in which public opinion is generated, reference publication, humanism, Latin and Mediterranean character. The concepts visually representative in the field of social sciences may be: books, the sphere of the world, political maps, newspapers, speakers…

Once all the variables of the matrix are completed, four sketches are produced, focused primarily on the image of the sphere from different interpretations. As regards typography, considering the symbolism of the various typographic categories (Loxley 2007; Nørgaard 2009; Subiela Hernández 2012), two possibilities are considered: maintaining a serif font to remember the previous design, and bet for a Roman one, associated with values such as correctness, reliability and credibility. It is particularly useful in this regard the synthetic table presented by Subiela Hernández (2012), in which a series of symbolic links to the main categories of type designs associated attributes.

The resulting sketches (Fig. 38.2) are posted on the journal's Facebook page, and the followers are asked—mainly from the scientific community in the area of social sciences and communication—to vote for the most identifying and attractive proposal, after describing the reasons why the editorial board has decided to renew the graphic mark.

Fig. 38.2 Sketches proposed. *Source* Compiled from matrix of visual metaphors and symbolism typographic table created by Subiela Hernández (2012)

Fig. 38.3 Final header. *Source* Designed using the font ʼAveriaʼ, by Dan Sayers

Sketches 2 and 4 are the most voted by the online community, but editors finally decide to choose option 4, since one of its early slogans did not maintain any connections to the previous graphic style.

Based on the sketch 4, they started working in the final proposal. Thus, the whole graphics are provided with a Mediterranean style, both in the printing and the iconic part. With regard to typography, that style is achieved by choosing a font whose design is the result of merging a set of popular fonts. It is the Averia font (modification of the English average), created by Dan Sayers and available on Google fonts with Open Font License. The unfocused contours from this source cause a liquid sensation that connects to the Mediterranean character sought for the header. Regarding the development of the polyhedron, technical drawing geometric shapes are abandoned and it is used a line of varying thickness, which provides a more humanistic style.

The final proposal is as follows (Fig. 38.3).

38.4   Discussion and Conclusions

Although creative and design processes have always been considered unstructured and random, here we show the usefulness of following a work plan to achieve orderly designs that meet the needs of the request. The phases in which we can summarize this process are: information gathering, creative process (based on the development of visual metaphors), making sketches, public exposure in social networks and voting by the online community and final selection for application of graphic style.

Thus, establishing a methodology for a co-creative process like the design process is not contradictory. This is because creativity is not incompatible with the processes and order. Tracking a series of steps and the use of certain creative techniques guarantees that the final design, beyond its aesthetic value, is an element of effective communication. We understand graphics efficacy capacity to identify and represent the organization to which it gives name, shape and colour. Moreover, the participation of the online community in the process also serves as a guarantee that the final brand will be quickly adopted in a nondramatic way for users.

The new era for "Sphera Publica" has a strong graphic reference with which being identified. And it meets the main requirements found in the briefing:

innovation, analysis, social science, humanism and Mediterranean character. And this statement is possible by monitoring the methodology presented in this text and the involvement of Facebook followers.

References

Alandete, D. (2010, Octubre 12). *GAP retira su nuevo logo ante la presión de las redes sociales.* Spain: El País.

Batey, M. (2014). *El significado de la marca, el cómo y por qué ponemos sentido a productos y servicios.* Buenos Aires: Granica.

Buttle, H., & Westoby, N. (2006). Brand Logo and Name Association: It's all in the name. *Applied Cognitive Psychology, 20,* 1181–1194. doi:10.1002/acp.1257

Chaves, N., y Belluccia, R. (2003). La marca corporativa. *Gestión y diseño de símbolos y logotipos.* Barcelona: Paidós.

Chesbrough, H., & Crowther, A. K. (2006). Beyond hight tech: Early adopters of open innovation in other industries. *R&D Management, 36*(3), 229–236.

García Jiménez, L. (2007). *Las teorías de la comunicación en España: Un mapa sobre el territorio de nuestra investigación (1980–2006).* Madrid: Tecnos.

Habermas, J. (1982). *Historia y crítica de la opinión pública.* Barcelona: Gustavo Gili.

Jardí, E. (2012). *Pensar con imágenes.* Barcelona: Gustavo Gili.

Levine, R., Locke, C., Searls, D., & Weinberger, D. (2000). *The Cluetrain Manifesto: The end of business as usual.* Cambridge (MA): Perseus.

Llorente-Barroso, C., & García-García, F. (2015). La Construcción Retórica de los Logos Corporativos. *Arte, Individuo y Sociedad, 27*(2) 289–309. Retrieved from: http://www.arteindividuoysociedad.es/articles/N27.2/LLORENTE_GARCIA.pdf. Accessed at October 22, 2015.

Loxley, S. (2007). *La historia secreta de las letras.* Valencia: Campgràfic.

Nørgaard, N. (2009). The semiotics of typography in literary texts. A multimodal approach. *Orbis Litterarum, 64*(2), 141–160.

Piller, F. T., Vossen, A., & Ihl, C. (2012). From social media to social product development: The impact of social media on co-creation of innovation. *Die Unternehmung, 65*(1).

Rivera Díaz, L. A. (2008). La retórica en el diseño gráfico. *Investigación y Ciencia, 41,* 33–37. Retrieved from: http://www.uaa.mx/investigacion/revista/archivo/revista41/Articulo5.pdf. Accessed at July 8, 2015.

Subiela Hernández, B. J. (2012). El simbolismo tipográfico en los nuevos dispositivos móviles: Hacia la reconciliación de letras y pantallas. *Icono14, 10*(2), 126–147.

Villafañe, J. (1999). *La gestión profesional de la imagen corporativa.* Madrid: Pirámide.

Chapter 39
Impact of Religious Tourism in Social Media in the Andean Region of Ecuador: The Case of the Pilgrimage of the Virgin of El Cisne and the Trade Fair of Loja

Eva Sánchez-Amboage, Alex-Paul Ludeña-Reyes and Christian Viñán-Merecí

Abstract The pilgrimage of the Virgin of El Cisne and the Trade Fair of Loja are the two most significant events in Southern Ecuador. This study makes a documentary analysis on the online activity and reputation of Facebook pages from the bodies responsible for the pilgrimage of the Virgin of El Cisne and the Trade Fair in Loja. The assessment tool employed for analysing Facebook is Fanpage Karma, since it simplifies the retrieval of data. The research is further reinforced by the information available from Google Trends in relation to the recent surveys on these events. Finally, it is done a contrast work by analysing the influence of social media in the decision-making process in relation to the participation in these events. The measurement is carried out through an online survey using Google Docs-Drive, which was shared on social networks.

Keywords Religious tourism · Pilgrimage of the Virgin of Cisne · Loja trade fair · Ecuador · Social media

39.1 Theoretical Framework

A key part of articles and books published in recent years, in which tourism is related with the Information and Communications Technology (ICT), explain how important are these latter for tourist industry (Baggio 2006).

E. Sánchez-Amboage (✉) · A.-P. Ludeña-Reyes · C. Viñán-Merecí
Technical Particular University of Loja (UTPL), Loja, Ecuador
e-mail: eva.amboaxe@gmail.com

A.-P. Ludeña-Reyes
e-mail: apludena@utpl.edu.ec

C. Viñán-Merecí
e-mail: csvinan@utpl.edu.ec

© Springer International Publishing Switzerland 2017
F.C. Freire et al. (eds.), *Media and Metamedia Management*,
Advances in Intelligent Systems and Computing 503,
DOI 10.1007/978-3-319-46068-0_39

303

Within the ICT, social media have become new platforms for companies and tourist destinations. Recent studies show an increase in the number of active tourists, which generate content in 2.0 platforms and use information of social media for planning trips. As Mich and Baggio (2015) note, Facebook has a dominant position in the world of social networks. Indeed, it has become a reference tool for online tourist industry.

The second main theme of this work is the religious tourism through the Virgin of El Cisne and the Trade Fair of Loja, events of major importance in Ecuador. Authors such as Raj and Morpeth (2007) see religious tourism as a sustainable and international phenomenon that is constantly growing. It is an increasingly important slice of the global tourist market, which is focused on an idea that has been present for thousands of years in the evolution of humanity and its relation with the divinity (Juárez et al. 2012).

Those religious landmarks where there is a growth and development of this type of tourism show a positive economic effect, but also social and cultural. This impact was reflected in the central area and the entire region (Lorenzo and Ramón 2011; Tobón and Tobón 2013).

On 28 July 1829, Venezuela's Liberator Simon Bolívar approved through a decree the pilgrimage of the Virgin of El Cisne from the temple to the city of Loja. Every 8 September, people pay tribute to the Virgin, and she heads the Trade and Religious chair in Loja. This has strengthened the strong devotional rooting to the Virgin of El Cisne, but it has also established a secular dimension to the event with the creation of the Trade Fair. This fair itself has a cultural component, reflected through the various homages to the Virgin, fireworks, and traditional dances and folk music. In short, it should be noted that, at present, the pilgrimage of the Virgin of El Cisne and the Trade Fair of Loja are the biggest tourism event of the city.

39.2 Methodology

The present study is exploratory and has the following general objectives (GO):

- GO1. To analyse the visibility and interactivity of the pilgrimage of the Virgin of El Cisne and the Trade Fair of Loja in Facebook.
- GO2. To understand the impact of both events on the search engine Google with the aim of knowing the online positioning.
- GO3. To study the influence of social media in the promotion of the Trade Fair of Loja and the Pilgrimage of the Virgin of El Cisne.

To accomplish the first objective, we employ the online free tool Fanpage Karma, http://www.fanpagekarma.com/, which is useful for analysing social media and monitoring various social networks. In order to determine the needed variables to analyse the visibility and interactivity, the study of Huertas et al. (2014) and the guidelines of Cavalganti and Sobejano (2011) has been used as a basis. The four selected Fanpages for the analysis correspond to the main official bodies responsible for the promotion of the pilgrimage of the Virgin of El Cisne and Trade Fair Loja on Facebook: *"Alcaldía de Loja"*, *"Feria de Loja"*, *"Perfectura de la provincia de Loja"* and *"Santuario de la Virgen de El Cisne"*. The analysis has been done on 15, 20, and 21 August 2015, on 8 and 13 September, and 1 November 2015, which means, during, before, and after the event.

Google Trends https://www.google.es/trends/ is used to reach the GO2. The use of this tool as the basis of a research project is a recent issue that, according to Matias et al. (2009), provide daily information on what people all over the world are searching.

The third objective is based on the implementation of surveys, which are designed for the purpose of gathering digital information and measuring variables (Punch 2003). In this way, a survey through Google Docs-Drive was undertaken: https://docs.google.com/forms/d/1L9-RVAvY0Vg3ug0TGay4fmd4w-CG5iTK0c_w61Rjmgl/edit. We take as a reference the study of the Chamber of Industries of Loja, which notes that the fair and the visit to the Virgin of El Cisne attract 514,159 visitors. With this universe, it is used a simple random sampling, since all potential visitors have the same probability of being selected in the sample. Also, this sample size has a 92 % confidence level, being the probability of happening (P) and not happening (Q) is 50 % each one; one sigma of 92 % and an estimated error of 6 %, resulting in 213 survey. Then, it was proceed to the distribution of the sample in each province according to the population weight and, finally, it was sent to a random person to complete the online survey.

39.3 Results

According to the data collected during this paper, results can be divided into three blocks related to each of the GO. The first objective, related to the visibility and interactivity of the above-mentioned Facebook pages, has shown that "Feria de Loja" and "Alcaldía de Loja" are the most visible in Facebook, followed by "Perfectura de Loja" and "Santuario de La Virgen de El Cisne", although to a minor extent. Over all, these Facebook pages reach the attention of 43,474 fans. It is also important to note that the four pages exceed by far the recommendation of 3 publications a week of Internet República (2012).

As regards the visibility of each publication, there is a clear deviation between the number of likes, comments, and shares in the analysed pages, since likes reach the highest figure. Furthermore, special mention should be made of likes related to the Virgin of El Cisne, which show an evident devotion to the Virgin in the online community. The most common format is the combination of text and pictures, even if there is also post with video and text, but is less frequent.

Besides, it is studied the interactivity between pages and users through the engagement. Data from Table 39.1 show that the number of fans is not important if the page is not well managed and if there is no participation of the online

Table 39.1 Interactivity in Facebook pages

Fanpage	Number of fans	Engagement (%)
Alcaldía de Loja	16,066	31.6
Feria de Loja	18,065	17.5
Perfectura de Loja	8599	44.5
Santuario de la Virgen de El Cisne	744	54.4

Source Prepared by the authors

community. The most striking case is the page "Santuario de la Virgen de El Cisne". Although it has the fewer number of fans, it has the best *engagement*, what means, commitment and interaction with the online community.

Regarding the second objective, the Pilgrimage of the Virgin of El Cisne has more impact on Google than the Trade Fair of Loja. It is noted that the pilgrimage has become well known in 2009 and the Virgin of El Cisne in 2007. Also, there is seasonality in searching processes. August is the most active month, since it is when these events are held. It is further recognized that Ecuadorians are those who become informed on these events via Google (Chart 39.1).

Finally, it is important to present the most significant results of the survey. Facebook and Google are the most used when gathering information on the pilgrimage of the Virgin of El Cisne and the Trade Fair Loja. Also, the vast majority of respondents agree with the promotion of the pilgrimage and the trade fair via social networks. It is further considered that comments on these events in online communities enjoy a high decree of credibility and are useful to increase visits to Loja. The types of contents that awaken most interest among respondents are pictures, followed by videos and text (Chart 39.2).

Lastly, as regards participation of respondents, the vast majority share pictures, videos, and events, and comment on their experiences at the fair and the pilgrimage. A small proportion interacts with likes in post related to these events. There is thus a contradiction with the information obtained in the section of visibility, where it was clear that the most common actions were likes in publications.

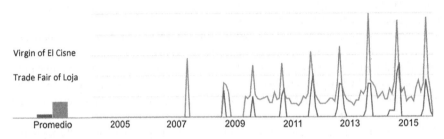

Chart 39.1 Popularity of the pilgrimage of the Virgin of El Cisne and the trade fair of Loja in Google. *Source* Prepared by the authors

Chart 39.2 Type of content that awakens interest for participating in the trade fair of Loja and the pilgrimage to the Virgin of El Cisne. *Source* Prepared by the authors

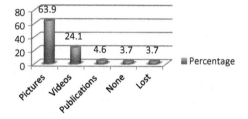

39.4 Discussion and Conclusions

According to the data collected in this work, we conclude that there is a segment of Facebook market interested in the Trade Fair of Loja and the Pilgrimage to the Virgin of El Cisne. This should be exploited both by responsible for the events and tourism administrations of Ecuador. It is noted once again the clear devotion of the online community to the Virgin of El Cisne.

Data accessible through Google Trends reinforces the idea that there is interest in analysed events through the Internet. Also, it is found that searches are mainly made in the Ecuadorian country, so it is advisable to take into account this information when undertaking advertising campaigns in order to make them known throughout the world. Finally, the survey allow us to conclude that social media are of great importance when it comes to promoting and report on the Trade Fair of Loja and the Virgin of el Cisne.

Acknowledgments The present work was sponsored by the Prometeo Project of the Ministry of Higher Education, Science, Technology and Innovation of the Republic of Ecuador—PROMETEO-CEB-008-2015-. It was also developed as part of the activities of the International Network of Communication Management—R2014/026 XESCOM-, supported by the Ministry of Culture, Education, and University Planning of the Xunta de Galicia. This network is formed by iMarka Research Group from the University of A Coruña and the Group of Innovation and New Enterprise from the Technical Particular University of Loja, among others.

References

Baggio, R. (2006). Complex systems, information technologies, and tourism: A network point of view. *Information Technology & Tourism, 8*(1), 15–29.

Cavalganti, J., & Sobejano, J. (2011). *Social Media IOR las relaciones como moneda de rentabilidad.* Madrid: Bubuk Publishing.

Huertas, A., Setó, D., & Miguez, M. (2014). Comunicación de Destinos Turísticos a ctravés de las Redes Sociales. *El Profesional de la Información, 24*(1), 15–21.

Juárez, J., Ramirez, B., Mota, J., César, F., & Valverde, G. (2012). Peregrinación y turismo religioso en los santuarios de México. *Revista Geográfica Valparaíso, 46*, 41.

Lorenzo, L., & Ramón, F. (2011). La ruta de los sagrados corporales de Llutxent (Valencia) como una nueva expresión de turismo religioso en España. *Estudios y Perspectivas en Turismo, 20*, 960.

Matias, Y., Niv, E., & Shimshoni, Y. (2009). On the predictability of search trends. Retrieved at October 28, 2015 from: http://googleresearch.blogspot.com.es/2009/08/on-predictability-of-search-trends.html

Mich, L., & Baggio, R. (2015). Evaluating Facebook pages for small hotels: A systematic approach. *Information Technology & Tourism*. Retrieved at October 28, 2015 from: http://www.iby.it/turismo/papers/baggio_mich-FB.pdf

Punch, K. (2003). *Survey research: The basic*. London: SAGE Publications Ltd.

Raj, R., & Morpeth, N. (2007). *Religious tourism and pilgrimage festivals management. An international perspective*. Cambridge: Cab Internacional.

Tobón, S., & Tobón, N. (2013). Turismo religioso: Fenómeno social y económico. *Anuario Turismo y Sociedad*. Retrieved at October 28, 2015 from: http://papers.ssrn.com/sol3/papers.cfm?abstract_id=2435380

Chapter 40
Social Networks and their Role in the Promotion of Emerging Tourist Destinations: The Case of the Area 7 of Ecuador

Clide Rodríguez-Vázquez, Valentín Alejandro Martínez Fernández, Ramiro Armijos-Valdivieso and María-Dolores Mahauad-Burneo

Abstract Digital social media have an essential role in the promotion of tourist destinations representing image and reputation where the visitant assumes an active role, as a prescriber. Destinations must be included in the social media as part of the strategy by well defining their communication strategies mostly for reaching efficiency. The management of promotion, through communication, is relevant for emerging destinations with a weak positioning but with a great potential, as it is the case of the destinations in the South of Ecuador. The tourism development is a priority for the Ecuadorian Government for economic and social policies, gathered within the PLANDETUR2020 and the National Plan for Good Living. This research project analyses the potential of tourist destinations from several towns located in the South Region (Part of the Region 7) of the country: Loja, Vilcabamba, Malacatos, Saraguro and Catamayo, as well as the Natural Park Podocarpus as a dynamic element. Official Facebook accounts were studied to define the projection of their identity in order to create a brand of image—place and the user interaction with the digital social media.

Keywords Digital social media · Facebook · Tourist destinations · Promotion management · South of Ecuador-Region 7

C. Rodríguez-Vázquez (✉) · V.A. Martínez Fernández
University of A Coruña (UDC), Coruña, Spain
e-mail: crodriguezv@udc.es

V.A. Martínez Fernández
e-mail: valejand@udc.es

R. Armijos-Valdivieso · M.-D. Mahauad-Burneo
Technical Particular University of Loja (UTPL), Loja, Ecuador
e-mail: prarmijos@utpl.edu.ec

M.-D. Mahauad-Burneo
e-mail: mdmahauad@utpl.edu.ec

© Springer International Publishing Switzerland 2017
F.C. Freire et al. (eds.), *Media and Metamedia Management*,
Advances in Intelligent Systems and Computing 503,
DOI 10.1007/978-3-319-46068-0_40

40.1 Theoretical Framework

In the 21st century, it has been supported that to be competitive in the market, the customer considered the main sponsor (Rosales Castillo 2010) and the relationship with the customer should be managed directly when introducing new strategies and proactive media.

The implementation of a new process must be built from the knowledge of the market segments; the new value chain for each of the tourist products; considering all those channels dominated by ICT such as; Internet and social media and the establishment a framework for collaboration to optimize efforts and generate a greater impact of the actions.

The use of catalysts, such as technologies, will allow a positive influence in the management of promotion and marketing; considering the traditional basic structure of promotional activities because it should be taken into consideration its relationship with other systems that configured the target structure. Even if it is proved that tools and technology have been strong developed (González Vázquez 1995), it is not enough, since they must merge from its basics to obtain good results.

In the tourist area the formula of traditional promotion has evolved. Nowadays, destinations have to look over the strategy to adapt it into the new reality, where difficulties and inherent cost of promotion activities are high. Some of the factors that affect this situation are a similarity and an overload of offered messages or an excess of advertising tools (Turgalicia 2006).

According to Bigné et al (2000) promotion of destinations concern the integration and coordination of the techniques used and the connection offline–online; as well as, the promotion planning through a global and general plan of destination that encourages the consumer or theirs analysis as a buyer.

One of the most influential aspects is the habits of using information by tourist consumers when planning their trips, especially in the relationship between the sources of information and the choice of destination (Um and Crompton 1992); as well as, the importance of the Internet use as impersonal source of decision (Collado Molina et al. 2007). There are also other elements that influence the intentions of tourists, such as; psychological motivations, demographic variables, the image or the attitude towards the destination (Court and Lupton 1997; Díaz 2002).

Kent (1991) and Goodall (1991) determine the need to classify the elements that affect decision making in Push and Pull actions, in which the image of the destination, drives the choice of it between the already preferred ones (Kang and Schuett 2013; Munar and Steen 2014; Hudson and Thal 2013). It is estimated that a significant part of the image, of the process decision and the evaluation of the current consumer destination comply through the network and in particular the social media. Being Facebook the preferred, the most used and visited (The Cocktail Analysis 2009; IAB 2015).

40.2   Methodology

The research, carried out between August 8th and October 24th, 2015, has consisted in the analysis of the official FanPage of the following communities located at the South of Ecuador (Table 40.1).

These accounts have been chosen for the analysis since they are official communities that manage destinations. The analysis of content, interactivity and visibility criteria were followed in accordance with the objective of the research (Huertas Roig et al. 2015). Their work has been of great importance to select the values and items: **Contents**—post frequency, format and type of information-; **Visibility**—number of fans, likes, shares, timing, post length and hashtags—and **Interactivity**—user response and engagement-.

40.3   Results

Content. The first item scanned was the format of the posts. This analysis has been exclusively done to the FanPage of Loja, Malacatos and Saraguro, as other destinations have not published during the period studied. The most commonly type of post used in three destinations was photography. As reflected in Table 40.2, Loja has done through this format 54.2 % posts (6373 pictures), Malacatos 57.1 % (524 pictures) and Saraguro 63.6 % (196 pictures). There were other meaningful type of post in the dissemination and visibility of content such as: video, Loja used 30 % of its posts and Saraguro 18.2 %. Other type of posts were links, Loja had 12.5 % and Saraguro 18.2 %; and finally the status type of posts used only by Malacatos from July 26th to August 25th, which meant 42.9 % total of its posts.

Regarding the frequency of posting (see Table 40.2), Loja is the only destination that published nearly a daily post (with an increase of 0.30 in July to 0.60 in October), between 3 and 5 per week. The rest of the destinations did not reach even one per week (0.55 Malacatos and 0.85 for Saraguro), which was clearly insufficient and did not help to encourage interest in these communities.

Table 40.1 FanPage analyzed

Destinations	FanPage
Loja	https://www.facebook.com/LojaEcuador
Malacatos	https://www.facebook.com/MALACATOSvalle?fref=ts
Saraguro	https://www.facebook.com/saraguros?fref=ts
Vilcabamba	https://www.facebook.com/vilcatour?nr
Catamayo	https://www.facebook.com/catamayocity
Parque Nacional de Podocarpus	https://www.facebook.com/pages/Parque-nacional-Podocarpus/637927456224286?fref=ts

Source Authors

Table 40.2 Post per day, per type and rate

Post per day

	Loja	Malacatos	Saraguro	Vilcabamba	Catamayo	Parque Nacional de Podocarpus
	0.52	0.07	0.1	0	0	0
Post per type and rate[a]						
Photos (%)	54.2	57.1	63.6	0	0	0
Video (%)	29.2	0	18.2	0	0	0
Link (%)	12.5	0	18.2	0	0	0
Status (%)	0	42.9	0	0	0	0
Others	0	0	0	0	0	0

Source Authors
[a]At the time of analysis 24/10/2015

Table 40.3 Fans, post per day, likes, comments and Shares per post

	Loja	Malacatos	Saraguro	Vilcabamba	Catamayo	Parque Nacional of Podocarpus
Fans[a]	26,019	2350	1637	3734	121	65
Post per day[a]	0.52	0.07	0.1	0	0	0
Likes,[b] comments and shares	126	88	34	0	0	0
Timing	Way off	Perfect timing	Way off	–	–	–
Length of posts[c]	Less than 100	Less than 100	Less than 100	–	–	–
Hashtags	Using	Not using	Using	Not using	Not using	Not using

Source Authors
[a]At the time of analysis 24/10/2015; [b]Average per post; [c]Characters

Finally, the category of information provided by these communities was analyzed. It was found that there were few questions in the case both of Loja and Saraguro. However, Malacatos used this form to get information from their fans, to encourage them, to be more visible, to create dialogue and to facilitate greater interactivity.

Visibility. The number of fans is the quantitative criterion that provides more information about the presence of these destinations on the net. In Table 40.3 is observed that Loja, with 26,019 fans was the destination that attracted more people, followed by Vilcabamba with 3734; Malacatos with 2350 and Saraguro with 1637.

Other rates were the average of "like", "comments" and "share" that each post produced. Although Loja attracts a higher value than the rest (126), especially with its photos, Malacatos and Saraguro, even with fewer number of fans and posts per day had a higher number of likes, comments, shares. Therefore, Malacatos and Saraguro had a higher proportional number, showing that the success of the publications of these two destinations is greater than Loja. In the same way, it is important to mention that Vilcabamba, with 3734 fans was not as visible as other destinations because of the absence of publications, this decreased the efficacy of the profile.

Another aspect was related to the timing, the length of the Post and Hashtags used in publications (see Table 40.3). Malacatos achieved the greatest coordination, to publish in the hours where the followers are active (from 00 to 03 h GMT). Loja and Saraguro should change the timeslot of their publications due that they did not adjust to their fans' interests. The length of the posting was right, it did not exceed 100 characters, this is an important aspect that fans appreciate and they reward it better and with more interaction. The analysis of hashtags with the posts showed that only Loja and Saraguro use them, giving them visibility into other social media as Twitter.

Interactivity. A primary index on this last parameter is the engagement (see Table 40.4), which measured the reactions, of "likes", "comments" and "share" (to get it the PTAT, People Talking About This) is divided between the number of 'like' and the result should be higher than 7 %).

There are certain anomalies in the analysis; even if Loja published more posts, Malacatos generated a greater engagement, 14.68 % (see Table 40.4) which means that not only the number of published post is essential, but also the promotion, interaction and monitoring, as the community of the Natural Park of Podocarpus has done, this activity has allowed to this community to achieve a degree of commitment to their fans 6.15 %.

Other important items were users responses related to published post or also called index of response. Figure 40.1 show that the posts made with photography's caused a better reaction in all the users, basically "like", followed by "share" and

Table 40.4 Fans, post per day, PTAT and engagement

	Loja	Malacatos	Saraguro	Vilcabamba	Catamayo	Parque Nacional de Podocarpus
Fans[a]	26,019	2350	1637	3734	121	65
Post per day[a]	0.52	0.07	0.1	0	0	0
PTAT[b] (peopletalking about this)	375	345	8	14	0	4
Engagement (%)	1.44	14.68	0.49	0.37	0	6.15

Source Authors
[a]At the time of analysis 24/10/2015; [b]Last month

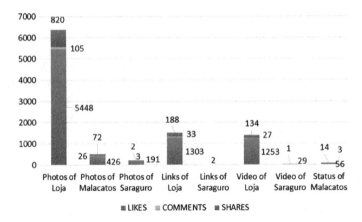

Fig. 40.1 Page's top posts. Top 5 posts. From 08/08/2015 to 24/10/2015. *Source* Authors

"comments". Also, videos and links generated responses, although to a lesser extent. Finally, we can find posts that were used only by Malacatos, so these posts generated reactions using "like" (56), comments (3) and shares (14).

40.4   Discussion and Conclusions

Social media is a key tool in the development of the tourist destinations due to the impact on communication of the brands and identity projection (María Munar 2011). Facebook is a likely way to connect with tourists, exchange content, or encourage viral posts and visibility. Analysis of the FanPage of Loja, Vilcabamba, Malacatos, Saraguro, Catamayo and the Natural Park Podocarpus, through the LikeAlyzer monitoring tool, showed the following results:

The FanPage gives weak data, since the FanPage only uses three formats, which allows to estimate that a greater variety and dialogue would create an attractive area to generate higher fidelity. The FanPage should encourage to ask more questions to enhance interactivity. In relation to the frequency of published posts, none of the tested pages performs daily publications therefore this stops an effective visibility and interactivity.

Regarding interactivity, posts with photographs got better and more reactions, followed by videos and links. The generation of a greater number of posts does not give a higher degree of engagement regarding to destination, in all cases. So, we advise to give a better treatment and monitoring to the publications.

If we want to attract more fans, some actions could be carried out to encourage them to follow the social media, such as: face-to-face discussions with the target; carry out contests; get ahead of the consumer; make the tourist feel part of the community or reward their loyalty (Merodio 2010).

Acknowledgments The present work was sponsored by the Prometeo Project of the Ministry of Higher Education, Science, Technology and Innovation of the Republic of Ecuador—PROMETEO-CEB-008-2015-. It was also developed as part of the activities of the International Network of Communication Management—R2014/026 XESCOM-, supported by the Ministry of Culture, Education, and University Planning of the Xunta de Galicia. This network are part of the iMarka Research Group from the University of A Coruña and the Group of Innovation and New Enterprise from the Technical Particular University of Loja, among others.

References

Bigné, J. E., Font, X., & Andreu, L. (2000). *Marketing de destinos turísticos. Análisis y estrategias de desarrollo.* Madrid: Esic.

Collado Molina, A., Esteban Talaya, A., & Martín Consuegra, D. (2007). Una aproximación al comportamiento del consumidor turístico y la importancia de las fuentes de información. *Papers de Turisme, 41,* 7–25.

Court, B., & Lupton, R. (1997). Customer portofolio development: Modeling destination adopters, inactives and rejecters. *Journal of Travel Research, 36*(1), 35–43.

Díaz, R. J. (2002). *Interrelación entre imagen y comunicación en destinos turísticos.* Juan Ramón Oreja Rodríguez e Isabel Montero Muradas (Dir.). La Laguna. Tesis doctoral.

González Vázquez, E. (1995). Aplicación de la tecnología de la información a las decisiones estratégicas de marketing: Análisis de la cartera de productos. *Investigaciones Europeas de Dirección y Economía de la Empresa, 1*(2), 137–146.

Goodall, B. (1991). Opportunity set concept: An application to tourist destination areas. In G. Asworth & B. Goodall (Eds.), *Marketing tourism places.* London: Routledge.

Hudson, S., & Thal, K. (2013). The impact of social media on the consumer decision process: Implications for tourism marketing. *Journal of Travel & Tourism Marketing, 30*(1–2), 156–160.

Huertas Roig, A., Setó-PàmieS, D., & Míguez-González, M. I. (2015). Comunicación de destinos turísticos a través de los medios sociales. *El profesional de la información, XXIV,* 1. Retrieved at August 8, 2015 from http://dx.doi.org/10.3145/epi.2015.ene.02

IAB. (2015). VI Estudio Redes Sociales de IAB. *IAB.* Retrieved at August 5, 2015 from http://www.iabspain.net

Kang, M., & Schuett, M. (2013). Determinants of sharing travel experiences in social media. *Journal of Travel & Tourism Marketing, 30*(1–2), 93–107.

Kent, P. (1991). People, places, and priorities: Opportunity sets and consumer's holiday choice. In G. Asworth & B. Goodall (Eds.), *Marketing tourism places.* London: Routledge.

María Munar, A. (2011). Tourist-created content: Rethinking destination branding. *International Journal of Culture, Tourism and Hospitality Research, 3*(5), 291–305. Retrieved at August 5, 2015 from http://dx.doi.org/10.1108/17506181111156989

Merodio, J. (2010). Marketing en redes sociales. Retrieved at August 25, 2015 from http://www.bubok.es/libros/191596/Marketing-en-Redes-Sociales-Mensajes-de-empresa-para-gente-selectiva

Munar, A., & Steen, J. (2014). Motivations for sharing tourism experiences through social media. *Tourism Managment, 43,* 46–54.

Rosales Castillo, P. (2010). *Estrategia digital.* Barcelona: Deusto.

The Cocktail Analysis. (2009). *I Oleada del Observatorio de Redes Sociales.* Retrieved at October 27, 2015 from http://tcanalysis.com/blog/post/informe-ebooks

Turgalicia. (2006). *Plan estratégico 2006–2010.* Dirección Xeral de Turismo, Xunta de Galicia.

Um, S., & Crompton, J. L. (1992). The roles of perceived inhibitors and facilitators in pleasure travel destination decisions. *Journal of Travel Research, 30*(3), 18–25.

Chapter 41
Eye Tracking: Methodological and Theoretical Review

Xosé Rúas Araújo, Iván Puentes-Rivera and Sabela Direito-Rebollal

Abstract Methods, techniques and tools in neuroscience offer new perspectives of analysis in social sciences, marketing and advertising when studying perception and attention. Eye tracking, which emerged in the late 19th and early 20th centuries, is one of these tools. At present, it has many applications in communication, including the experiential contrast to improve usability and web design. From the theoretical review and description of the historical evolution of eye tracking, this exploratory work presents a methodology for the analysis and contrast of the websites from the University of Vigo (UVigo) and Santiago de Compostela (USC). The sample is composed by a group of students from the faculties of communication, aiming at checking whether the result of students´ viewing are in line with the recommendations of experts in composition, content, images perception and web texts.

Keywords Eye tracking · Attention · Persuasion · Communication · Web design

41.1 Introduction

41.1.1 Visual Attention

The application of traditional methods and techniques in neurosciences is reaching social sciences and, particularly, marketing and communication, which are using them to analyse attention and persuasion processes and to contrast the efficiency of messages. The most used techniques are the functional magnetic resonance imaging

X. Rúas Araújo (✉) · I. Puentes-Rivera
University of Vigo (UVigo), Vigo, Spain
e-mail: joseruas@uvigo.es

I. Puentes-Rivera
e-mail: ivanpuentes@uvigo.es

S. Direito-Rebollal
University of Santiago de Compostela (USC), Santiago de Compostela, Spain
e-mail: sabeladireito@hotmail.com

© Springer International Publishing Switzerland 2017   317
F.C. Freire et al. (eds.), *Media and Metamedia Management*,
Advances in Intelligent Systems and Computing 503,
DOI 10.1007/978-3-319-46068-0_41

of the brain (fMRI); the electroencephalogram (EEG); the analysis of galvanic response or thermal conductance; the heart rate variability (HRV); and the facial recognition of emotions for assessing responses to aural and visual stimuli (Rúas Araújo et al. 2015).

Eye-tracking is one of the techniques for analysing perception and the connection between our brain and the outside world. The term is related to the Human Visual System (HVS) and the visual attention (immediate or derivate, passive, reflexive, active and voluntary), where associations between vision and action and the control system over gaze come into play.

The gaze is used to locate, and the motor system to operate. All that remains is to decide how to use the action. Hence the importance of predisposition, as noted by one of the first researchers on visual attention (Von Helmholtz 1925); and focusing, concentration and conscience—it is evident that humans cannot pay attention to several things at once. These three last elements are used by James (1890) to define attention in terms of meaning and associated expectation that, at the same time, give us clues about memory.

Accordingly, the analysis of information flows in visual attention processes cannot be addressed in isolation. There is also a need to talk about the concepts of system and scheme, indispensable in describing the internal representation of a task and the instructions associated with its execution. Hence the approach of the so-called "supervisory attentional system" (Norman and Shallice 1986) and the "working memory model" (Baddeley 2007), which provide information on where, how and what to look for.

Also, two elements come into play as regards visual psychology: the relevance and the prominence of images (what is calling attention on them), as regards details related to brightness, colour and movement. It is a very important factor for the organization of a "map of relevance" (Itti and Koch 2001), in order to discover the hot sports on which people focus their attention. The setting, duration and location are also decisive.

The eyes are also an important element in our social interactions and relations as a mode of non-verbal communication and a complement to language, to the extent that mutual gaze can affect other person's behaviour (Argyle 1988) and reveal aspects of personality (introverted, extroverted), and the expression of emotions and social skills (Kleinke 1986).

In short, from a perspective of communication analysis, all above described lead us to move away from isolated approaches based solely on the use of point tools. It should be noted that vision does not necessarily imply action and a casual relation; it depends on multiple factors.

41.1.2 Historical Evolution

The record of eye movement has been done for more than a century, as Duchowski (2002), Wade and Tlatler (2005) note. The first experiments were carried out by

Dodge and Cline in 1901 and were based on photographic records; after them, Guy Buswell used the record of routes followed by the eyes when observing images. In 1948, Hartridge and Thompson designed the first eye tracker, which was installed on the head; in the 50s, the Russian physiologist Alfred Yarbus set out recording techniques; in the 60s, Mackworth and Thomas used TV cameras; in the 70s, Merchant and Morrisette invented the eye-tracker (Jacob and Karn 2003); and present techniques, which are using infrared lights.

The first studies, conducted by James and Helmholtz, were focused on WHAT attracted attention and WHERE was it produced (issues related to the foveal and parafoveal vision and the saccadic movements of the eye). Later, in the 40s and 50s, research was centred in HOW the attention was produced. The maximum exponent of this topic was the North-American psychologist Jerome James Gibson, who wrote *The Perception of the Visual World*, a book focused on the visual perception of space that resulted in the emergence of the ecological psychology. Contrary to the theory of Gestalt, Gibson considered that perception was not part of the body structure, but of the environment in which the organism is set.

In 1953, Deutsch and Deutsch reported on the existence of a selective filter—the bottleneck model—in the sensory memory. Thus, studies focused on WHERE and HOW is the attention attracted were combined with the search for answers to studies on expectation—WHAT, as a basis for the creation of computational models of visual attention.

Specifically, there are two strategies for information processing: the "top-down" and the "bottom-up", which reproduce the neurobiological behaviour. These models, established by Treisman (1988), are based on the assumption that the various properties of the space are encoded in maps whose characteristics are reproduced in different brain regions.

On this basis, there is also discussion on the encoding process, which does not start until a sufficient level of detail is reached; and also on the understanding, which is achieved through a sequential and hierarchical learning of a series of visual discriminations and perceptive, lexical, syntactic and semantic processes.

41.2 Eye-Tracking on Marketing and Advertising

Consumer actions widely combine internal and external elements in decision-making. Among the external ones, there are some related to the 4Ps of marketing: product, placement, price and promotion.

The most advanced eye-tracking devices allow monitoring consumers visual attention while they are on the move. For instance, eye-tracking glasses allow observation of consumers' activity, their stroll down the supermarket aisles and sections, as well as their choices between the different brands and products. In short, they make it possible to describe consumers' experiences.

The Eye-tracking also contribute, from an internal point of view, to carry out perceptual and cognitive assessments. It all depends, essentially, on whether we want to study the effect or the cause. And, although a comprehensive research should be the result of the combination and contrast with other qualitative techniques and tools, the eye-tracking provides information on how users focus their visual attention on products and ads.

Foremost among the first experiments is the one carried out by Lohse (1997), who studied the ads on yellow pages and analysed their location, colour and size. He concluded that these factors had an impact on attention. Pieters et al. (1997) also conducted studies to check the effectiveness of print ads as regards the length and beginning of attention, the movement of the gaze between different parts of the ad through the analysis of saccadic movements.

The Poynter Institute was one of the first organisms focused on analysing journals, news and press photos (García et al. 1991). Also, the National Association for Stock Car Auto Racing (NASCAR) worked together with students from informatics, marketing and industrial engineering to test perception according to the different areas of specialisation, with the aim of describing where students look whilst using racing cars and how their fixations were distributed (Duchowski 2002).

Since then, there appeared many studies on perception using eye-tracking tools. Some of the most relevant and recent studies were carried out with the objective of analysing the difference in purchase decision behaviour of compression sportswear and the observation of their promotional pictures (Rahulan et al. 2015); examining the effects of smiling models on consumer joy and attitudes (Berg et al. 2015); and analysing the efficiency and effects of personalized banner ads on visual attention and recognition memory (Koster et al. 2015).

Also, there are recent studies focused on the analysis of new devices, such as the Kinect sensor, a tool used in various console games (Wii, Xbox). These works aim at analysing viewers' behaviour when watching different TV channels and programmes within the family environment (Takahashi et al. 2015); and understanding iPads' uses and their applications for people with disabilities and cerebral palsy (López et al. 2015).

In Spain, research on eye-tracking has been focused on the impact of press photos and social networks (Arrazola and Marcos 2014); digital press (Rovira et al. 2014); and TV advertising on young and older generations (Añaños and Valli 2012; Añaños 2015).

In short, there are many works confined to the study of Human Visual Systems, which conditioned theories and studies on the measurement of psychological (and physiologic) responses to the media (paraphrasing the title of Lang's work (2011). These studies influenced on the media and their terminology, standing at the objective of the media's frame and priming.

41.3 Conclusions

The present works aimed at establishing a theoretical and methodological framework, based on the defence of neurosciences' interdisciplinarity and their application to social sciences and communication through the combination of different methods, techniques and research tools.

Any visual and aural detection process, and any memory reconstruction and social construction of reality using new technologies, require a multifactorial analysis and an observation of the constant information flow.

In this regard, eye tracking is an efficient tool for undertaking direct and objective monitoring, but it needs to be combined with other methods and techniques, escaping from isolated approaches and, in consequence, providing different views on the same subject of study. Therefore, the application of these techniques from neurosciences to social sciences is considered appropriate and relevant when studying human behaviour and social relationships.

References

Argyle, M. (1988). *Bodily communication*. London: Methuen.

Arrazola, V., & Marcos, M. C. (2014). Fotografía de prensa y redes sociales: la técnica de Eye-Tracking. *Ámbitos, 27*, 31–40.

Añaños, E. (2015). La tecnología del "Eye-tracker" en adultos mayores: cómo se atienden y procesan los contenidos integrados de televisión. *Comunicar, 45*, 75–83.

Añaños, E., & Valli, A. (2012). La publicidad integrada en el contenido TV: Atención visual y reconocimiento cognitivo en los jóvenes en los adultos mayores. *Pensar la publicidad, 6*(1), 139–162.

Baddeley, A. (2007). *Working memory. Thought and action*. Oxford: Oxford University Press.

Berg, H., Soderlund, M., & Lindstrom, A. (2015). Spreading joy: Examining the effects of smiling models on consumer joy and attitudes. *Journal of Consumer Marketing, 32*(6), 459–469.

Campbell, D. T., & Fiske, D. W. (1959). Convergent and discriminant validation by multitrait-multimethod matrix. *Psychological Bulletin, 56*, 81–105.

Denzin, N. (1989). *Strategies of multiple triangulation. The research act: A theoretical introduction to sociological methods*. New York: McGraw Hill.

Denzin, N. K. (1970). *Sociological methods: A source book*. Chicago: Aldine Publishing Company.

Deutsch, J. A., & Deutsch, D. (1963). Attention: Some theoretical considerations. *Psychological Review, 70*(1), 80–90.

Duchowski, A. T. (2002). *Eye tracking methodology. Theory and practice*. Londres: Springer.

García, M. R., Stark, M. M., & Miller, E. (1991). *Eyes on the news*. Florida: Poynter Institute.

Gibson, J. J. (1974). *La percepción del mundo visual*. Buenos Aires: Infinito.

Hassan, Y., & Herrero, H. (2007). Eye tracking en Interacción Persona-Ordenador, *No sólo Usabilidad*. Retrieved from: http://www.nosolousbilidad.com/articulos/eye-tracking.htm. Accessed at October 15, 2015.

Itti, L., & Koch, C. (2001). Computational modelling of visual attention. *Nature Review of Neuroscience, 2*, 194–203.

Instone, K. (2010). *Navigation stress test*. Retrieved from: http://instone.org/navstress. Accessed at October 15, 2015.

Jacob, R., & Karn, K. (2003). Eye-tracking in human-computer interaction and usability research. *Ready to deliver the promises. Mind, 2*(3), 4.

Kleinke, C. L. (1986). Gaze and eye contact: A research review. *Psychological Bulletin, 100*, 78–100.

Koster, M., Ruth, M., Hamborg, K Ch., & Kaspar, K. (2015). Effects of personalized banner ads on visual attention and recognition memory. *Applied Cognitive Psychology, 29*, 181–192.

Krueger, R. (1988). *Focus group. A practical guide for applied research.* Newbury Park: Sage.

James, W. (1890). *The principles of psychology.* New York: Henry Holt & Co.

Lang, A. (Ed.), *Measuring psychological responses to media.* New York: Routledge.

Lohse, G. L. (1997). Consumer eye movement patterns on yellow pages advertising. *Journal of Advertising, 26*(1), 61–71.

López Gil, J. M., et al. (2010). Análisis de la arquitectura de webs mediante tests de estrés de navegación, de usabilidad y eye tracking. *El Profesional de la Información, 19*(4), 359–367.

López, A., Méndez, A., & García, B. (2015). Eye/head tracking technology to improve HCI with iPad applications. *Sensors, 15*(2), 2244–2264.

Merton, R., Fiske, P., & Kendall, P. (1956). *The focused interview.* Glencoe: The Free Press.

Morgan, D. (1988). *Focus group as qualitative research.* Newbury Park: Sage.

Nielsen, J., & Pernice, K. (2009). *Eyetracking web usability.* Berkeley: New Riders Press.

Nielsen, J. (2002). Top ten guidelines for homepage usability. Retrieved from: http://www.useit.com/alertbox/20020512.html. Accessed at October 13, 2015.

Nielsen, J. (1994). Estimating the number of subjects needed for a thinking aloud test. *Journal of Human-Computer Studies, 41*(3), 385–397.

Norman, D. A., & Shallice, T. (1986). Attention to action: Willed and automatic control of behaviour. In R. J. Davidson, G. E. Schwarts & D. Shapiro (Eds.), *Consciousness and self-regulation. Advances in research and theory* (Vol. 4, pp. 1–18).

Pieters, R., Rosbergen, E., & Wedel, M. (1997). Visual attention to repeated print advertising: A test of scanpath theory. *Journal of Marketing Research, 36*(4), 424–438.

Rahulan, M., Troynikov, O., Watson, C., Janta, M., & Senner, V. (2015). Consumer behaviour of generational cohorts for compression sportswear. *Journal of Fashion Marketing and Management, 19*(1), 87–104.

Rovira, C., Capdevila, J., & Marcos, M. C. (2014). La importancia de las fuentes en la selección de artículos de prensa en línea: un estudio de Google Noticias mediante seguimiento ocular (eye-tracking). *Investigación Bibliotecológica, 28*, 63.

Rúas Araújo, J., Punín Larrea, M. I., Gómez Alvarado, H., Cuesta Morales, P., & Ratté, S. (2015). Neurociencias aplicadas al análisis de la percepción: Corazón y emoción ante el Himno de Ecuador. *Revista Latina de Comunicación Social, 70*, 401–422.

Shapiro, M. A. (2011). Think-Aloud and Thought-List Procedures in Investigating Mental Processes. In A. LANG, *Measuring Psychological Responses to Media* (1–14). New Jersey: Routledge.

Takahashi, M., Clippingdale, S., Naemura, M., & Shibata, M. (2015). Estimation of viewer´s ratings of TV programs based on behaviours in home environments. *Multimed Tools Applied, 74*, 8669–8684.

Treisman, A. (1988). Features and objects: The fourteenth bartlett memorial lecture. *The Quarterly Journal of Experimental Psychology, 40*(2), 201–237.

Von Helmholtz, H. (1925). *Treatise on physiological optics.* Rochester, NY: The Optical Society of America.

Wade, N. J., & Tatler, B. W. (2005). *The moving tablet of the eye: The origins of modern eye movement research.* Oxford: Oxford University Press.

Chapter 42
Impact and Positioning in Social Media of Events in Southern Ecuador

Eva Sánchez-Amboage, Verónica Mora-Jácome
and Estefanía Sánchez-Cevallos

Abstract In the southern region of Ecuador, the framework for developing event tourism is quite good. According to data from the Universidad Técnica Particular de Loja (Ecuador), in 2012 there were 123 national and international events attended by 26 thousand people. That generated an annual income of almost $3 million. The present research aims at identifying what is the positioning of the southern region of Ecuador—through the city Loja—in events organizations and their impact on social networks.

Keywords Tourism · Events · Conferences · Ecuador · Social networks

42.1 Theoretical Framework

According to the OMT (2002) and Del Valle (2008), Conferences and Event tourism is part of the so-called Business Tourism, which mobilises essentially opinion leaders in different branches of knowledge—science, technology, finances, and commerce. Business Tourism may be also called MICE Tourism (*Meeting, Incentives, Conventions and Exhibitions*), which includes conferences, seminars, workshops, symposiums, fairs and exhibitions, incentive travels and events. This new form of practising tourism is also based on traditional activities, additional leisure and recreation activities.

As López and Osácar (2008) note, the availability of large convention centres and the city's image are key factors for developing the Tourism of Congress and Conventions. The construction of a MICE destination's image takes time and

E. Sánchez-Amboage (✉) · V. Mora-Jácome · E. Sánchez-Cevallos
Technical Particular University of Loja (UTPL), Loja, Ecuador
e-mail: eva.amboaxe@gmail.com

V. Mora-Jácome
e-mail: vlmora@utpl.edu.ec

E. Sánchez-Cevallos
e-mail: resanchez@utpl.edu.ec

© Springer International Publishing Switzerland 2017
F.C. Freire et al. (eds.), *Media and Metamedia Management*,
Advances in Intelligent Systems and Computing 503,
DOI 10.1007/978-3-319-46068-0_42

requires knowledge of the meaning it has for tourists. The image of a destination is formed comparatively; when you talk about a city, leads you inevitably to think of similar destinations. Accordingly, it is important to pay attention to one of the more significant elements as regards promotion of Information and Communications Technologies (ICT), which have a direct impact on a destination's image: social media, also known as social networks.

Authors such as Álvarez et al. (2011); Heerschap et al. (2014); Theodosiou and Katsikea (2012); Hudson and Thal (2013) support that there is a clear transformation of sectors—including tourism—because of the online evolution and activity. Wichels (2014) explains that there is a change in the process by which the tourist makes an information query, receives influences, and makes the decision to book a destination, on the basis of recommendations and opinions of other travellers.

The city of Loja is one of the MICE destinations in Ecuador. While it is only starting to practice this king of tourism, it has several features that leave it in a good position. Loja is the cultural and musical capital of the southern region. It has various tourist attractions and it is considered one of the most productive counties of the country, due to its great wealth of natural, cultural, gastronomic, and commercial resources. In September 2004, Loja begins to get into the MICE Tourism and opens the UTPL—Universidad Técnica Particular de Loja—Convention Centre. From the time it was established to the end of 2009, the economy in Loja was clearly benefitted from this activity.

On the basis of the results obtained in the study "Analysis and value of MICE Tourism in Loja" (2014), 54 % of survey respondents would like to receive this kind of information via a website, 22 % via email, and 4 % via press releases. What is perhaps more significant is that 20 % of participants would like to be informed by social media. This lead us to analyse the promotion of the sector of Conferences and Convention sector in social networks, in order to understand its situation and determine, if necessary, any proposals for improvement.

42.2 Methodology

The present study is exploratory and is based on the documentary analysis undertaken to know the presence of Loja MICE Tourism in Social Media. Corona (2013) notes that the Internet as a subject of study is difficult to determine, not only for its ongoing changes, but also for its blurred boundaries and its influence in the social sphere. This situation poses a significant challenge when trying to define exactly what and how to investigate the Internet.

The assessment tool used in this study is Fanpage Karma http://www.fanpagekarma.com/, an online free service fro social media analysis and monitoring of social networks such as Facebook, Twitter, YouTube, Google+ and Instagram. It has been already used in other works on tourism, as that presented by Huertas et al. (2014). The analysis was carried out for three months—from August 1, 2015 to November 1, 2015, both for the study of the official page in Facebook and the Twitter account of the *Buró de Convenciones e Incentivos de Quito (BCIQ)*. It is the minimum period of time so that the results are representative (Internet República 2012). This paper makes an approach to state of Conferences and Event Tourism in Loja and checks its presence in social media. The aim is to draw conclusions that could lead to an increase in events promotion via online.

42.3 Results

In order to understand the activity of the Conferences and Event Tourism in social networks, it is sufficient to analyse the presence of the main events of the city in the most used in social networks.

Conferences of greater impact are divided into two different topics: technological conferences (II *Congreso Internacional de Innovación y Desarrollo*) and academic conferences, organized by the UTPL. In the former case, there is no representativeness of the International Conference in social networks. Secondly, the UTPL, as a university, is present in various social media, such as Facebook, Twitter, Google+, Instagram, Flickr and YouTube, and is applying sound online management. However, there has not a profile and website dedicated to the promotion of its events.

The lack of data forces us to analyse the case of Quito, through the study of Quito Convention Bureau. Convention Bureaux are non-profit organizations whose main objective is to position their cities, regions and countries in the market, as places for holding meetings. This paper tries to report on a new management model for tourism that can be developed to improve promotion and communication of MICE Tourism in Loja.

The *Buró de Convenciones e Incentivos de Quito* (BCIQ) promotes the capital of the Ecuador in the region and in the global markets as an ideal destination for conferences, fairs, incentive trips and national and international events. As a non-profit organization, the BCIQ brings together private companies related to MICE Tourism. Also, the entity has the support of official sector bodies, such as: the Ministry of Tourism, the Ministry of Foreign Affairs, the Municipality of Quito, and the Chamber of Tourism. The *Buró* is member of the International Congress and Convention Association (ICCA).

The BCIQ has a website (http://www.eventosquitoecuador.com/), which presents clear and updated information, and guides the visitor through all the events in Quito. Also, users can find information on accommodation, places to visit, restaurants, travel agencies, and mobile operators. It has thus become a site to promote MICE tourism throughout the city, which allows it to have presence in social media using an only brand: *Buró Convenciones de Quito*.

The Facebook page of BCIQ has more than 5653 likes. The results obtained through Fanpage Karma show that the average activity per week in Facebook is 2.4 dairy posts, so it exceeds recommendations proposed by Internet República (2012). Mondays are the most active days, but best results are obtained on Fridays between 20:00 and 22:00. When it comes to the content, Fanpage Karma notes that, in many cases, BCIQ uses external links. The programme recommends including more pictures in its posts in order to improve results. With regard to the topics, the Convention Bureaus reports—as it does in the website—on events in Quito and attractions and services. The programme detects that the Facebook page has a yield rate of 6 per cent, so its interactivity with the online community may be improved.

BCIQ's Twitter account has 1489 followers and 1887 published tweets. The average activity per day is 2.5 posts, which is a low figure when compared with the indications of Internet República (2012). When it comes to the content, it could be said that its behaviour is similar that in Facebook. It combines text and external links, and provides information on events and attractions of the city. Fanpage Karma estimated that there is a yield rate of 25 %, so it is advisable to take advantage offered by this social network.

42.4 Discussion and Conclusions

This exploratory study serves as a starting point for approaching the MICE Tourism in southern Ecuador as a whole and Loja in particular. While this city is only starting its activity in the field of MICE Tourism, it has several features that put them in a good position on the sector. The results of the investigation allow for the conclusion that the UTPL serves as an academic events creator in Loja. The city is backed up by all the tourist attractions in the southern region of Ecuador, which understands MICE Tourism as a way to foster economic and social development. The increasing interest of the government in developing projects of MICE Tourism is also seen as an opportunity.

The data obtained from the survey of the study "Analysis and value of MICE Tourism in Loja" (2014) show that there is a potential market for MICE Tourism, especially taking into account that 85 % of respondents attend one event per month, and 6 % twice a month. Also, it is detected that 20 % of respondents would prefer to receive information on events via social media.

The emergence of Convention Bureaux helped to improve the management model for MICE Tourism, whereby Loja could focus all their events. The main goal is to attain greater online representativeness, offering not only information on conferences and events, but also on the city itself. That could be of great help to consolidate the image of Loja as a MICE destination with an efficient market.

Acknowledgments The present work was developed with the support of researcher Valentín Alejandro Martínez Fernández and it was sponsored by the Prometeo Project of the Ministry of Higher Education, Science, Technology and Innovation of the Republic of Ecuador—PROMETEO—CEB-008-2015. It was also developed as part of the activities of the International Network of Communication Management—R2014/026 XESCOM, supported by the Ministry of Culture, Education, and University Planning of the Xunta de Galicia. This network is formed by iMarka Research Group from the University of A Coruña and the Group of Innovation and New Enterprise from the Universidad Técnica Particular de Loja, among others.

References

Álvarez, I., Benamou, J., Fernández, J., & Solé, C. (2011). *España conecta: cómo transforma internet la economía española*. Boston, MA: The Boston Consulting Group. Retrieved at October 23, 2015 from: http://www.espanaconecta.es/pdf/Spanish_Executive_Summary.pdf

Corona, J. (2013). *Etnografía de lo virtual: Experiencias y Aprendizaje de una Propuesta Metodológica para Investigar Internet*. Retrieved at October 29, 2015 from: http://www.razonypalabra.org.mx/N/N82/V82/36_Corona_V82.pdf

Del Valle, E. (2008): *El turismo de negocios y motivos profesionales. Marco de análisis y reflexion*. Retrieved at November 8, 2015 from: https://dialnet.unirioja.es/servlet/articulo?codigo=2719980

Heerschap, N., Ortega, S., Priem, A., & Offermans, M. (2014). Innovation of tourism statatistics through the use of new big data sources. *Statistics Netherlands*. Retrieved at October 28, 2015 from: http://tsf2014prague.cz/assets/downloads/Paper%201.2_Nicolaes%20Heerschap_NL.pdf

Hudson, S., & Thal, K. (2013). The impact of social media on the consumer decision process: Implications for tourism marketing. *Journal of Travel & Tourism Marketing, 30*(1–2), 156–160.

Huertas, A., Setó, D., & Miguez, M. (2014). Comunicación de Destinos Turísticos a ctravés de las Redes Sociales. *El Profesional de la Información, 24*(1), 15–21.

Internet República. (2012). *Estudio la Banca a Examen en las Redes Sociales*. Retrieved at November 8, 2015 from: http://www.slideshare.net/slideshow/embed_code/11256128

López, E., & Osácar, E. (2008). *Tourism Destination Placement: La imagen de los destinos turísticos a través de los largometrajes. El caso Barcelona: la web Barcelona de película*. Elche: CityMarketing.

Ruiz, A., & Astudillo, F. (2014). *Análisis y puesta en valor del Turismo de Congresos y Convenciones en la ciudad de Loja, año 2014*. Estudio de pregrado de la titulación de Administración de Empresas Turísticas y Hoteleras de la Universidad Técnica Particular de Loja, Ecuador.

Theodosiou, M., & Katsikea, E. (2012). Antecedents and performance of electronic business adoption in the hotel industry. *European Journal of Marketing, 46*(1/2), 258–283.

Wichels, S. (2014). Nuevos desafíos en Relaciones Públicas 2.0: La creciente influencia de las plataformas de online review en Turismo. *Revista Internacional de Relaciones Públicas, 4*(7), 197–216.

Chapter 43
Galician Spas in Facebook

**María-Magdalena Rodríguez-Fernández, Eva Sánchez-Amboage,
Clide Rodríguez-Vázquez and María-Dolores Mahauad-Burneo**

Abstract In Spain, Galicia is considered one of the best-rated regions in reference to Thermal Tourism. This is because of its medicinal and mineral waters and the quality of its thermal places. In this regard, the ICT have contributed positively to improve the online presence of spas. The research tries to bring together two elements: four Galician spas—one for each region—, and social media—Facebook —, in order to determine how those could improve their online visibility. *LikeAlyzer* is the online tool used to make the present analysis, since it allows assessing a series of interesting values. Also, it can analyse the effectiveness of websites, detect problems, and provide solutions for covering gaps.

Keywords Thermal tourism · Social media · Galicia · *Likealyzer*

43.1 Theoretical Framework

Information and Communications Technologies (ICT) have had a significant impact on the tourism industry, changing the way travellers access information, plan a trip, share experiences (Senecal and Nantel 2004; Buhalis and Law 2008; Xiang and Gretzel 2012), as well as compare offers and destinations (Avila and Barrado 2005; Prat Forga and Cànoves Valiente 2013).

M.-M. Rodríguez-Fernández (✉) · C. Rodríguez-Vázquez
University of A Coruña (UDC), A Coruña, Spain
e-mail: mmrodriguez@udc.es

C. Rodríguez-Vázquez
e-mail: crodriguezv@udc.es

E. Sánchez-Amboage · M.-D. Mahauad-Burneo
Technical Particular University of Loja (UTPL), Loja, Ecuador
e-mail: eva.amboaxe@gmail.com

M.-D. Mahauad-Burneo
e-mail: mdmahauad@utpl.edu.ec

© Springer International Publishing Switzerland 2017
F.C. Freire et al. (eds.), *Media and Metamedia Management*,
Advances in Intelligent Systems and Computing 503,
DOI 10.1007/978-3-319-46068-0_43

Within the ICT, social media—also known as social networks—have become key platforms for social change (Agarwal et al. 2011), and some of the most advanced channels for the development of tourist promotion. As Hudson and Thal (2013) note, social media have changed fundamentally the purchase decision process. At present, these networks are elements for evaluating products, taking decisions and providing recommendations to other users.

Among all social networks, Facebook is the most used worldwide (Teles 2014). This tool allows users to publish individual messages, share photo albums, follow their friends' experiences, and organize events, among others (Acar 2009). Llorens and Capdeferro (2011) note that the power of Facebook to share resources and to link online contents to profiles and pages, enable it to be a medium for complex and ongoing experiences of interaction, as well as to structure processes of collaborative learning.

Against this background, it was considered appropriate to analyse Galician spas on Facebook, due to the positioning of the Autonomous Region in Spa Tourism. Spa hotels are a kind of tourism capable of revitalising the local economy, since they are in the hinterlands, provide work for a large proportion of the population, and act as deseasonalised elements. In Galicia, spas are present in the whole region and represent an attraction for tourists and Galician people. Also, these places have modern equipment, qualified professionals and, especially, quality waters, features that give Galician spas an indisputable reputation in the market.

The research brings together two elements: four Galician spas—one for each region, and social media, particularly Facebook, in order to determine whether they optimise the use of this tool and if not, to make recommendations for improving their profitability.

Amongst all Galician spas, there are major differences as to how these companies make tourist promotions online. In this regard, it should be noted that the selection of spas was not accidental, but chosen by an online tool: *LikeAlyzer* (http://www.likealyzer.com/). This allowed identifying all the Galician spas with Facebook accounts, the ranking of spas selected by province, and the best positioned. These were the results: Hotel Balneario de Compostela, Oca Augas Santas Balneario and Golf Resort, Caldaria Balnearios y el Gran Hotel La Toja.

43.2 Methodology

The methodology is based on the observation of Facebook official accounts related to the four selected spas. It is used LikeAlyzer because it allows the measurement of interesting parameters for this study; the analysis of the efficiency of each fanpage; and the identification of problems and possible solutions. We have relied on the study of Huertas et al. (2014) with the aim of analysing information offered by Facebook. Their work has been of great importance to select the values and items: **Contents**—post frequency and format and type of information; **Interactivity**—user response and engagement; and **Visibility**—number of fans, likes, shares, favourites, timing, and post length. The study was carried out from 15 September to 6 November 2015, with three data collections—these two dates above-mentioned and the 6 October 2015.

43.3 Results

As regards **Content**, it was first analysed the post frequency (Table 43.1). All pages, with the exception of the Hotel Balneario de Compostela, publish every day, which promotes interactivity and dissemination of contents. On the other hand, the average daily number of post in the fanpage Oca Augas Santas Balneario and Golf Resort is higher than 2 daily posts, which is a decisive factor for maintaining and improving relationships with followers. However, the data obtained in September— 15 September and 6 October—showed that both the Oca Augas Santas Balneario and Golf Resort and the Gran Hotel La Toja did not publish more than once a day, which point out a decrease in interactivity.

Photography is the most used format in all cases. Besides, both the Hotel Balneario de Compostela and the Gran Hotel la Toja, only use photos when publishing, which reduces their visibility and creativity. Even though it is undeniable that contents in all analysed fanpages help to interact with users, there is a need of asking more questions to users. There are two main benefits of posing questions to fans: they provide greater visibility of the website and answers that allow companies to better understand and know customers.

As regards **Interactivity**, it should be noted that one of the measured items in all fanpages was the user response in relation to the sort of published post. The five most successful publications are photos and links.

Table 43.1 Daily post, post per type and rate

	Daily post[a]			
	Hotel Balneario de Compostela	Oca Augas Santas Balneario and Golf Resort	Caldaria Balnearios	Gran Hotel La Toja
	0.34	2.27	1.02	2.09
Post per type and rate[b]				
Photos (%)	100	91.7	87.5	100
Link (%)	0	4.2	12.5	0
Status	0	0	0	0
Video (%)	0	4.2	0	0

Source Prepared by the authors
[a]Average daily post at the time of study 06/11/2015; [b]Last month 06/10/2015–06/11/2015

Another analysed aspect that is closely linked to response is the engagement rate (see Tables 43.2 and 43.3), which measures users reactions, likes, comments, and shares. The method for measuring engagement consists of dividing the PTAT (People Talking About This) between likes. According to the *LikeAlyzer* analysis, it would be necessary to reach an engagement rate higher than 7 % if they are to be successful in Facebook. In this way, only Oca Augas Santas Balneario and Golf Resort—8.3 % and Caldaria Balnearios—almost 15 %—reached that figure. The case of the Gran Hotel de La Toja is special, since it does not comply with the items, it has more fans than Augas Santas and has more than two daily publications, but users engagement does not reach 6 %—5.5. Overall, it can be concluded that the increasing number of post is not directly related to the engagement.

Table 43.2 Number of fans, daily post and *engagement*

	Number of fans, daily post and *engagement*			
	Hotel Balneario de Compostela	Oca Augas Santas Balneario and Golf Resort	Caldaria Balnearios	Gran Hotel La Toja
Number of fans[a]	1.422	2.024	7.330	2.746
Daily post[a]	0.34	2.27	1.02	2.09
Engagement[b] (%)	0.98	8.3	14.71	5.5

Source Prepared by the authors
[a]At the time of study 06/10/2015; [b]Last month

Table 43.3 Number of fans, daily post, average likes, average comments and average shares

	Number of fans, daily post, average likes, average comments and average shares			
	Hotel Balneario de Compostela	Oca Augas Santas Balneario and Golf Resort	Caldaria Balnearios	Gran Hotel La Toja
Number of fans[a]	1.422	2.024	7.330	2.746
Daily post[a]	0.34	2.27	1.02	2.09
Likes[b] Comments Shares per post	11	21	42	28
Timing	Perfect coordination	Improvable	Perfect coordination	Bad
Post length[c]	From 100 to 500	Less than 100	From 100 to 500	From 100 to 500

Source Prepared by the authors
[a]At the time of study 06/10/2015; [b]Average *per post*; [c]Characters

The number of fans is an essential element for assessing **Visibility**, since it shows how important is the presence of the site and the whole tourist company.

The most noteworthy results show that the account Caldaria Hotel Balneario has the greatest number of fans—more than 7300 followers, 2000 more than in the first data collection, followed by the Gran Hotel of La Toja—more than 2700 fans. Other interesting item is the average of likes, comments and shares per post. It is observed that, in proportion, the Gran Hotel La Toja understand how to encourage its followers, but the Hotel Balneario Compostela should involve people with questions if it is to become more active.

The timing and the post length are also important when talking about **visibility**. Timing is the moment of the day when publications are more effective and encourage fans to participate. The tool identifies that, in relation to the timing, fanpages of Hotel Balneario Compostela and the Group Caldaria have a perfect coordination, since they publish many post between 12 and 15 h—GMT. Nevertheless, Oca Augas Santas Balneario and Golf Resort and the Gran Hotel La Toja move away from that time slot, the moment when there are more active users. Regarding the length of the post, it is preferable to write messages with less than 100 characters, since in such a way that generates more responses.

43.4 Discussion and Conclusions

Results show that, as regards content, it is appropriate to publish every day, since that encourages interactivity and ensures contents dissemination. In this regard, posts including photo, followed by links and videos, are the most popular formats amongst Galician spas.

When it comes to interactivity, publications including photos are getting more reactions between users, followed by links. It should be noted that there is no evidence of correlation between the number of posts and the engagement with the company, so it is advisable to increase interaction and to do live monitoring of publications. Otherwise, pages with a high number of daily posts could have a low level of engagement. It is important to be aware of the relevance of social networks for achieving interaction with clients and engagement. In this regard, social media create emotional ties: destinations inspire feelings, bring value, and create forum and contents.

Finally, it should be noted that the number of fans of the analysed pages show how important is their presence, but companies should not forget to improve participation and conversion ratios, and to make grow the unit value of each client over time, which would increase the whole value of the social network. Against this background, the best and worst times to publish on social networks are important factors in relationships between users and companies.

Acknowledgments The present work was sponsored by the Prometeo Project of the Ministry of Higher Education, Science, Technology and Innovation of the Republic of Ecuador— PROMETEO—CEB-008-2015. It was also developed as part of the activities of the International Network of Communication Management—R2014/026 XESCOM, supported by the Ministry of Culture, Education, and University Planning of the Xunta de Galicia. This network is formed by iMarka Research Group from the University of A Coruña and the Group of Innovation and New Enterprise from the Technical Particular University of Loja, among others.

References

Acar, A. (2009). Antecedents and consequences of online social networking behaviour: The case of *Facebook*. *Journal of Website Promotion, 3*(1), 62–83.

Agarwal, S., Mondal, A., & Nath, A. (2011). Social media. The new corporate playground. *International journal of research and reviews in computer science, 2*(3), 696–700.

Avila, R., & Barrado, D. (2005). Nuevas tendencias en el desarrollo de destinos turísticos: marcos conceptuales y operativos para su planificación y gestión. *Cuadernos de Turismo, 15*, 27–43.

Buhalis, D., & Law, R. (2008). Progress in information technology and tourism management: 20 years on and 10 years after the internet—the state of eTourism research. *Tourism Management, 29*, 607–623.

Hudson, S., & Thal, K. (2013). The impact of social media on the consumer decision process: Implications for tourism marketing. *Journal of Travel & Tourism Marketing, 30*(1–2), 156–160.

Huertas, A., Setó, D., & Miguez, M. (2014). Comunicación de Destinos Turísticos a través de las Redes Sociales. *El Profesional de la Información, 24*(1), 15–21.

Llorens, F., & Capdeferro, N. (2011). Posibilidades de la plataforma *Facebook* para el aprendizaje colaborativo en línea. *RUSC. Universities and Knowledge Society Journal, 8*(2), 31–45.

Prat Forga, J., & Cànoves Valiente, G. (2013). La participación en redes sociales y su incidencia sobre el comportamiento y satisfacción de los consumidores de turismo. Un estudio comparativo en diferentes recursos de turismo industrial en Cataluña. *Alsacia y Escocia. Investigaciones turísticas, 5*, 29–59.

Senecal, S., & Nantel, J. (2004). The influence of online product recommendations on consumers' online choices. *Journal of Retailing, 80*, 159–169.

Teles, V. (2014). *Comunidades de práticas nas redes sociais: atos de discurso em interação e estratégias discursivas principais*. Lisboa: Universidade Aberta. Retrieved at October 27, 2015 from https://repositorioaberto.uab.pt/bitstream/10400.2/

Xiang, Z., & Gretzel, U. (2012). Role of social media in online travel information search. *Tourism Management, 31*, 179–188.

Chapter 44
Gastronomy as a Part of the Ecuadorian Identity: Positioning on the Internet and Social Networks

María-Magdalena Rodríguez-Fernández,
Patricio-Mauricio Artieda-Ponce,
Patricia-Marisol Chango-Cañaveral
and Fabián-Mauricio Gaibor-Monar

Abstract Gastronomy is one of the distinguishing marks of greater strength to build the image of a destination, since it creates new motivations for travellers: culinary tourism. This is raising the interest of tourist managers for stressing the culinary singularities of regions, countries, and specific areas. In Ecuador, the Government has launched a specific action to give value to traditional and innovative dishes. It promotes gastronomy through competitions among local chefs and around four specific dishes: the *encebollado*; the *hornado*; the *colada morada* and the *fanesca*, since they are the most representative and consumed of Ecuador. This research aims to identify the positioning in social media of Ecuadorian gastronomy and the above-mentioned dishes. In such a way, it has been studied the positioning of these dishes in the main search engine—Google—and analysed the presence on Facebook and YouTube.

Keywords Gastronomy · Tourism · Internet · Social networks · Ecuador

M.-M. Rodríguez-Fernández (✉)
University of A Coruña (UDC), A Coruña, Spain
e-mail: magdalena.rodriguez@udc.es

P.-M. Artieda-Ponce · P.-M. Chango-Cañaveral · F.-M. Gaibor-Monar
Technical Particular University of Loja (UTPL), Loja, Ecuador
e-mail: mpartieda@utpl.edu.ec

P.-M. Chango-Cañaveral
e-mail: pmchango@utpl.edu.ec

F.-M. Gaibor-Monar
e-mail: fmgaibor@utpl.edu.ec

© Springer International Publishing Switzerland 2017   335
F.C. Freire et al. (eds.), *Media and Metamedia Management*,
Advances in Intelligent Systems and Computing 503,
DOI 10.1007/978-3-319-46068-0_44

44.1   Theoretical Framework

Culinary tourism is an important incentive to give value to destinations, since gastronomy is a key element for diversifying tourist supply and stimulating the economic development of a country.

When tourism began to grow, people did not care for local cultures when travelling. In the last half of the 20th century, there was a structural change in the way people appreciated food. Since the 80s, there started the travels with exclusively gastronomic purposes and the heritage status of local gastronomy (Schlüter and Thiel 2008).

In 2000, this type of tourism began to draw the Academia's attention because of the International Gastronomy Tourism Congress, "*Local Food and Tourism*", conducted by the World Tourism Organization in Larnaca (Chipre). From then on, there appeared scientific literature on gastronomy tourism (López Guzmán and Margarida Jesus 2011).

Gastronomy tourism is the "activity of tourists who are planning their trips for tasting the local cuisine and carrying out activities related to it" (Flavián Blanco and Fandos Herrera 2011). As (Hall and Sharples 2003) note, it is "an experiential journey to a gastronomic region with recreational purposes, and whose main reason includes a visit to primary and secondary food manufacturers, food festivals and fairs, farmer markets, shows and cooking demonstrations, organic food tastings and any activity related to the world of food".

There are many destinations with strategic positioning as regards culinary tourism. In Europe: Spain, France, Italy and Portugal; in Asia: Japan and India; in Latin America: Argentina, Mexico, Peru, Uruguay and Ecuador, the most recent case. As mentioned above, the Government has allocated resources to encourage Ecuadorian gastronomy—including four typical dishes: *encebollado*, *hornado*, *fanesca* and *colada morada*—turn it into a key element of the country.

Organizations and tourist destinations should consider the advantages of the Information and Communication Technologies (ICT). Consumers see ICT as key elements when planning trips, and comparing offers, products and destinations (Ávila Bercial and Barrado Timón 2005). The Internet and social networks are essential platforms for social change (Agarwal et al. 2011).

In October 1994, the Internet emerged as a new media for channelling advertising. Gomis (2000) stresses the importance of social networks as advertising channels and define them as "media that, in different formats, provide access to users segments that can be classified according to their interests, concerns, feelings, and ideologies". For Hudson and Thal (2013) the consumer's purchase decision process has radically changed.

The Internet and social networks have revolutionized the way to communicate, by allowing tourism organizations to interact anytime and directly with their target audiences at a relatively low price (Kaplan and Haenlein 2010). For this reason, these two elements have been identified as essential for future tourists when making decisions (Prat Forga and Cànoves Valiente 2013), and for destinations when promoting tourist products.

44.2   Methodology

The methodology is based on the use of Google Trends—to understand the evolution and future of web searches, YouTube, and Facebook, which is used to analyse the impact of the four dishes. Matias et al. (2009) note that *Google Trends* provides daily information on what the world is searching; has predictability; and shows the most popular terms in Google, given de possibility of filtering by many fields.

The first step in our research was the introduction of the four dishes—*encebollado, hornado, colada morada* y *fanesca*—on Google Trends, in order to analyse the their evolution and foreseeable future in the web and in YouTube. It was carried out on 30 October 2015. The second step, the study of the impact of dishes on Facebook, consisted of analysing if they had official fanpages, but they did not. We therefore studied the fanpage of the Ecuadorian Ministry of Tourism, considering that there could be found related publications. This information was collected on 28 and 29 October 2015.

44.3   Results

Firstly, it is presented the evolution of web searches from 2004 to 2015—in this last year, from January to October (Fig. 44.1).

The most searched dish is the *encebollado*, followed by the *fanesca* and *colada morada*—same level-m and the *hornado*, which has little interest. When analysing them separately, the *encebollado* awakens the most interest. The search trend does not follow a logic; anytime is great to gather information. But summer is the time when more surveys are received, especially during August 2015. The *fanesca* generates stronger tracking in Easter, since it is a typical dish. That was on April, with the exception of the years 2005, 2008 and 2013, where Easter fell on March.

The *colada morada* is a typical beverage of the Day of the Dead. In the analysed years, there is stronger interest in October, while in November 2006, 2007, 2008, 2011, 2012 and 2013 it also outnumbered the rest of dishes. The *hornado* is less

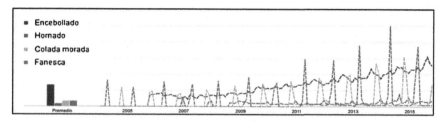

Fig. 44.1 Searches in Google Trends on the online interest in the four dishes. *Source* Prepared by the authors

searched; the only time it reached some visibility was December 2006. From then on, it was that month when it receives more attention, with the exception of 2014, when August is the most active.

Google Trends also provides information on the location of searches. Data obtained show that the vast majority are from Ecuador. As regards the *encebollado*, Ecuadorian people, followed by these countries, make 100 % of surveys: Puerto Rico (47 %), Guatemala (27 %), Spain (15 %), Venezuela (7 %), Peru (5 %) and Mexico (3 %). 100 % of searches on the *hornado* are from Ecuador, as on the *colada morada* and the *fanesca*, even though in this last dish 1 % is from Spain (Fig. 44.2).

As regards future trends, the *encebollado* is expected to be the most searched dish, followed at some distance by the *colada morada*. When analysing data individually, forecasting online surveys in 2015 show that the *encebollado* will continue at the top, and the *colada morada* a will be the second, even though over the first months will not awaken great interest.

When looking at searches on YouTube—from 2008 to 2015- the *encebollado* is the focus of attention, followed at some distance by the *fanesca* and the *colada morada*. The *hornado* arouses less interest. In a detailed analysis, the trend shows that the most searched dish until 2012 is the *encebollado*, while in April of that year, as well as March 2013, 2014 and 2015 the *fanesca* was ranked first. The *colada morada* is barely shown in Fig. 44.3; video searches only increased in December 2013 and August 2014, but it did not exceed the *encebollado*. Ecuador is

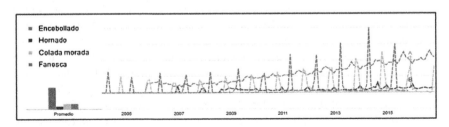

Fig. 44.2 Searches in Google Trends on the online future interest in the dishes. *Source* Prepared by the authors

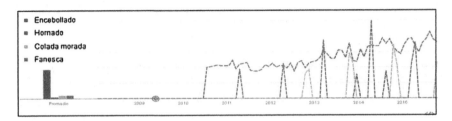

Fig. 44.3 Searches in Google Trends on the interest in the dishes on YouTube. *Source* Prepared by the authors

Table 44.1 Impact on *Facebook* of posts related to the analysed dishes

Dish	2012	2013	2014	2015
	Colada Morada	*Fanesca*	*Hornado*	*Encebollado*
Type of publication	1 photo	1 Link	10 photos 1 Video 1 Link	2 photos 1 Video
Likes	39	31	Photos: 936 Videos: 77 Link: 45	Photos: 64 Videos: 17
Comments	3	2	Photos: 32 Videos: 5 Link: 1	Photos: 1 Videos: 2
Shares	34	12	Photos: 215 Videos: 70 Link: 13	Photos: 15 Videos: 2

Source Prepared by the authors

the main origin of searches; in the case of the *encebollado*, 100 % of searches come from Ecuador, but no data is available on the rest of the dishes.

As regards the impact on Facebook, the fanpage of the Ecuadorian Ministry of Tourism publish many posts related to the Ecuadorian gastronomy in general. Thus, in order to not deviate so much from the subject of study, reference is made only to our four dishes. The analysis was carried out from 2012 to 2015 (Table 44.1).

Results respond to three items: **Contents**—Type of publication (photos, videos and links); **Interactivity**—likes, comments, and shares; and **Visibility**—likes and shares.

It is striking that the *colada morada* is only mentioned in 2012, the *fanesca* in 2013, the *hornado* in 2014 and the *encebollado* in 2015. The reason was the celebration in these two last years of the World Championship of the *Hornado* and the *Encebollado*, respectively.

The fanpage has few posts (17) and the most used types are photos (13), followed by videos (2) and links (2). Photos cause more reactions in terms of interactivity and visibility, although they are not in line with the number of posts, which are used for reporting rather than interacting.

44.4 Discussion and Conclusions

Culinary tourism is an important driver of economic growth, since it generates employment and has a cross-multiplier effect on other sectors. Tourist managers and promoters should benefit from the advantages offered by the ICT. The Internet and social networks as communication channels are essential for increasing visibility and interactivity.

Ecuador has culinary resources to encourage this type of tourism. There are many typical dishes, nationally and internationally recognized. These include the *encebollado*, *hornado*, *colada morada* and the *fanesca*.

According to Google Trends, web and YouTube searches show that the *encebollado* is the most recognized dish. As regards the *colada morada* and the *fanesca*, they are season products, so their interest is limited to specific dates: the Day of the Dead and Easter. The *hornado* is of little interest. In YouTube, videos related to the *encebollado* are the most popular, in contrast to the *hornado*, which is far less searched than the other dishes.

Forecast on the future web searches, the *encebollado* will occupy the first position in 2016, while the *colada morada* will also attract interest. As regards Facebook, results show that the analysed fanpage could make better use of the social network. There is a need for more attractive and continuous posts if it is to be more visible, interactive and known, especially in relation to the *fanesca* and the *colada morada*.

In order to improve the positioning of the Ecuadorian gastronomy—particularly the four analysed dishes, it is recommended to create specific websites for each one, containing links to social networks—Facebook, YouTube and Twitter, and providing information on events, products, chefs, and products.

Acknowledgments The present work was sponsored by the Prometeo Project of the Ministry of Higher Education, Science, Technology and Innovation of the Republic of Ecuador—PROMETEO-CEB-008-2015. It was also developed as part of the activities of the International Network of Communication Management—R2014/026 XESCOM, supported by the Ministry of Culture, Education, and University Planning of the Xunta de Galicia. This network is formed by iMarka Research Group from the University of A Coruña and the Group of Innovation and New Enterprise from the Technical Particular University of Loja, among others.

References

Agarwal, S., Mondal, A., & Nath, A. (2011). Social media—The new corporate playground. *International Journal of Research and Reviews in Computer Science, 2*(3), 696–700.

Ávila Bercial, R., & Barrado Timón, D. (2005). Nuevas tendencias en el desarrollo de destinos turísticos: marcos conceptuales y operativos para su planificación y gestión. *Cuadernos de Turismo, 15,* 27–43.

Flavián Blanco, C., & Fandos Herrera, C. (2011). *Turismo gastronómico.* Estrategias de marketing y experiencias de éxito. Zaragoza: Prensas Universitarias.

Gomis, J. M. (2000). *La información turística: del papel a la Red en Manual de Comunicación Turística. De la información a la persuasión, de la promoción a la emoción.* Girona: Editorial Documenta Universitaria.

Hall, C. M., & Sharples, L. (2003). The consumption of experiences or the experience of consumption? An introduction to the tourism of taste. In C. M. Hall et al. (Eds.), *Food tourism around the world* (pp. 1–24). Elsevier, Oxford.

Hudson, S., & Thal, K. (2013). The impact of social media on the consumer decision process: Implications for tourism marketing. *Journal of Travel & Tourism Marketing, 30*(1–2), 156–160.

Kaplan, A. M., & Haenlein, M. (2010). Users of the world, unite! The challenges and opportunities of social media. *Business Horizons, 53*, 59–68.

López Guzmán, T., & Margarida Jesus, M. (2011). Turismo, cultura y gastronomía. Una aproximación a las rutas culinarias. *Tourism & Management Studies, 1*, 915–922.

Matias, Y., Niv, E., & Shimshoni, Y. (2009). *On the predictability of search trends*. Retrieved from http://googleresearch.blogspot.com.es/2009/08/on-predictability-of-search-trends.html. Accessed at September 30, 2015.

Prat Forga, J., & Cànoves Valiente, G. (2013). La participación en redes sociales y su incidencia sobre el comportamiento y satisfacción de los consumidores de turismo. Un estudio comparativo en diferentes recursos de turismo industrial en Cataluña. *Alsacia y Escocia. Investigaciones turísticas, 5*, 29–59.

Schlüter, R., & Thiel, D. (2008). Gastronomía y turismo en Argentina polo gastronómico Tomas Jofré. *Pasos Revista de Turismo y Patrimonio Cultural, 6*(2), 249–268.

Chapter 45
New Digital Metrics in Marketing: A Comparative Study on Social Media Use

Joan Francesc Fondevila Gascón, Pedro Mir Bernal,
Eva Santana López and Josep Rom Rodríguez

Abstract Analysis metrics of social media are multiplying in order to reflect the performance that companies can get from the Internet activity. It is an essential formula for the information professional as embodying the efforts made in communication. After analyzing different metrics to be used depending on the type of target to be attained with them, the article deals with the analysis of feeling. From the classification of Lovett's performance, we compared VuelingPeople! and EasyJet. We observe alternate domain in the various parameters.

Keywords Marketing · Social media · Metrics · Reputation · Sentiment analysis

45.1 Theoretical Framework

The relevance that social media has acquired over the past years has made participating in them an imperative for each brand. Social media offers both opportunities but also risks. Thus, it is an imperative to design a clear and well-defined strategy. This strategy must have objectives along with metrics in order to

J.F. Fondevila Gascón (✉) · P. Mir Bernal
Pompeu Fabra University (UPF), Barcelona, Spain
e-mail: joanfrancescfg@blanquerna.url.edu

P. Mir Bernal
e-mail: pmir@unav.edu

J.F. Fondevila Gascón ·
E. Santana López · J. Rom Rodríguez
Ramon Llull Blanquerna-University (URL), Barcelona, Spain
e-mail: Evasll@blanquerna.url.edu

J. Rom Rodríguez
e-mail: joseprr@blanquerna.url.edu

J.F. Fondevila Gascón
Mediterrani University School of the University of Girona (UdG),
Barcelona, Spain

© Springer International Publishing Switzerland 2017
F.C. Freire et al. (eds.), *Media and Metamedia Management*,
Advances in Intelligent Systems and Computing 503,
DOI 10.1007/978-3-319-46068-0_45

determine if the objectives have been accomplished. However, sentiment analysis and ROI (Return on Investment) metrics have to be taken into account when performing this evaluation.

Reichheld (2003) found a direct correlation among the percentage of clients that would recommend a brand and the growth indexes among competitors for most of the industries about which he did research. Consumer loyalty implies certain commitment since, when recommending, they risk their own reputation. Firms must focus on increasing the number of advocates or promoters of their brand in order to turn their loyal clients, figuratively, into their marketing department.

East et al. (2008) claim that Reichheld's proposition does not work when predicting the behavior or yield of a brand. They claim that his proposition uses self-prediction, since the surveyed population cannot anticipate easily the circumstances in which they would make a recommendation. They also claim that the word-of-mouth phenomenon affects the outcomes when choosing a brand, but does not have an effect when making recommendations. They propose a combination of experiments based on role-play games and surveys, but they also recognize the limitations of both methods.

Social media has turned consumers into active participants on the creation, spread and content search (Hunter and Soberman 2010; Bonsón and Ratkai 2013). Nowadays, brands have to give control to the consumers and start a dialogue with them (Bonsón and Ratkai 2013). Information experts and executives need accurate metrics for each objective, since the information provided by each one will vary significantly and will affect the way decisions are made. Metrics are proposed for journalism (Fondevila Gascón 2012, 2014).

Information and public relations professionals need to measure the impact of social media beyond simply counting the number of fans and followers. They are pondering the way to conceptualize and measure user experience towards the brand content on social media and the way to measure their interactions (Smith 2013).

One of the challenges (Table 45.1) is to choose of how many social media networks a firm should be a part. Opportunity cost exists and the fact of

J.F. Fondevila Gascón
Open University of Catalonia (UOC), Barcelona, Spain

J.F. Fondevila Gascón
University of Barcelona (UB), Barcelona, Spain

J.F. Fondevila Gascón
Autonomous University of Barcelona (UAB), Barcelona, Spain

J.F. Fondevila Gascón · P. Mir Bernal
CECABLE (Cable Studies Center), Barcelona, Spain

P. Mir Bernal
University of Navarra (UNAV), Pamplona, Spain

P. Mir Bernal
AQUAMMM (Asymetric Qualitative Understanding and Modelling
for Marketing and Management), Boston, USA

Table 45.1 Challenges of social media metrics

Author	Challenge
Smith	Measure the social media user' experience related to the brand content
Smith	Measure the value of social media user's interactions.
Barger and Labrecque	Number of social media presences of a firm
Barger and Labrecque	Active role of social media presence
Barger and Labrecque	Synergies among different social media channels

Source The authors

participating could not only result in a loss of clients but also could expose the brand to the danger of an identity theft by a third party. Another aspect to consider is how active this presence in social media should be. There are both inactive and false followers (you can buy fake Twitter followers for less than a penny per follower) (Barger and Labrecque 2013).

Barger and Labrecque (2013) propose metrics based on volume (number of mentions received by a brand on social media during a specified period of time), share of voice (number of mentions received by a brand as a percentage of the total mentions of the brands that belong to a specified product category), engagement (interaction: consumers who carry out an some kind of action beyond seeing or reading; for instance, give a "Like", commenting or replying to the content), advocates (supporters with active participation by giving opinion and sharing), return on investment or ROI (revenue generated by a marketing campaign carried out on social media minus the cost of the campaign, all divided by the cost of the campaign), leads generated (through social media), and response time (32 % of consumers who contact this way expect a less-than-thirty minutes response).

Metrics proposed by Bonsón and Ratkai (2013) aim for measuring the interaction of all stakeholders and the brand in terms of popularity, commitment and "virality". They also take into account and try to measure the mood of those same stakeholders (positive feedback, negative or neutral). The validity of the "Like" and followers used as metrics is a controversial topic in the academic literature. A "Like" is always more than just a number (Gerlitz and Helmond 2013), because its value lies in the potential "Like", comments or other additional responses it can generate.

45.2 Methodology

The used methodology is quantitative; data were collected and analyzed from the Internet websites VuelingPeople! and EasyJet in July 2015 with the aim of analyzing the creation of promoters (Reichheld 2003) using the sentiment analysis.

This study focused on the social network Facebook and the metrics provided by SimilarWeb.com addressed VuelingPeople! and EasyJet websites. It can be considered that both companies are competing and conducting the same kind of service, so the analysis will show the reader two different strategies.

To perform this analysis, the interactions of these two companies were reviewed and analyzed during the month of July 2015. Lovett (2012) criteria were used as a basis to discuss some important aspects of the study.

The sentiment analysis performed considered positive feedback to every comment that implied good reviews about the services provided by the brand or the neutral participation in the content they shared on Facebook. That is, if VuelingPeople! raised a question in order to generate engagement, active participation of users was considered positive.

45.3   Results

Lovett (2012) introduces a series of social media metrics for the analysis that consists of the following: be interaction (understanding and interaction only activity that ends in conversion), commitment (which does include visits, comments, shared content and time devoted to these activities), influence, and impact defenders (identified with the ROI). Progress metrics and ROI metrics are differentiated (Table 45.2).

Opinions play an essential role when determining the success or failure of brands and products because people trust them (Wright 2009; Kennedy 2012). Therefore,

Table 45.2 Social media results and associated metrics results

Results	Metrics in progress	ROI metrics
Exposure	Scope, media mentions, spread and posting activity	Cost per campaign
Dialogue	New visitors, referral sources, relevant links, active users, user growth rate, brand trending topics and brand keywords	Cost per new visitor, cost per qualified prospective client
Interaction	Time spent, visited pages, games played, contests presented, downloaded apps, sent messages and posted comments	Cost per interaction, cost per engaged visitor
Support	Returning visitors, recent visits, frequency of visit, satisfaction rating and positive reviews percentage	Cost per satisfied customer
Defense	Content variety, likes, shared content, bookmarks, community status, influence, weight and important commentators	Defense campaign cost
Revenues	Total customers, average order value, average orders per customer and customer lifetime value	Cost per campaign revenue per total clients revenue

Source Lovett (2012)

various software applications have been developed in order to analyze the sentiment (Wright 2009).

The information through surveys and focus groups is now immediately available and free of charge on the Internet (Mostafa 2013). Moreover, sentiment analysis tools can identify the subject under discussion and even classify the opinions according to different points of view (Jacobson 2009).

Determining the ruling sentiment in a text consists of dividing it and categorizing it into three possible categories: positive, negative and neutral. There are three possible approaches to perform sentiment analysis (Haddi et al. 2013): based on machine learning methods; relying on the use of lexical methods; and linguistic analysis. To obtain the opinions, sentiment analysis involves two consecutive tasks: identify which parts of the text contain some sort of opinion or feelings; determine the polarity and the intensity level of this feeling (Yu et al. 2013). The categorization of words is a key step in applying sentiment analysis technique (Mostafa 2013), which has evolved from a narrow approach to the analysis of the words individually to a consideration that also takes into account the context and the way the words are used (Rao et al. 2014).

Most of the accuracy problems vitiating the analysis of sentiment arise because of the peculiarities and complexities of language. Direct opinion expressions such as "I hate this movie," are easy to spot, but there are many ways to state an opinion. In addition, the use of irony, sarcasm, humor and, the use of abbreviations (specifically in the social media) make this task more complicated.

The amount of feeling expressed in social media is relatively small (Kennedy 2012). For example, in the case of Twitter, approximately 19 % of the tweets sent refer to a brand, but of that 19 %, only 20 % contain some manifestation of feeling or opinion (Bae and Lee 2012). Finally, about data "mining", Kennedy (2012) refers to the fact that most of the opinions expressed via Internet are favorable to brands (about 65 % would be positive, while only 8 % would be negative), which raises suspicion that brands remove negative comments about them on the Internet and that they write many positive reviews. Moreover, approximately 10 % of the negative reviews are false (Kennedy 2012; Mostafa 2013). For all this, Kennedy (2012) wonders if a real sentiment analysis exists.

The sentiment analysis has also received some criticism of moral order by the academy, as some consider it as a monetization of privacy (Kennedy 2012). In spite of this, it also allows the brands to take into account the preferences and opinions of the users, allowing them to develop more personalized and relevant services (Rao et al. 2014).

In the comparative study (Fig. 45.1), the first point to highlight is the difference between the number of users who regularly visit the site (scope), the time spent on it, the content they read and the bounce rate. In these four aspects, EasyJet is the winner, since it beats VuelingPeople! in all four categories, especially in the scope.

As for Facebook content, the brand Vueling People! shows a greater activity and also generates content that is most interesting to their target audience. Although the number of new visitors per day is higher in EasyJet (consider that EasyJet has a more international scope), VuelingPeople! shows a better management of its

Exposición		easyJet		vueling	
	Alcance	13M	✓	5.1M	✗
	Menciones en google (buscador)	14.500.000	✓	7.250.000	✗
	Difusión y actividad de publicación (frecuencia posts por día)	0,58	✗	1,10	✓
Diálogo					
	Nuevos visitantes (nuevos likes al día)	2611	✓	1687,25	✗
	Enlaces relevantes	Página propia	✓	Página propia	✓
	Usuarios activos (Promedio likes)	155,72	✗	1982,47	✓
	Tasa de crecimiento de usuario	43%	✗	555%	✓
	Temas de tendencia de la marca	Pregunta, Oferta, Disculpa, Información, App.	✓	Promoción, Información, RSC, Oferta, Concurso.	✓
	Palabras clave de la marca	Cheap, travel, flight, airline tickets, search travel ideas	✓	españa, cost, vacaciones, buscadores	✓
	Renovación de contenidos (en horas)	41,94	✗	21,33	✓
Interacción					
	Tasa de rebote (bounce rate)	21,79%	✓	21,16%	✗
	Tiempo pasado	0:10:22	✓	0:06:58	✗
	Páginas visualizadas	6,34	✓	5,61	✗
	Concursos presentados (en un mes)	0,00	✗	3	✓
	¿Tiene APP?	APP	✓	APP	✓
	Mensajes enviados (Tasa respuesta)	37,4%	✓	12,0%	✗
	Comentarios emitidos (promedio)	51,88	✗	152,44	✓
Soporte					
	Puntuación de satisfacción	NO DISPONIBLE		NO DISPONIBLE	
	Porcentaje de revisiones positivas	22,4%	✗	51%	✓
Defensa					
	Distribución de contenidos	Pregunta, Oferta, Disculpa, Información, App.	✓	Promoción, Información, RSC, Oferta, Concurso.	✓
	Gustos (dato obtenido 02/08/15 13:22)	358648	✗	622013	✓
	Comparticiones	8,83	✗	135,18	✓
	Estado de la comunidad	ACTIVO	✓	ACTIVO	✓
	Comentaristas importantes				
Ingresos					
	Clientes totales	No vende por Facebook			
	Valor medio de un pedido				
	Media de pedidos por cliente				
	Valor de vida del cliente				

Fig. 45.1 Comparative study between VuelingPeople! and EasyJet. *Source* The authors

content, since content renovation rate is higher (1.10 posts per day against 0.58 posts per day). This increases the growth rate of users (in terms of participation, since it has been taken into account the increase of users that give "Like" to each new publication). Moreover, the average number of comments is also greater in VuelingPeople! and the number of comments received are growing at a much faster pace than that of EasyJet.

Another property that should be taken into account is the percentage of reviews obtained, which is prepared by analyzing the comments received. At this point VuelingPeople! scored higher than EasyJet. However, it is necessary to mention that the response rate is significantly lower. As demonstrated throughout the study, good communication between the customer and the brand is a key factor when it comes to consolidating a good reputation. As stated Barger and Labrecque (2013), the short-term objectives can be used to obtain consideration and encourage first and subsequent purchases. This is reflected in the type of content that each of the brands post. EasyJet focuses primarily on promoting its App, reporting flight incidents and information and offering promotions, as well as releasing and making open questions about their services. VuelingPeople!, however, tends to release more 'neutral' content less subject to complaints, such as broad questions about travel or consumer interests. Although both brands are investing resources to improve customer communication in order to strengthen the brand image and reach new markets, they should keep focusing its efforts on complaints management and the development of quality content, as it is the form to consolidate its position.

"*Marketing mix modelling*" technique can be used here in order to determine the effects of each of the marketing actions and the external circumstances happening that can affect the strategy at the time of generating the revenues (Powell et al. 2011), leading to different allocation models.

45.4 Discussion and Conclusions

In order to determine the impact on the final user, social media plays a crucial role that is of high interest for the professionals in this area. The diagnosed challenges consist of measuring the user experience when using the brand content on the social media as well as determining the number of social media the brand should use. They also want to measure if there is an active role within the social media presence and propose possible synergies between different social media channels.

In the studied case, Easyjet has better exposure and interaction while VuelingPeople! outscores Easyjet on dialogue, support and advocacy. Comments help to design solutions that are more accurate and improve the consumer experience. Increasing the sample and analyzing companies from other sectors are future research projects to address.

References

Bae, Y., & Lee, H. (2012). Sentiment analysis of Twitter audiences: Measuring the positive or negative influence of popular Twitterers. *Journal of the American Society for Information and Technology, 63*(12), 2521–2535.

Barger, V. A., & Labrecque, L. (2013). An integrated marketing communications perspective on social media metrics. *International Journal of Integrated Marketing Communications, Spring 2013*.

Bonsón, E., & Ratkai, M. (2013). A set of metrics to assess stakeholder engagement and social legitimacy on a corporate Facebook page. *Online Information Review, 37*(5), 787–803.

East, R., Hammond, K., & Lomax, W. (2008). Measuring the impact of positive and negative word of mouth on brand purchase probability. *International Journal of Research in Marketing, 25*, 215–224.

Fondevila Gascón, J. F. (2012). El uso de recursos del periodismo digital en la prensa del Reino Unido, Francia, Estados Unidos y España. *Estudios sobre el Mensaje Periodístico (EMP), 18* (1), 73–87.

Fondevila Gascón, J. F. (2014). El uso de hipertexto, multimedia e interactividad en periodismo digital: propuesta metodológica de ranking de calidad. *ZER, Revista de Estudios de Comunicación (Journal of Communication Studies), 19*(36), 55–76.

Gerlitz, C., & Helmond, A. (2013). The like economy: Social buttons and the data-intensive web. *New Media & Society, 15*(8), 1348–1365.

Haddi, E., Liu, X., & Shi, Y. (2013). The role of text pre-processing in sentiment analysis. *Procedia Computer Science, 17*, 26–32.

Hunter, M. L., & Soberman, D. A. (2010). The equalizer': Measuring and explaining the impact of online communities on consumer markets. *Corporate Reputation View, 13*(4), 225–247.

Jacobson, L. (2009). Take a sentimental journey: What sentiment analysis means for PR professionals. *Public Relations Tactics, 18*.

Kennedy, H. (2012). Perspectives on sentiment analysis. *Journal of Broadcasting & Electronic Media, 56*(4), 435–450.

Lovett, J. (2012). *Social media. Métricas y análisis.* Madrid: Ediciones Anaya Multimedia.

Mostafa, M. M. (2013). More than words: Social networks' text mining for consumer brand sentiments. *Expert Systems with Applications, 40*, 4241–4251.

Powell, G., Groves, S., & Dimos, J. (2011). *ROI of Social Media: How to improve the return on your socialmarketing investment.* Singapore: Wiley

Rao, Y., Li, Q., Mao, X., & Wenyin, L. (2014). Sentiment topic models for social emotion mining. *Information Sciences, 266*, 90–100.

Reichheld, F. F. (2003). The one number you need to grow. *Harvard Business Review, 81*(12), 46–55.

Smith, S. (2013). Conceptualising an evaluating experiences with brands on Facebook. *International Journal of Market Research, 55*(3), 357–374.

Wright, A. (2009). Our sentiments. Exactly. *Communications of the ACM (Association for Computing Machinery), 52*(4), 14–15.

Yu, Y., Duan, W., & Cao, Q. (2013). The impact of social and conventional media on firm equity value: A sentiment analysis approach. *Decision Support Systems, 55*, 919–926.

Chapter 46
The Sport as an Element of Appreciation for the Cities. The Case of Pontevedra and Its Treatment in the Digital Media

Montse Vázquez-Gestal and Ana Belén Fernández-Souto

Abstract Pontevedra is a capital of a province in Spain with more than 80,000 habitants that is looking for a different position in the Galician ambience for tourism. In this city, sport have always been important, like soccer (the team of the city has climbed this year to 2ªB), and other sports that obtain less visibility in media, for example handball, basketball, indoor soccer, judo. In this way, the city has thought about how to be reinvented across town-planning, cultural and social actions, and even with sports, what it has been allowed to place the city itself in a place privileged in activities of this type, celebrated throughout the whole year in very different disciplines but that leads to the city receiving national and international participants in events of a very big importance, like tests qualifying rounds for world and European in triatlon, championship of the world of duatlon, concentrations of judo, championship of Spain of rhythmic gym. It has meant a big change and an important implication of the city, not only of its institutions but also of its inhabitants, who accompany all these activities on its presence and support.

Keywords Sport · City · Tourism · Participation

46.1 Prelude

There are a lot of cities that they bet for realizing campaigns for tourist promotion based on its diverse local references. Most of them emphasize its sceneries (well be of interior or of sea), in its nature (rivers, seas, mountains, lakes…), or in a tourism already settled down (tourism of the sun and beach, rural tourism, green tourism, gastronomic tourism…)… but there are less cities that bet for attracting specific

M. Vázquez-Gestal (✉) · A.B. Fernández-Souto
University of Vigo (UVigo), Vigo, Spain
e-mail: mvgestal@uvigo.es

A.B. Fernández-Souto
e-mail: abfsouto@uvigo.es

© Springer International Publishing Switzerland 2017 351
F.C. Freire et al. (eds.), *Media and Metamedia Management*,
Advances in Intelligent Systems and Computing 503,
DOI 10.1007/978-3-319-46068-0_46

publics with sport actions, as there can be the congresses, the professional fairs or the sports meetings.

There are many difficulties to find a good position or an element of really interesting differentiation between the territories, so we believe in the celebration of sport activities and the development of all this type of actions destined to improve the promotion of the cities, in our case, the small capital of Pontevedra, in the Galicia region. This is a city that has elected the sport as a revitalizing element and, of course, like an economic revulsive for the local activity. In the year 2012 more than 120 sports events were organized. We emphasize this year, because of the large number of sport activities celebrated there and because in all of them the city was involved. Since then the number of sports events has not stopped growing.

46.2 Theoretical Context

46.2.1 Historical and Sociocultural Context

The city of Pontevedra is the capital of the Rías Baixas in the Galicia region. It is a city between the sea and the river, and in whose waters and streets have developed centuries of sports activities of very diverse nature. The legend says that the city was founded by Teucro, son of Telemón and Hersione, who after the end of the war of Troy, went to Galicia and baptized the city where he stayed with the name of Helenes.

On one stone of the building of the town hall exists an inscription recorded of unknown author which alludes to the origin of the city: "Fúndote Teucro valiente/de aqueste rio en la orilla/para que de España fueres/de villas, la maravilla". Leaving its legend origin, traditionally, the historiography has affirmed that there exist studies that relate the foundation of Pontevedra to the establishment of Turoqua, mansion of the roman route XIX founded on the south shore of the river Lérez, after the Gallaecia integration in the Roman Empire. The name of the city derives from the latin and its meaning is an old bridge. In 1833 it turns in the capital of the province of the same name. It transforms the city in an administrative metropolis that attracts bureaucrats, bourgeoises, professionals and all kinds craftsmen. The social and cultural splendor made possible the town planning, commercial and industrial development of Pontevedra.

46.3 Methodology

The sport is one of the elements about the government of the city which has thought about how to turn into revitalizing, not only to announce the city, but also like a form of civil implication in the urban development. To get it, we analyze de aftereffect that the different sports event meant for the city, basing our study in the local and autonomic media and their social media. We select the three most important newspapers of the city, Diario de Pontevedra, La Voz de Galicia (Pontevedra edition) and Faro de Vigo (also Pontevedra edition). We have prepared the study in its digital formats, after the access of the contents turned out to be simpler, from the year 2007 up to the actuality. We complete the information with the web of the Instituto Municipal de Deportes (IMD), entity dependent on the town hall and that organize the sport activity of the city. We also analyzed their social media: Twitter and Facebook.

46.4   Pontevedra and the Triatlon

The bet of the city of Pontevedra for the triatlon was in 2008, like one of the test that were taking part of the European Cup (Diario de Pontevedra 2011).

Its aftereffect went so far as to compare Pontevedra with the city of Hamburg, as for services, resources, ambience and popular support. It was the first Galician city that was lodging a test of the continental tournament, entering the European circuit: twelve test in ten countries. It was celebrated on April 19, with more than 150 participants, a champion from Galicia—Gómez Noya—, and approximately 20,000 followers in the streets of the city. In January, 2009 it is confirmed that Pontevedra will receive a test of the European Cup in March: "we try to show that Pontevedra has capacity for the organization of the European Cup of Triatlon, and we will keep on asking for it with the support of the Headquarters of Sports of the Galician autonomous government before the pertinent instances" (La Voz de Galicia 2009). More than 2000 people in different categories and more than 40,000 visitors for these dates, they force three administrations "Galician autonomous government, Delegation of Pontevedra and Concello de Pontevedra" to invest to obtain high comeback profitability, especially for hotel trade, restoration or tourism.

In July, 2009, the Executive board of the European Union of Triatlon (ETU) designates Pontevedra head office of the Championship of Europe of 2011, as the second Spanish city doing this important sports competition, after Valencia. In the year 2010 on of the test is organizated again for the Championship of the World, and the champion Gómez Noya talks about the possibility of the Championship of Europe of the specialty in 2011, which predicted and investment of close to a million euros (Diario de Pontevedra 2010).

Pontevedra received the test basing on this geographical place, next to the airports of Vigo, Santiago and Oporto, its hotel disposition (city and surroundings), its experience in the organization of sports events and the support of the champion Gómez Noya: "… for me is a pleasure to be present in this act, and it is a pride to compete in Pontevedra, the city where I live and it he one that I train every day, and with a test of this level. To compete at home motivates me specially and I would like a repeating victory in Pontevedra" (Diario de Pontevedra 2009).

The appointment of 2011 attracted more than 40,000 visitors and 3000 sportsmen. Activities and events were added for a major participation and activity in the city. The inaugural day was reserved for the careers junior, both feminine and masculine. One day later, it was the time for the sportsmen in the categories, feminine and masculine elite and paratriatlon. The third day was for the athletes that they were competing in junior reliefs and reliefs of elite. It there joined the celebration of the University Championship of Spain. With an entire full house as form accommodation it refers, the organization had to freight buses towards nearby towns, like Vigo or Sanxenxo to lodge all the participants and teams. The triatlon allowed to feel as for the locomotion means, since the closest airports (Santiago, Vigo and Oporto, the last one more focused in the international flights) without forgetting of the restoration.

46.5 Results

46.5.1 *The Sports Activity as New Position for the City*

In the year 2012 Pontevedra celebrated 120 sports events, like the Championship of Spain of Route, or the stages of the Vuelta Ciclista a España the championship of Spain of swimming for persons with disability, the Average Marathon Pontevedra Serviocio of different test of boating, sport especially loved by the habitants of the city, where the Olympic champions David Cal have trained. And other activities and events of sports like the handball, the fencing, the tumbling, the shot with arch of the PonteRaid, inside the Galician calendar in the league organized by the Galician Association of Clus of Orientation. At the end of the year, the councillorship of Sport of Town Hall was getting the award of the Galician Association of Sports agents for the development of the activities programmed (Instituto Municipal dos Deportes 2013).

The Vuelta Ciclista of Spain had an important presence in the year 2013 in Galicia where it started its journey, specifically in Vilanova de Arousa, receiving the exit of its second state the city of Pontevedra, which was repeating presence in the Spanish round, in 2012, was a protagonist for double entry, the squad was crossing the city in its stage disputed between Puenteareas and Sanxenxo. On the following day, the arrival in the city or Pontevedra. And in the round of 2016, Pontevedra will be part of the trip again. 2013 began with the celebration of the Championship of Spain of Tennis of Table, with more than a thousand sportsman and woman. In March, the celebration of the Open of Spain of Swimming required the coordination for human resources with a staff around 300 people (umpires, representation of the Spanish federation, timekeepers, personnel of assembly, control of accesses, kitchen personnel or secretariat). More than 200 swimmers came from France, England, Portugal, Hungary or Romania, along with its accompanists and that, therefore, allowed to add in the economy of the city. In April there was celebrated the Championship of Spain of Duatlon, with more than 2000 sportsmen and woman in all its categories. It is necessary to remember that it was allowing the punctuation for the Duatlon world cup in the categories paralimpic and age groups. In July we emphasize the I International Tournament of Chess City of Pontevedra, with a hundred of participants in with Nikita Meskovs won. It is necessary to add to these activities events of less importance but that the sport activities maintain in the city the whole year with the celebration of test in disciplines like the Taekwondo, Rugby, Waterpolo, Fitness, Tennis, Fencing, in Sync Swimming, Bicycles BTT... In 2013 there started also the Circuit of Popular careers of Pontevedra, with is provided every year with a hundred of sportsmen and woman. To emphasize that in this year there is celebrated the II Edition of the Galician league of Canicross and Bikejoning, sport with dogs, the participants are running and cycling, accompanied by the dogs.

The year 2014 has kept on consolidating this position in the ambience of the sports activities. So the Mayor of the city has emphasized it in different occasions. Fernández Lores has pointed out that the town hall wants Pontevedra as a regarding city, so much in practice of sport as in the organization of events and sports competitions: "Non creo que haxa en Galicia unha cidade comparable a nós nese aspecto; e só unhas poucas no conxunto do Estado" (http://imdpontevedra.blogspot.com.es/2013/04/o-concello-de-pontevedra-vive-o.html). Throughout the year 2014 the Galician Shot championship has been celebrated with arch of the Championship of Spain Junior of Taekwondo, or the winter Challenge of boating in March, the feminine Championship of Infantile Spain of Basketball, the Championship of Spain of Springboard or the Championship in Sync Swimming. But if we emphasizes something in 2014 it is the celebration of the Championship of the World of Duatlon in May and Jun, an international event with an economic impact superior to 2 million euros. On it, there took part a whole of 1412 sportsmen from more than 30 countries, more than 500 volunteers and Pontevedra took a hotel occupation of 100 %.

Pontevedra maintains a special relation with boating, because some international champions are training in the waters of the Lérez, case of David Cal or Paula Portela. In this sense, the city will receive in the year 2016 the championship of Europe of Boating Marathon. In the year 2015, the sports events increased as for disciplines, the innovation has come from the hand of the rhythmic gymnastic that celebrated in Pontevedra the Championship of Spain of Rhythmic Gymnastic of Clubs and regions, with more than 300 gymnasts and 400 sports clubs and that continued with the Championship of Orientation has been celebrated also for the first time with important success of participation.

46.6   Conclusions

The celebration of sport events brings to Pontevedra the discovery of an option of urban position, placing it like a model in Spain in the celebration of sports test.

These activities have provided an important economic propulsion because it maintains high hotel occupation, provides through out the year tourist arrival and has influenced sector like restoration and services.

Another achievement to be emphasized is the urban participation, like volunteers, participants and quizmasters in the streets to give fortitude and support to the sportsmen.

We understand that the celebration of test of less importance in the previous years served as experience with this type of events, being the definitive recognition to turn them into engines of the economic and tourist revival of the city in a new position.

The role of the sportsmen was fundamental. There was clear support in the declarations of many of them. The increase of sports activities have been progressive since 2011, year of the gunshot of exit for the achievement of sports test. Since the Championship of the World of Duatlon in the year 2014 and the Championship of Europe of boating Marathon in summer 2016, the confidence of the sports authorities in this city was increased.

The town hall of the city is the main promoter and organizer of the sports events, although the clubs and schools of the city help them a lot. The promotion and communication of these sports events is managed straight from the Council and usually bases its communications on the relation with the mass media and, especially its digital editions, but also giving a big importance to the social media and in particular to Face, Twitter and Instagram. These accounts are managed by the Council itself or the Instituto Municipal dos Deportes.

References

Diario de Pontevedra. (2009, July 03). *Triatlón. Pontevedra acogerá el Campeonato de Europa de Triatlón de 2011.* Retrieved from http://diariodepontevedra.galiciae.com/nova/94783.html. Accessed at December 12, 2015.

Diario de Pontevedra. (2010, June 12). *Copa de Europa de Triatlón Premium. Espectacular triunfo de Gómez Noya en Pontevedra.* Retrieved from http://diariodepontevedra.galiciae.com/nova/55927.html. Accessed at January 20, 2016.

Diario de Pontevedra. (2011, June 11). *La gran cita de Pontevedra.* Retrieved from http://diariodepontevedra.galiciae.com/nova/94783.html. Accessed at January 23, 2016.

La Voz de Galicia. (2009, January 09). *Pontevedra tendrá en 2009 una prueba puntuable de la Copa de Europa.* Retrieved from http://www.lavozdegalicia.es/deportes/2009/01/09/0003123152446726532452.htm?utm_source=buscavoz&utm_medium=buscavoz. Accessed at January 22, 2016.

Municipal Sports Institute. (2013, April 25). *O Concello de Pontevedra vive o Campionato de España de Dúatlon como un ensaio xeral para o Mundial de 2014.* Retrieved from http://imdpontevedra.blogspot.com.es/2013/04/o-concello-de-pontevedra-vive-o.html. Accessed at February 02, 2016.

Chapter 47
Treatment in the Spanish Digital Press of the Brazilian Carnival and Its Dissemination in Social Media

Jaime Álvarez de la Torre and Diego Rodríguez-Toubes

Abstract Mainstream press reaches the greatest number of people throughout the consumer decision-making process despite the fact of being less credible than specialized tourist press. However, tourism is a transversal sector and news related to tourism is susceptible of appearing in any one of the media sections. Thus it may gain presence or make a greater impact on the public who will subsequently share or disseminate this information through social media. The mass media may either enhance or attenuate stereotype images associated to Brazil, even throughout seasons like Carnival. This paper aims to analyse the information about Brazil provided by mainstream media and assess its impact and diffusion throughout internet and social networks. It conducts a context and content analysis of the digital editions of the main Spanish newspapers by measuring the interaction of these editions with Facebook. Our results show the presence of certain stereotypes associated to the image of Brazil as well as a certain emphasis on the negative aspects of events that are either directly or indirectly related to Carnival. However, social media seem to make no elevated diffusion of these events.

Keywords Carnival Brazil · Negative stereotypes · Image · Digital press · Social media

47.1 Conceptual Framework

Nowadays no activity may be understood without establishing its relation to mass media. Journalism, particularly press, plays a key role in conveying concise information about people, destinations and events (Ruibal 2009). With the recent

J.Á. de la Torre (✉)
University of a Coruña (UDC), Corunna, Spain
e-mail: jaime.delatorre@udc.es

D. Rodríguez-Toubes
University of Vigo (UVigo), Vigo, Spain
e-mail: drtoubes@uvigo.es

© Springer International Publishing Switzerland 2017
F.C. Freire et al. (eds.), *Media and Metamedia Management*,
Advances in Intelligent Systems and Computing 503,
DOI 10.1007/978-3-319-46068-0_47

advances in ICT's, users have also become diffusers of information. The mass media must now decide how they want to be represented on social networks and exactly what type of information they want to convey.

Destination images consist in several attributes represented in the main tourist attractions of a place (Stabler 1988) that shape the image individuals form of it. Baloglu and McClearly (1999) considered the image as an attitudinal construct composed of individual mental representations of beliefs, feelings and global impressions surrounding a destination. The subjective component of this term seems to be one of its few constants throughout the literature review (Beerli and Martin 2004; Bigné and Sánchez 2001; Gallarza et al. 2002). Additionally, elements like *impressions*, *perceptions* or *representations* are also commonly used to define it. Echtner and Ritchie (1991) point out that individuals may form an image even without having previously visited the destination or having had any direct contact with promotional commercial sources.

Gunn (1972) distinguishes between two types of information sources: induced sources, based on promotions and communications that come from tourism organizations; and organic sources, messages unrelated to a tourist destination that are informative and have no clear commercial intention. Following Gunn, Gartner (1994) constructed a continuum of separate agents and sources of information that either act independently or in combination to form an image of the destination. For our study we have focused on the so-called autonomous agents (which include the mass media). Their importance is explained by their wide diffusion and high level of credibility (Gartner 1994). In addition to this, the a priori lack of specific commercial interest implies that they are impartial. Depending on the type of event, these agents are capable of generating immediate changes in image as was the case in China shortly after the Tiananmen Square events (Gartner and Shen 1992).

Nowadays, tourists dispose of a wide range of sources, and information is crucial to choose a destination. Tourism requires greater access to information; travellers no longer only selects a destination but also search for information on additional complementary tourist services (Fodness and Murray 1997). Specialized journalism is a source of interest to tourists, but most of them choose mainstream media.

47.1.1   Digital Press

The media is in charge of collecting, selecting and processing information that is later diffused by public opinion both within their closer circles of influence as well as in th social media. The development of ICTs has changed the way we access and read information. Cañigral (2015) notes two aspects of news on Internet. On the one hand, its narrative component intends to call attention. On the other hand, its structural component seeks to integrate elements into the news that may make the information reach a greater number of people via social media and/or search engines.

Searching information process may often lead us to unfamiliar media despite the fact that our selection is based on preferences, interests and ideologies. Much the same happens with the information shared in social networks. The digital media offers the chance to establish links to other texts or web sites (Sandoval 2003) which will later be diffused through social media.

47.2   The Brazilian Carnival

Celebrations, folklore or music are essential in describing the main attributes of Brazil (Bignami 2002; Leal 2004). Carnival may be the main tourist attraction that collectively comes to mind no sooner Brazil is mentioned. This celebration is the most internationally broadcast event related to Brazilian culture apart from football. With it, the arrival of tourists rises across all over the country.

The Brazilian Carnival has become a national ritual that contributes to Brazil's national identity. Even Da Matta (2002) compares it to National Day given its appeal and the way society commits to it. The celebration reflects contrasts that define Brazilian society: the event may be understood as being either formal or more informal and festive. Carnival represents the moment in which society has no hierarchy; there is no apparent order and everything is allowed. This celebration is marked by happiness and positive values transcending the limits of the everyday world (Da Matta 2002). Madness and passion mark this cultural expression in which the Brazilians feel care-free and they welcome visitors to behave the same way.

Besides the organization of mega sports events, the image of Brazil has always been biased or stereotyped. The publication of garish or sensationalist news about a period of festivity as peculiar as the Carnival may come hand in hand with subsequent interaction and diffusion through social media.

This main aim of this paper is to analyse the way digital editions of leading Spanish newspapers deal with news concerning Brazil throughout the Carnival season and how this news is later diffused in social media. Our main hypothesis seeks to either confirm or reject the low interaction and diffusion of news related to the Brazilian Carnival in a scenario in which the treatment of matters specifically related to Carnival and to tourism in general is markedly sensationalist, alarmist and/or stereotyped.

47.3   Methodology

This study is based on the analysis of the descriptive content present in messages coming from leading Spanish journals. The interaction analysis is based on specific indicators provided by the websites of these journals as well as those provided by Fanpage karma, a tool used to specifically analyse the social network Facebook.

The media was selected according to the news compiled by the Embassy of Brazil in Spain which was taken from the digital editions of the following newspapers: El Pais, El Mundo, ABC, La Razon, La Vanguardia, Expansion and Cinco Dias. The selection was conducted from 13 to 24 February 2015. We analysed all the news bearing either a direct or indirect relation to tourism or the Brazilian image. The thematic fields under study were: Carnival, mega-events, sport and security. We identified the following subtopics under the topic *Carnival*: touristic dimension, sensationalism, incidents and image. As a result of this process, we identified and subsequently analysed a total of 40 units.

We used an analysis card to register and analyse the news. On it we recorded a series of variables based on other journalistic content analyses (Ruibal 2009; Piñuel et al. 2013). We then selected a total of 9 variables: newspaper, weekday, working day, section, genre, relation to tourism, type of headline, type of expression, and assessment of the news. We followed the same procedure, using the Karma Fanpage tool, to analyse the presence of this news on Facebook.

47.4   Results

The thematic distribution of the news shows that the impact of Brazil on the media is extremely heterogeneous (see Table 47.1). Seven articles were sensationalist; they exclusively referred to the problem arising from the apparent sponsorship of a parade by the government of Equatorial Guinea. Six articles were touristic, containing news merely referring to matters like tourist arrivals and expenditures per tourist. This is clearly indicative of how much interest is aroused by these matters.

The *Tourism and leisure* section has the most news, 13 articles. *Culture* and *International* follow with 11 and 8, respectively. It seems rather surprising that in such a short period Brazil is mentioned 3 times in the *Incidents* section. This reinforces the stereotyped image of Brazil as an unsafe destination. Yet few comments or shares about this were detected on the web or even published on the Facebook pages of the newspapers, e.g., the news 'One dead, nine wounded in a shootout at the Brazilian carnival' from El Mundo on 15 February 2015 only received 6 comments on the web and was shared on Facebook 68 times. Another piece of news from *ABC* newspaper on 15 February 2015 entitled 'One dead and

Table 47.1 News topics

Categories		Number of articles
Carnival	Touristic dimension	6
	Sensationalism	7
	Incidents	3
	Image	17
Mega-events		3
Security		4

Source Prepared by the authors

ten wounded in a shootout at Paraty Carnival in Rio de Janeiro' received only 3 comments and 46 shares.

Three pieces of news unrelated to Carnival referred to security as generally being one of the country's main concerns without making reference to specific facts. The text 'Why is Rio so violent?' published in El País on 19 February 2015 is the story with the largest number of interactions on Facebook throughout the period under study: 508 likes, 70 comments and 219 shares. On the same social network, we also found the news 'Silenced deaths' from the 24 February 2015 edition of El Mundo with 47 likes and 11 shares.

The most widely used journalistic genre by far was news (27 times) followed by chronicles (9 times). It should be noted that a chronicle is extensively used as a travel narrative because it recounts first-hand travel experiences. Thus, we find references to Carnival as explicit as 'Raving up in Rio (just like every February)' published in El País on 17 February 2015. The website edition of this news only received six comments but on Facebook it received 149 likes, 4 comments and 25 shares.

Focusing on headlines, almost half of the news analysed (19) were of informative style, followed by expressive style (10) and appellative style (6). If we consider that the period under study is characterized by joy and debauchery, the use of appellative and expressive headlines seems logical. The distinguishing element is the way this expression is revealed. That is to say, it may have a positive, negative or stereotypical connotation. In any case, a slightly tendentious intent was detected concerning the controversy surrounding Equatorial Guinea and its sponsorship of the winner of the Rio Sambodromo parade. A reference to this fact entitled 'A Carnival dictator', posted in El País on 13 February, was one of the most widespread stories on the web and social media with 311 likes, 45 comments and 218 shares.

Table 47.2 Most commented and shared articles during Rio Carnival

| Headline | Genre | Fanpage Karma (Facebook) | | | Digital edition | |
		Like	Comment	Shares	Comment	Shared in Facebook
1. Why is Rio so violent?	Press report	508	70	219	18	–
2. A carnival dictator	News	311	45	218	4	–
3. Raving up in Rio (just like every February)	Chronicle	149	4	25	6	–
4. King Momo opens the five Carnival Rio holidays	News	119	6	19	–	17
5. Silenced deaths	Op-ed	47	–	11	4	4

Note Compiled from the digital editions of seven Spanish newspapers through Fanpage karma
Source Prepared by the authors

As may be seen in the Fangpage karma analysis of Facebook (Table 47.2), digital editions lean towards more personal genres like chronicles and press reports. Paradoxically, however these contents are not among the most shared or commented on Facebook (see Table 47.2).

47.5   Conclusions

Mass media should avoid sensationalist approaches when narrating facts surrounding a certain event. It is advisable to not fall into an excessive use of stereotypical or flashy tactics which clearly call more social attention. Mass media plays a role close to that of a public service (Costa-Sánchez 2011), so society trusts it and gives credit to the information it conveys. Adopting an alarmist stance either in terms of the way the news is presented, or by simply just doing so in the headlines may make a major difference on how visitors may perceive a tourist destination. Moreover, the news selected to be published in the social media is expressive or alarmist rather than informative because it attracts the most attention.

Even though this analysis covers a very short period of time, we observe the latent presence of certain ideas rooted in our society which describe Brazil as a backward and dangerous destination. Yet at the same time Brazil is also described as a destination full of energy and festivity. These are precisely the topics that have the greatest impact on social media. Thus it seems clear that the mass media must increasingly pay attention and dedicate more time to social networks.

References

Baloglu, S., & Mcclearly, K. W. (1999). A model of destination image formation. *Annals of Tourism Research, 26*(4), 868–897.

Beerli, A., & Martin, J. D. (2004). Factors influencing destination image. *Annals of Tourism Research, 31*(3), 657–681.

Bignami, R. (2002). *A imagem do Brasil no turismo: construção, desafios e vantagem competitiva.* Sao Paulo: Aleph.

Bigné, J. E., & Sánchez, I. (2001). Evaluación de la imagen de destinos turísticos: Una aplicación metodológica en la Comunidad Valenciana. *Revista Europea de Dirección y Economía de la Empresa, 10*(3), 189–200.

Cañigral Giner, M. (2015). *La influencia de Google en la redacción de contenidos periodísticos.* PhD thesis, CEU Cardenal Herrera University, Faculty of Humanities and Communication Sciences, Valencia. Department of Audiovisual Communication, Advertising and Information Technology.

Costa-Sánchez, C. (2011). Tratamiento informativo de una crisis de salud pública: Los titulares sobre gripe A en la prensa española. *Revista de la SEECI, año, 14*(25), 43–62.

Da Matta, R. (2002). *Carnavales, Malandros y héroes: hacia una sociología del dilema brasileño.* México: Fondo de Cultura Económica.

Echtner, C. M., & Ritchie, J. B. (1991). The meaning and measurement of destination image. *Journal of Tourism Studies, 2*(2), 2–12.

Fodness, D., & Murray, B. (1997). Tourist information search. *Annals of Tourism Research, 24*(3), 503–523.

Gallarza, M. G., Saura, I. G., & García, H. C. (2002). Destination image: Towards a conceptual framework. *Annals of Tourism Research, 29*(1), 56–78.

Gartner, W. C. (1994). Image formation process. *Journal of Travel & Tourism Marketing, 2*(2–3), 191–216.

Gartner, W. C., & Shen, J. (1992). The impact of Tiananmen Square on China's tourism image. *Journal of Travel Research, 30*(4), 47–52.

Gunn, C. (1972). *Vacationscape: Designing tourist environments.* Austin: University of Texas.

Leal, S. (2004). A imagem de destinações turísticas: um estudo de caso do Brasil na percepção de alunos baseados na Austrália. *Retur, 2*(2), 1–8.

Piñuel, J. L., Gaitán, J. A. & Lozano, C. (2013). *Confiar en la prensa o no. Un método para el estudio de la construcción mediática de la realidad.* Colección Metodologías Iberoamericanas de la Comunicación. Salamanca: Comunicación social.

Ruibal, A. R. (2009). *Periodismo turístico: análisis del turismo a través de las portadas* (Vol. 144). Barcelona: Editorial UOC.

Sandoval, M. T. (2003). Géneros informativos: la noticia. En J. Díaz Noci y R. Salaverría, (Coords.), *Manual de redacción ciberperiodística* (425–448). Barcelona: Ariel.

Stabler, M. J. (1988). The image of destination regions: Theoretical and empirical aspects. In B. Goodall & G. Ashworth (Eds.), *Marketing in the tourism industry; promotional destination regions* (pp. 133–161). London: Croom Helm.

Chapter 48
Web Communication: Tourist Content Management in Web Portals for City Councils in the Province of Badajoz

María del Rosario Luna and Guadalupe Meléndez González-Haba

Abstract In the last few years the technological revolution has substantially changed the manner in which tourist destinations are advertised. Due to the fact that the internet has become the preferred guide for tourists, designing an appealing website is indispensable. The purpose of this paper is to explore whether the city councils in the province of Badajoz are effectively providing tourist information. We will evaluate the homepages of every city hall's tourism website, focusing on how the information is presented, and what resources are employed. Our research includes interviews with the personnel responsible for managing the content displayed in the region's corporate websites. Some of the conclusions reached in this study reveal the ways in which the city council websites in the province of Badajoz are flawed: (1) They are geared towards the region's residents, and not tourists; (2) The information provided is unappealing to travelers; (3) The staff in charge is unqualified.

Keywords Communication · Tourism · Corporate websites · Province of Badajoz

48.1 Theoretical Framework

In Spain, tourism is a vital source of economic growth and development. However, its strength varies considerably in each region. The Autonomous Community of Extremadura is not one of the more popular tourist destinations among non-residents, although domestic tourism has increased (The Institute of Tourism Studies of Spain, Egatur 2012). TICs have revolutionized trade patterns in the sector and it is increasingly rare to find tourists who travel without prior online

M.d.R. Luna (✉)
University of Extremadura (UNEX), Badajos, Spain
e-mail: mariadelrosarioluna@yahoo.com.ar

G. Meléndez González-Haba
University of Cadiz (UCA), Cadiz, Spain
e-mail: guadalupe.melendez@uca.es

© Springer International Publishing Switzerland 2017
F.C. Freire et al. (eds.), *Media and Metamedia Management*,
Advances in Intelligent Systems and Computing 503,
DOI 10.1007/978-3-319-46068-0_48

research: in the year 2009, 60 % of the travelers in Spain planned their trips using the internet (The Institute of Tourism Studies of Spain, Egatur 2012). Websites are therefore crucial mediums for a region's development. Internet content in the new global economy is a source of production. The utilization of websites does not guarantee a competitive advantage in and of itself, rather it is a requirement for participating in the labor market (Martínez et al. 2012).

Osorio et al. (2009) agree that, given the need to increase the competitiveness of tourist destinations, it is required that they adapt to the market, using the internet as a key tool for promoting tourism. This has inspired tourism companies to invest in consolidating their web presence. Customers find a wide range of options, and their needs and requirements tally with the predominating market demands. Web content plays a decisive role in the consumers' decision (Davis 2007). Official tourism websites are therefore key to both choosing a destination and planning a trip. We can conclude that, "it is now essential to design, develop and maintain websites that inform, persuade and commercialize their tourist brands efficiently" (Fernández-Cavia 2010: 3).

48.2 Methodology

This paper is part of a larger research project called *Las Webs turísticas de la Provincia de Badajoz: estrategias para la comunicación de una experiencia de marca*, (Tourist Websites in the Province of Badajoz: Brand Experience Communication Strategies), carried out with the support of the 2013 Research Grants for Regional Studies from the provincial council of Badajoz. Our research was conducted using the methodology proposed by Fernández-Cavia et al. (2010), introduced in the VIII TURITEC Congress, Information *and Communication Technologies in Tourism 2010* where the project "Nuevas estrategias de publicidad y promoción de las marcas turísticas españolas en la web" (CSO 2008-02627) was shown. To obtain empirical evidence, we studied the official tourism websites of the 180 city councils in the province of Badajoz. We analyzed the tourist content on their homepages according to the following parameters: design, content, resources, maintenance, and information sources.

In addition, in order to provide a qualitative analysis of the situation, we conducted twelve in-depth interviews. The interviewees were members of the tourism staff from different city councils in the province of Badajoz, as well as members of the tourism board in the region's provincial council. The heterogeneity of the region's tourism websites was taken into account when selecting the interviewees, and a wide range of testimonies were gathered to provide a fuller account of the kind of work being done.

48.3 Results

48.3.1 Web Portal Design

Official tourism websites for the councils of Badajoz employ different frameworks for displaying web content. Most frequently used is the template provided by the Badajoz provincial council (36.1 %; 65 city councils). A similar figure (33.3 %) is reached by a group of sixty municipalities, each of which has its own website with an exclusive format designed by private companies. The remaining cities and towns in the province have no corporate portals (30.5 %; 55 city councils).

The template from the Provincial Council Office of Badajoz was created more than fifteen years ago with the ".es" web domain, available for free to all city councils in the province. Each municipality in the province creates its own web address, adding the aforesaid domain after the name of the municipality itself. This feature is important to show that the portal is open to the public (which is valued by average users, who usually look for official tourist information because of its reliability). The template comprises a series of basic standardized items, which can be useful for local governments when informing their residents about city hall related matters.

Take for instance the homepage of the Ribera del Fresno City Council (http://www.riberadelfresno.es/); it is organized by links to City Hall, Municipal Services, Tourism and Enterprises. Content related to tourism are displayed in the following semantic links: History, Monuments, Historical Figures, Village Festivals, Gastronomy, and Tourist Routes. The template's main flaw is that it is designed for the town residents, and the section for tourist information is not the main focus.

It is important to highlight the increasing percentage of city councils lacking a website of their own (30.5 %). In practice, the provincial council offers the required infrastructure (domain, template, rights, technical advice), but it is each city or town council's responsibility to provide functioning web content. In an interview with one of the province's tourism supervisors, we were told: "We don't intervene, city halls and local governments are free to do as they please. At first we wanted them to develop their websites, and we gave them lots of encouragement, but the truth is that the mayors didn't even sign on to read a single email" (Director of the Board of Tourism, Badajoz Provincial Council, 2014). Regarding the traditional templates, the main complaint dealt with the size limitations when adding photos and videos: "That limits us a bit, yes, because we have to adapt to the template, especially with regards to the size of the files and formats. That's the main setback, we are really constrained" (Tourist Office Manager of the Badajoz provincial council, 2013).

One of the main advantages of most websites is the possibility to include multimedia, especially images. Websites with fewer images are the result of a technological restraint: "I have some very beautiful photographs of the region, but on the website they don't look good. The quality of the pictures worsens considerably after uploading them" (Manager of the city hall tourist office in the province of Badajoz, 2013). As for videos, some city councils have found a solution by creating a link to other websites such as YouTube, where their videos can be located.

During our research, we have come across special cases in which municipalities using the provincial council's basic framework, link their tourist network node to a website dedicated exclusively to tourist information. Here are the municipalities in which tourist heritage is of upmost importance: Mérida (http://www.turismomerida. org/), Zafra (http://www.visitazafra.com/), Badajoz (http://www.turismobadajoz.es/), La codosera (http://www.turismolacodosera.es/). The provincial council's improved official templates reflect a considerable effort to display tourist information. This can be observed by the inclusion of new links for content organization, quality content, enhancement of expressive webpage resources, etc.

However, a second group of city councils manage their content without employing the official template; instead they rely on a template developed by private companies. For example, the design for the town hall homepage of Frenegal de la Sierra (http://www.fregenaldelasierra.es/) has veered away from the traditional template. Not only does it provide information in several languages, it offers different types of tourism (cultural tourism, nature tourism and business tourism), special tourist packages, and various mediums for displaying information (videos, photographs, audio-guides, etc.). "We have worked hard and had in-depth discussions about what we wanted to offer on our website. We wanted to attract people by emphasizing our heritage and by making maximum use of available resources" (Council tourism technician 2014). Nonetheless, more than half of all the municipalities using templates from private companies suffer from the same constraints found in the official templates (e.g. the city halls of Azuaga, Campanario, Campillo de Llerena, El Carrascalejo). Currently there is no explanation for the decision to switch from the official format to one with similar attributes but from a private company.

48.3.2 Content

The analyzed data shows that some city councils have made an effort to modify the appearance of their websites, albeit without achieving any significant changes. In certain cases, switching to innovative designs actually resulted in the use of outdated information from tourist brochures and books. In this manner, content quality has been unattended to while the actual mediums have been heavily relied upon: "Web content is the same as it was in the previous version of the websites. We thought that the critical issue had to do with the format since the actual tourist information hasn't changed over time" (Tourism technician, 2014).

It is now worth asking about the people who determine what tourist information is used in a city hall's official website, and how it is presented. In some city councils, the content is chosen by general consensus, whereas in others it is the task of one or two people. The following testimony exemplifies the former: "The information we include in the web is compiled by taking into account citizen feedback. For instance, with content related to our heritage we rely on a teacher with in-depth knowledge of our village's history," (Tourism technician, 2014). In other municipalities, content is defined without any consensus; information comes from one point of view: "As mayor and as someone well acquainted with the area, I am the one who decides what tourist information to use on the website" (A mayor in the province of Badajoz, 2014). Regardless of how the information is chosen, each town hall has full autonomy when carrying out the selection process. The regional board of tourism does not play a role: "It would be impossible for us to intervene because it is not our area of expertise. The provincial council can't possibly provide guidelines about what each town hall should do with its website" (Tourism manager at the Provincial Council of Badajoz, 2014).

48.3.3   Maintenance

Keeping web content up to date is crucial for maintaining the website's credibility among users. In more than half of the cases analyzed we found outdated information (52.78 %), which leads us to question how content is renewed and updated on a city hall's website, and who is in charge of it. Staff shortages, a lack of skills, or an inefficient procedure for generating web content all lead to information not being properly updated: "The web was launched by a computer technician who had been employed for just a few months. I am the only tourism technician here, and I have plenty of tasks to carry out, so if I had to keep the tourism website up to date, it would require all my time" (Tourism technician, 2014).

During our visits to the town halls we discerned that a lot depends on the skills and attitude of the staff at the tourist offices. Destinations with a high influx of tourism will have properly managed websites that are up to date, whereas destinations with a much lower influx of tourists are not. Regarding web content management and display, different organizational methods have been observed: sometimes the town hall supervisor for tourism or even the technician himself is in charge of uploading web content. In other cases, the task is the sole responsibility of the local computer specialist. Be that as it may, experience has shown us that website maintenance depends largely on a region's organization protocols. Clearly, it is not being given neither the commitment nor the resources required for efficient web management.

48.3.4   Professionals

The professional profile of those managing tourist information on the web is varied, from those without the relevant qualifications (administrative assistants, IT Specialists), those who do, and those who have undergone postgraduate studies or training courses in tourism. Nevertheless, communication strategies vary from one city council to the next, and they depend more on the way professionals approach their work rather than on what they studied. The favorable results achieved by an employee are usually owed more to a strong work ethic than to the kind of training. That's not to say that adequate training is not necessary, but for our interests it is more important to highlight how, in the absence of an established procedure for supervising tourism websites, members of a community will make a joint effort to provide tourist information about their hometown: "The best thing about this job is that little by little I have been able to make the locals aware that the tourism office belongs to us all, that in one way or another tourism enriches us" (Tourism Manager, 2014). All in all, good work is owed more to a community's strong work ethic than it is to a team of qualified professionals following a specific protocol.

48.4 Discussion and Conclusions

The weakest aspects of the official city hall websites for the province of Badajoz are the following: (1) They are designed for the region's residents, and not for tourists. (2) They provide information that is unappealing to travelers. (3) The web content of several city councils is managed by unqualified staff. Therefore, our main proposals focus on the relationship between structure, staff, and content: (1) The professionalization of tourism services, investing in human and technological resources to improve the platforms; (2) Make a joint effort between public and private sectors, including small entrepreneurs, with the aim of promoting and developing of the region. (3) Provide information appealing for travelers. (4) Content must be based on feedback from travelers, tourism staff and local entrepreneurs.

Acknowledgments This work has been funded by the government of Extremadura and The European Regional Development Fund (ERDF).

GOBIERNO DE EXTREMADURA

EUROPEAN UNION
EUROPEAN REGIONAL
DEVELOPMENT FUND

References

Davis, T. H. (2007). Are websites like MySpace, YouTube and TripAdvisor relevant to today's business travellers? *Times Online*, 6 de marzo de 2007. Retrieved November 10, 2015 from http://travel.timesonline.co.uk/tol/life_and_style/travel/business/article1477541

Fernández-Cavia, J. (2010). Marcas de territorio y comunicación a través de la Web: un proyecto de investigación. In the *II Congreso Internacional AEIC "Comunicación y desarrollo en la era digital"*. Málaga (pp. 3–5).

Fernández-Cavia, J., & Rovira, C., et al. (2010). Propuesta de diseño de una plantilla multidisciplinar para el análisis y evaluación de webs de destinos turísticos. In *TURITEC 2010, VIII Congreso Nacional Turismo y Tecnologías de la Información y las Comunicaciones*.

Instituto de Estudios Turísticos (IET). (2012). *Turismo en cifras*. Retrieved November 10, 2015 from http://www.iet.turismoencifras.es/turismoporccaa/item/79-extremadura.html

Martínez, V., Penelas-Cortés, M. & Rodríguez, C. (2012). Análisis y Balance de un nuevo paradigma de comunicación y comercialización aplicado a destinos turísticos. In Turitec 2012*: IX Congreso Nacional Turismo y Tecnologías de la información y las Comunicaciones.*

Osorio, J., Gallego, J., & Murgui, S. (2009). *Turismo, planificación y gestión estratégica.* Valencia: Polytechnic University of Valencia.

Chapter 49
Neuromarketing: Current Situation and Future Trends

María del Mar Lozano Cortés and María García García

Abstract To understand the usefulness of neuromarketing in business terms, one needs to understand how consumers behave during the purchasing process. Thanks to brain imaging, one can better understand what determines this behaviour, and the opinions and preferences of consumers (Lindstrom in Buyology: verdades y mentiras de por qué compramos. Gestión, Barcelona, 2010). However, according to Zaltman (Cómo piensan los consumidores: lo que nuestros clientes no pueden decirnos y nuestros competidores no saben. Empresa Activa, Madrid, 2003), the ethical and moral implications underlying this activity require a judicious and socially responsible use of the information obtained. The method used is based on semi-structured interviews with seven Spanish neuromarketing firms. Thus, in order to respond to the current situation and to predict the future scenario in the short term, they were asked about the knowledge of both the population and Spanish firms, about the situation of the discipline versus traditional market research, and about the lack of specific regulations, among other things. The results demonstrate the incipient state of the discipline, although extensive growth is expected in the coming years. With greater current application in multinationals than in SMEs, the discipline is complemented with traditional methods, which in no case should be replaced.

Keywords Neuromarketing · Current situation · Market research

M. del Mar Lozano Cortés (✉) · M. García García
University of Extremadura (UNEX), Badajoz, Spain
e-mail: mlozanoz@alumnos.unex.es

M. García García
e-mail: mgargar@unex.es

© Springer International Publishing Switzerland 2017
F.C. Freire et al. (eds.), *Media and Metamedia Management*,
Advances in Intelligent Systems and Computing 503,
DOI 10.1007/978-3-319-46068-0_49

49.1   From Neuroscience to Neuromarketing

Neuroscience is an emerging interdisciplinary science. It is defined by Braidot (2005: 11) as "a merger, fairly recent, across disciplines, including: molecular biology, electrophysiology, neurophysiology, anatomy, embryology and developmental biology, cell biology, biology, behavioural biology, neurology, cognitive neuropsychology, and cognitive science."

One of the fundamental objectives of the neurosciences lies in understanding the biological mechanisms responsible for man's mental activity, allowing knowledge to be gained on the processes occurring within the human brain, and thus find deeper explanations about the behaviour of people and their decision making.

But also the emotions are an essential part of neuromarketing. For Álvarez del Blanco (2011), it is the most important motivating force known in humans. Zaltman (2003) agrees that it is the primary force acting on the mental processes and behaviour.

Likewise, marketing is also a multifaceted and multidisciplinary discipline. Since its inception, it has drawn on knowledge from other disciplines such as economics, philosophy, psychology, or sociology. Braidot (2005) noted the connection between these two sciences, and that one of the greatest challenges facing the economy and modern administrations at present is to understand how the human brain works.

According to Zaltman (2003), most consumers make decisions based on irrational processes. The current homogenization of products, globalization of the economy, saturation of commercial information, and even online purchases require firms to advance in their understanding of consumer tastes so as to bring to market products that are more relevant. Numerous studies have confirmed that over 80 % of new products coming to market end up failing in their first six months of life, or simply do not reach the objectives presupposed for them (Lindstrom 2010).

As a fast growing topic, more and more scholars are taking interest in neuromarketing, and consequently the number of definitions has also grown. Álvarez del Blanco (2011: 9) defines it as "the use of neuroscientific methods to analyse and understand human behaviour and emotions in relation to the market and its exchanges."

Neuromarketing is characterized by the use of tools midway between medicine and marketing. Thus, from a neurological point of view, there stand out functional nuclear magnetic resonance imaging and electroencephalography as the most advanced techniques today. However, also used are such tools as diffuse functional optical tomography, stable state typography, or magnetoencephalography, among others, with satisfactory results.

49.1.1 Ethics in Neuromarketing

In general, there is concern that advertisers have access to sophisticated tools capable of understanding and even directing consumer behaviour with the ultimate goal of selling a product (Andreu-Sánchez et al. 2014), and will use that knowledge to convert the consumer into a "prisoner", threatening their independence, health, and psychological balance (Álvarez del Blanco 2011). According to Zaltman (2003: 180), "the possibility of misuse should not pull us away from the discoveries, although it should sharpen our vigilance."

Although there are bodies at the European level concerned about the regulation of neuromarketing, there is no official body to ensure the legal regulation of the discipline. Only a few organizations, such as the Neuromarketing Science and Business Association (2016), are trying to respond to the existing lack of global regulation.

49.2 Objectives and Hypotheses

The main objective of this research was to analyse the current state of the discipline of neuromarketing in Spanish territory, and to try to predict the future situation in the short term. To this end, the following specific objectives were set:

OBJ1: To investigate the level of demand for services related to the discipline that exists in the Spanish population and Spanish firms.
OBJ2: To determine the implementation of the tools of neuromarketing today.
OBJ3: To establish the relationship between neuromarketing and traditional market research.

Similarly, on the basis of the initial data, the following hypotheses were posited:

H1: Neuromarketing is an unknown discipline for Spanish firms, so there is little demand for its services.
H2: Derived from this limited knowledge, the firms apply few neuromarketing tools to reach their public.
H3: Because of its potential, Consumer neuroscience will erode traditional market research.

49.3 Methods

To achieve the above objectives, a qualitative approach was made to the topic through an exploratory inquiry that was not intended to establish statistical generalizations, but to serve as a basis for future studies.

Thus, virtual semi-structured interviews were conducted with a sample of 7 firms professionally connected with neuromarketing, with activity in Spain, and available for participation in the study.

The interview consisted of twenty questions grouped into three distinct blocks:

Demand: Including questions on the current knowledge about the discipline of both the population and businesses in Spain, on the level of implementation, and on the profile of the firms that most demand these services.

Ethical and legal aspects: This block was aimed at inquiring into the level of knowledge that the firms have about the regulatory norms, and into the most important aspects that they believe should be regulated.

Results: Questions inquiring into the specific tools of the discipline, and their coexistence with traditional market research methods.

49.4 Results

After a literature review, we present the results for their subsequent analysis and interpretation. To facilitate their reading, we shall split the presentation into the same blocks as in the interview.

49.4.1 Demand

First, and to give a reasoned picture of the implementation of the discipline today, the study's participants were asked about what knowledge they thought the population (citizens) and firms of Spain have about neuromarketing. The results, which indicate a marked difference between the two objects of study, fall into three categories based on the observed level of knowledge (generalized ignorance, approach to the concept, and basic notions).

Thus, the responses indicate a generalized ignorance of the discipline both among the firms (28.5 %) and the general population (42.9 %) (Chart 49.1).

Chart 49.1 Comparison of the level of knowledge of the population and of firms, according to the interview responses. *Source* The authors

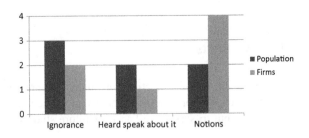

Another of the questions posed alluded to how much firms expected from consumer neuroscience. This time, 100 % of the responses were that, today, there is not enough interest in applying the techniques of neuromarketing. The main reasons the respondents gave can be summarized in the generalized ignorance noted above, and the firms' capacity to apply the discipline to their business.

Regarding the potential benefits of the discipline, 42.9 % of the respondents say that it improves marketing strategies, and the same percentage that it gives a better understanding of the brain. That it provides a better measure of stimuli is noted by 14.3 %.

The profile of the firm that tends to request neuromarketing studies reflects a clear demand on the part of large firms, and more specifically of multinational corporations, because of their great production capacity. Only one of the responses highlights the current participation of SMEs in the field of neuroscience applied to consumer behaviour.

When asked which are the sectors that dominate the implementation and application of the discipline, the respondents agree that it is the service sector targeted at mass consumption. More specifically, they highlight telecommunications, banking, and the automotive sector.

With regard to the national implementation, Madrid and Barcelona stand out as the areas with greatest activity (according to 42.8 % of the respondents).

In general, the purpose of any neuromarketing study is to analyse the responses of subjects when exposed to certain stimuli. However, the interviewees were also asked about what kind of study firms tend to request, and with what specific goals. Their responses pointed mainly towards three goals: an effective business strategy design, improved communication with customers, and evaluation of new products or services, all with identical percentages (33.3 %).

As for the approximate price of a standard study, only 1 of the 7 participating firms (i.e., 14.2 %) answered: "They are usually around €20,000" (Firm n° 2).

Another question inquired into the viability of a possible integration of neuromarketing by way of its own department in the firm. In this case, 71.4 % of the respondents would not be in agreement with such a proposal.

49.4.2 Ethical and Legal Aspects

The study's participants were asked to describe the evolution that neuromarketing has undergone up to today. Because the history of neuromarketing is still short, one can say that the discipline is still in a "real introductory phase, via testing and the expansion of its use" (Firm n° 6). Therefore, it is considered that the development and assimilation process will be somewhat slow and costly. This was highlighted by 33.3 % of the respondents. Finally, also 33.3 % of the respondents appreciate very positively that neuromarketing is being taken up by academia, in reference both to the formation of professionals and to the opening of new lines of research.

With respect to ethical and legal aspects, 42.8 % of the respondents acknowledge their ignorance of regulations existing at present.

They were also asked whether there existed a specific ethical commitment on the part of firms dedicated to providing services in neuromarketing. In responses to this question, 71.4 % were in favour of firms in which there is a solid commitment with the results of their studies, i.e., 5 out of the 7 participants. Just one respondent stated that there is no procedure to ensure confidentiality, "but there should be" (Firm n° 7).

49.4.3   Effectiveness and Tools

As for the tools used in neuromarketing, the participants were asked about the most effective technique for an average study (Chart 49.2). Notably, Firm n° 1 split itself off from this block since it does not use tools in its research, but bases it on the data provided by scientific studies. Also, 28.5 % of the respondents could not provide a valid response due to the wide range of possible studies, and to the specific application of the techniques in each case.

However, according to the other responses analysed, the most effective tools are nuclear magnetic resonance imaging and encephalography. We obtained 28.5 % for each technique.

With respect to the technique that is most demanded, eye tracking and encephalography stand out over the rest (Chart 49.2). One appreciates an increase in demand for biometric or physiological techniques relative to the responses to the previous question.

For 28.5 %, scientific and comparative studies could be the key to effective measurement, as well as "scientific publications in academic journals with high impact" (Firm n° 5). For Firm n° 6, the best way to measure the effectiveness is to check with the customer whether the behaviour predicted to the stimulus was verified in actual use. The other responses do not give a specific method, but limit themselves to stating that there exist different applicable theories and metrics.

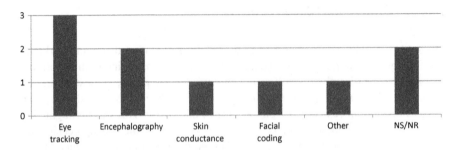

Chart 49.2 Demand for neuromarketing tools. *Source* The authors

We also inquired into whether there is some benefit for firms offering neuro-marketing services. To this question, 100 % of the responses are positive, affirming that it creates greater understanding of the consumer, it encourages curiosity and the possibility of observing new effects and learning from them.

Finally, there were questions dealing with the relationship between traditional market research and neuromarketing techniques. All the respondents (100 %) consider traditional market research to be in a situation of complementarity with neuromarketing. They are therefore in full agreement with a coexistence between the two, in some cases perceived as essential.

49.5 Conclusions

The results allow us to draw a picture of the current state of the discipline, and to put forward proposals for future development. Thus, in relation to the degree of knowledge of the discipline (H1), it was verified that it is still an unknown science for the vast majority of businesses and for citizens in general.

Especially noteworthy is the association of neuromarketing with large firms, partly due to the cost of the studies which is regarded as a brake on the current implementation of the technique in SMEs.

In line with this, and in relation with the scarcity of applied tools with which to reach their public (H2), the cost of the technology is also confirmed as being the main obstacle to a science that is still "in a real introductory phase", since it represents a disbursement that not all firms can afford.

Thus, among the most used tools, it is concluded that encephalography is the technique which in principle is best placed in both aspects, taking a standard study as the basis.

As for the relationship between traditional market research and neuromarketing (H3), everything made one expect that market research would be relegated to a second place. However, the responses indicate an equal coexistence of the two research methods, with that coexistence even becoming indispensable.

Thus, the theoretical consolidation of the discipline faces two key issues. On the one hand, there is the development of a committed normative regulation understood by all as a guarantee against possible malpractice and misuse, ensuring the confidentiality and privacy of the results, and guaranteeing the rights of individuals. And on the other, there is the need to incorporate multidisciplinary professionals specialized in neuromarketing into companies' marketing and/or communications departments. The academic development of the subject in universities which already offer postgraduate courses focused on the discipline will give a major impulse to this process.

It would be interesting to go deeper into the study of the current state of neuromarketing by expanding the sample of firms interviewed.

Acknowledgments This work was supported by the Regional Government of Extremadura and FEDER funds.

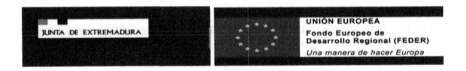

References

Álvarez Del Blanco, R. (2011). *Neuromarketing, fusión perfecta: seducir al cerebro con inteligencia para ganar en tiempos exigentes*. Madrid: Pearson Educación.

Andreu-Sánchez, C., Contreras-Gracia, A., & Martín-Pascual, M. Á. (2014). Situación del neuromarketing en España. *El profesional de la información, 23*(2), 151–157.

Braidot, N. (2005). *Neuromarketing: neuroeconomía y negocios*. Madrid: Puerto Norte-Sur.

Lindstrom, M. (2010). *Buyology: verdades y mentiras de por qué compramos*. Barcelona: Gestión.

Neuromarketing Science and Business Association (NMSBA). Ethical code. Retrieved from http://www.nmsba.com/Ethics. Accessed at February 08, 2016.

Zaltman, G. (2003). *Cómo piensan los consumidores: lo que nuestros clientes no pueden decirnos y nuestros competidores no saben*. Madrid: Empresa Activa.

Part VII
The Internet and Social Networks

Chapter 50
Online Communication and Galician Top Bloggers

Julia Fontenla Pedreira

Abstract The emergence of fashion blogs and bloggers has become a communication phenomenon that has originated new forms of consumption in the textile industry, with brands using them as a showcase to ensure that products reach the consumer more effectively. This study aims to be an approach to Galician bloggers communication. Particularly, it focuses on those considered as the most international: Lovely Pepa (Alexandra Pereira), Amlul (Gala González) and Bartabac (Silvia García). For this purpose, it is carried out a quantitative analysis of their activity, i.e. their interactions, messages and daily posts. This research highlights changes performed in communication strategies in order to reach users in a more effective, interactive and participative way. This has led to an increase in designers' and brands´ positioning as well as to a strong impact in society in terms of communication.

Keywords Blogs · Communication 2.0 · Fashion communication · Online communication · Fashion blogs

50.1 Fashion and Communication

Communication is a phenomenon that is present in all areas, and fashion industry is no exception. Theorists have contemplated it from different approaches. Paricio (1998) distinguishes between communication with the fashion, fashion's communication and communication in fashion. Both terms—communication and fashion —are complementary realities and according to Paricio (2000: 1), "there's no fashion without communication". This perspective is shared by Del Olmo (2005: 231), for whom "in order to make fashion accepted and consumed, it has to be known and followed in the first place, and that is only possible if it is known and

J. Fontenla Pedreira (✉)
Pontificia Universidad Católica de Ecuador—Sede Ibarra, Ibarra, Ecuador
e-mail: julia.fontenla@gmail.com

© Springer International Publishing Switzerland 2017   383
F.C. Freire et al. (eds.), *Media and Metamedia Management*,
Advances in Intelligent Systems and Computing 503,
DOI 10.1007/978-3-319-46068-0_50

communicated". There is where their total dependence appears, since there is no fashion without communication.

Fashion has a communicative dimension: it is a non-verbal communication system, a concept defended by theorists such as Davis (1992: 3), who refers to fashion as a code. Davis continues with the line established by Umberto Eco, for whom clothing is communication (Eco 2000: 103). For both authors, dressing styles and fashion has constituted something similar to a code, dependant from a context. From that perspective, the person who is wearing the clothes, the setting, the moment and the intention are the elements that will give meaning to the whole. This is why the meaning will be different depending on the given time and environmental conditions. Davis describes the features of this code focusing also on the fact that fashion has a changing nature and thus its meanings will constantly and imperatively change. This proves that fashion is indeed communication and it does not know of cultural and economic boundaries.

50.1.1 Fashion Communication Online by Blogs and Bloggers

New technologies are reconfiguring practically all the areas of knowledge, fashion included. It is present in the daily life of all people—everybody understands and talks about it, independently of their style. This is the result of all the information spread on the Internet and on those blogs that have fashion as a central topic. Therefore, the access to information is gaining adepts, and the updates about tendencies and accessories cause a homogenisation in consumption habits.

Fashion and beauty blogs are not going unnoticed and are influencing potential consumers, particularly young female. The starting point is that blogs have an impact on clothes people get or on styles chosen, becoming a stronger reference point than specialized magazines such as *Vogue*, *Glamour* and *Telva*. On the other hand, these specialized magazines have changed their working dynamics and they have been forced to introduce the blog method in their online edition. This is how fashion bloggers become a communication tool able to get into consumers' minds.

The subject of this study are personal blogs, *egoblogs*—characterized by being highly subjective depending on the test of their creators, who show their wardrobe or their physical appearance, most of them without linguistic or journalistic rigor. A number of them have undergone a process of professionalization due to the amount of people they move. Besides, there are faithful followers who buy the products showed in the posts. Hence, it is not surprising that this personality have a seat in the front line of fashion shows and important events. In fact, Giro (2012: 3) writes about fashion blogs addicts and considers it a justified addiction, since blogs represent a type of communication "that allows visualizing different fashion styles simultaneously and at a record-breaking speed, accelerating their spread all around the planet".

50.1.1.1 Galician Bloggers with International Style

Speaking about fashion communication online implies mentioning international referents such as the Galician bloggers Alexandra Pereira (Lovely Pepa), Gala González (Amlul) and Silvia García (Bartabac). According to the fashion site *Trendencias*, they lead the list of the most influential bloggers, ranking 3rd, 4th and 9th, respectively. They even compete with celebrities such as Blanca Suárez and Paula Echevarría. Appart from them, there are other Galician bloggers worth mentioning such as Alba Cuesta (Descalza por el parque), Carmela Lastres (Qui est in qui est out), Irene Medín (La Gordita Presumida), Casanova (Ropa Low Cost, Marta Fernández (Candies Closet), Nerea Vilela (Nery Poppins) Paloma Hernández (Chocolate & Lemon), Paola Ameigeiras (Love Cadessi), Paula Boado (Mi mundo by Paula Boado), Patricia G. Lema (Donkeycool) or Rebeca Rodríguez (El armario de Silvia).

The success of fashion bloggers in Galicia led to the organization in 2010 of the 1st Gathering of Fashion Bloggers in Pontus Veteris fashion show (Pontevedra). Galiciademoda, a fashion agency and an online specialized magazine, organized the event, just when great Galician bloggers were starting to emerge.

50.2 Methods and Results

The three blogs from the internationally successful bloggers Lovely Pepa (Alexandra Pereira), Amlul (Gala González) and Bartabac (Silvia García) were analysed following a quantitative methodology. The analysis was conducted during the week of 5–15 October 2015. The activity generated by each of them was studied by following their interactions, messages and daily posts. The last registered figure was noted at the end of each day, but these data might have increased due to the fact that the information is still present in the networks. Furthermore, a survey was conducted among 100 people, 50 men and 50 women between the ages of 16 and 45, this being the sector of Galician population that uses Internet the most according to Spanish Statistical Office. It aimed (i) to study their attitude towards consulting blogs, (ii) to determine how blogs may influence shopping choices; (iii) to detect advantages and disadvantages of online shopping.

The results obtained from the analysis of the blogs are shown in Tables 50.1, 50.2 and 50.3. The following data is analysed: number of post items per day and weekly average of posts; number of likes given by users and weekly average; answers to post, presence of advertising and social networks followers.

The results obtained in the programmed survey with the key question "Do you read fashion blogs?" are shown in Table 50.4.

Table 50.1 Data analysis from lovely pepa blog (Alexandra Pereira)

	Monday	Tuesday	Wednesday	Thursday	Friday	Saturday	Sunday
Posts	2	2	1	1	2	0	1
Weekly average of posts	1.4						
Comments on Facebook	230	299	176	189	196	0	155
Weekly average of comments	177						
Likes	2036	1657	874	1354	1101	0	3486
Weekly average of likes	1501						
Answers to posts	No						
Followers	Facebook: 385.911						
	Twitter: 31.366						
	Instagram: 67.9k						
Presence of advertising	Yes						

Source Prepared by the author

Table 50.2 Analysis data from Amlul blog (Gala González)

	Monday	Tuesday	Wednesday	Thursday	Friday	Saturday	Sunday
Post	1	1	1	1	2	1	0
Weekly average of posts	1						
Comments on Facebook	129	154	111	128	131	122	0
Weekly average of comments	110						
Likes	581	311	278	566	429	332	0
Weekly average of likes	356						
Answers to posts	No						
Followers	Facebook: 245.415						
	Twitter: 81.2k						
	Instagram: 508K						
Presence of advertising	Yes						

Source Prepared by the author

Table 50.3 Analysis data from Barbatac blog (Silvia García)

	Monday	Tuesday	Wednesday	Thursday	Friday	Saturday	Sunday
Post	1	1	1	1	1	1	1
Weekly average of posts	1						
Comments on Facebook	69	23	41	32	26	52	23
Weekly average of comments	38						
Likes	126	77	87	76	58	115	55
Weekly average of likes	78						
Answers to posts	No						
Followers	Facebook: 40.121						
	Twitter: 25K						
	Instagram: 296k						
Presence of advertising	Yes						

Source Prepared by the author

Table 50.4 Fashion blogs users

Do you read fashion blogs? Yes/No			
15 % NO			
Point out the reasons why you do not read these publications (such as lack of interest, you think they are trivial, you do not master social networks, other…)		Do you read fashion magazines in digital format?	
90 % think fashion blogs are trivial	10 % are not interested in them	100 % Yes	0 % No
85 % Yes			
The factors gender and age are considered		With what purpose do you use them? (Informative, professional, for entertainment, other…)	
5 % Men	Age 18–27: 62 %	95 % informative and for entertainment	5 % professional
95 % Woman	Age 27–35: 29 %		
	Age 35–4: 9 %		
What are the fashion blogs you follow the most closely? (including social networks, Lovely Pepa, Amlul, Bartabac, other)			
97 % Lovely Pepa		3 % other	
Do you read fashion magazines in digital format?			
100 % Yes		0 % No	
Have your consumption habits become more abusive since you started reading these blogs?			
98 % Yes		2 % No	

(continued)

Table 50.4 (continued)

Do you read fashion blogs? Yes/No		
Have your consumption habits become more abusive since you started reading these blogs?		
88 % Yes		12 % No
What product do you buy the most influenced by fashion blogs? Shoes, complements, underwear, young and adult's fashion, children's fashion		
Young/adult's fashion	67 %	
Complements	19 %	
Shoes	10 %	
Underwear	4 %	
Children's fashion	0 %	
What are the main reasons for buying products? Price, access to brands, exclusive items, comfort, avoiding crowds		
Comfort	75 %	
Access to brands	19 %	
Exclusive items	5 %	
Avoiding crowds	1 %	
What deters you from buying a product? Being worry for the size, shipping costs, not trusting the payment method, not trusting the shipping method, not liking the return system		
Being worry for the size	90 %	
Not trusting the payment method	10 %	

Source Prepared by the author

50.3 Discussion and Conclusions

The results indicate that number of posts is similar in the three blogs analysed by this study, the daily average of posts ranging from one to two. All of them give priority to photography. They are constantly being updated, particularly Lovely Pepa. An in-depth analysis shows that this results from the fact that the contents in Lovely Pepa cover a wider range of subjects, such as decoration, beauty products and daily life products. On the other hand, Amlul's contents are more based on a professional background, denoting the author's qualification. Conversely, Barbatac Mode's contents do not have the same impact as those of the other two blogs, probably because the other two include more advertising.

Furthermore, there is a massive participation both in blogs and in all social networks. These bloggers have numerous fans, although some of them follow these blogs not only on Facebook, but also on Twitter and Instagram. Therefore, some part of the public is accessing contents from different sources. Moreover, even if these bloggers claim to be close to their followers, they never answer any of their questions about clothes or style, which goes against the main purpose of blogs: having a greater contact with the public.

A longitudinal analysis proves that the three bloggers publish similar contents: all of them have the "look of the day" section, information on tendencies, key events related to fashion, etc.

Besides, the dynamic of posts and likes creates a relationship between fashion bloggers and their readers. Bloggers, for their part, talk to other bloggers, creating a huge network that is constantly increasing. This is a power relationship, since it is always stimulated by an advertising based on readers' subjectivity and seeking stimulate consumption.

Moreover, considering who the target audience is and analysing the results obtained in the surveys, it is clear that as a general rule women read more fashion and beauty blogs (95 % age 18–27) than men. However, the proportion of men reading these blogs is increasing, partly thanks to homosexual males. Regarding the effects of fashion blogs on the population, the respondents admit they have acquired products influenced by this type of blogs, sometimes making unnecessary purchases. They also claim consumption habits are changing towards online shopping.

Practically all the respondents (98 %) have bought influenced by fashion blogs. This proves bloggers have actually modified consumption habits. Indeed, among this group only 12 % have not bought over the Internet yet, although they have not ruled it out.

If we consider sectors, young fashion clothes are the most demanded, while consumption of children's clothes is low. Among main reasons that move consumers to buy products one of the priorities is having them delivered at home. Others—mostly fashionistas—want to find products to create a personal style that shows their personalities, and at the same time meet their requirements. These are the ones who usually become opinion consumers, those who first adopt a garment and can be followed afterwards. This led companies to stablish segmentation strategies for their communication campaigns to meet their goals. For example, the Italian brand Doce and Gabanna invited some of the most influential bloggers in the United Estates, such as Tommy Tom, Brian Boy, Constance Doré or Scott Shuman to sit in the first row of its fashion show and provided them with the necessary equipment so they could post about it in their blogs. This measure was also taken in Madrid Fashion Week.

Many bloggers have reached a level of success they could not imagine when they started their web pages. Lovely Pepa is a significant example of this, since now she is creating her own lines and collections for the world's top brands, such as Krack and so is Amlul designing for Adolfo Domínguez. They are the inspiration for thousands of people that look them up online; they are opinion formers and advisors for big events (Lovely Pepa or Bartabac in *Las Cancelas*, a shopping centre in the Spanish city of Santiago de Compostela). This has granted them great privileges among fashion companies and they have become referents who have privileged sits in fashion shows and are paid for their presence there.

The obtained data indicate that the main advantage of blogs is the form of access: you just need a computer connected to the Internet and it is charge free. Besides, there is no limit in terms of space and time, there is more freedom of opinion and information is handled in a light-hearted and fun way.

It is a real community, with a huge potential for cohesion and levels of participation that are unreachable for unidirectional mass print media. Therefore, these blogs autonomously made by users or readers are a reality that deserves attention, since their creators have a personal and independent initiative. Moreover, they are not subjected to the influence of advertising and brands, which are increasingly intervening in the thematic agenda of written publications or commercial blogs.

References

Del Olmo, J. L. (2005). *Marketing de la moda*. Madrid: ediciones Internacionales Universitarias.

Martínez Barreiro, A. (2006). *La moda en las sociedades modernas*. Madrid: Tecnos.

Mora, E. (2004). Globalización y cultura de la moda. In M. Codina & M. Herrero (Eds.), *Mirando la moda. Once reflexiones*. Madrid: Ed. Internacionales Universitarias.

Paricio, P. (1998). *Una aproximación a las dimensiones comunicativas de la moda: análisis de la comunicación de la moda en la prensa de información general española durante el S. XX*. (Tesis doctoral). Madrid: Universidad Complutense de Madrid.

Paricio Esteban, P. (2000). El encuadre de la moda en los diarios españoles de información general de ámbito nacional (1900–1994). *Revista Latina de Comunicación Social, 28*.

Rivière, M. (1977). *La moda, ¿comunicación o incomunicación?*. Barcelona: Gustavo Gili.

Squicciarino, N. (1990). *El vestido habla: Consideraciones psico-sociológicas sobrela indumentaria*. Madrid: Cátedra.

Chapter 51
Additional Barriers to Access to Labour Market for Prisoners Due to Digital Isolation

Isabel Novo-Corti and María Barreiro-Gen

Abstract Digital communication and social networks have become crucial elements in socialization. The deprivation of freedom for people in prison, is characterized not only by physical isolation, but also by a digital isolation. A sense of orphan hood or deprivation of communication appears, especially significant for "digital natives", who are also the youngest part of the prison population. Through the collection of opinions of inmates of all prisons in the region of Galicia (northwest Spain), it has been found that in addition to digital isolation, the possibilities of staying up-to-date in terms of finding a job are reduced due to the loss of "digital skills".

Keywords ICTs · Prison population · Labour market · Social networks

51.1 Digital Divide and Prison Population

Digital communication has become a crucial element in socialization, not only through social and professional networks, but also just sharing ideas via mobile devices.

Living in jail, without freedom and far from our "information society", implies a strong isolation from digital advances that may occur during the period of serving the sentence. Security requirements constitute the main reason for this isolation. This digital isolation can involve an irreversible digital divide. The digital divide leads to social disintegration caused by unequal ownership of ICTs and unequally distributed access to the online world, with consequences, for instance, in labour market.

I. Novo-Corti (✉) · M. Barreiro-Gen
Economic Development and Social Sustainability Unit (EDaSS),
University of a Coruña (UDC), A Coruña, Spain
e-mail: isabel.novo.corti@udc.es

M. Barreiro-Gen
e-mail: maria.gen@udc.es

© Springer International Publishing Switzerland 2017
F.C. Freire et al. (eds.), *Media and Metamedia Management*,
Advances in Intelligent Systems and Computing 503,
DOI 10.1007/978-3-319-46068-0_51

Helsper's model (2012) includes this new factor of social exclusion, along the same lines as the programs against exclusion designed by national and international organisms (Wong et al. 2009). This situation occurs due to the fact that the growing importance of information and communications technology (ICT) is a key factor in gaining employment. For example, the European Commission has developed the Digital Agenda for Europe, and consequently, each Member State has had to develop their own Agenda (European Commission 2013). However, a specific program for prison settings has not yet been established (Novo-Corti and Barreiro-Gen 2014). Thus, an in-depth study and analysis is necessary to examine of how convicts can be helped in avoiding their social exclusion and, more specifically, to prepare them for the time they leave prison and can properly face the requirements for labour market and for social relations. The aim of this paper is to analyse the key factors involved in explaining skills in the digital environment, applied to the case of prison inmates. We have relied on a study conducted in the autonomous community of Galicia (northwest Spain).

Theoretical models of social exclusion include variables such as income, employment, education and health (Gallie et al. 2003; Naraine and Lindsay 2011; Barreiro et al. 2013; Novo-Corti and Barreiro-Gen 2015). Helsper (2012) drew on the key variables included in research on social exclusion for developing a model linking this concept, in its traditional form (off-line), to digital exclusion (on-line). There are four key areas of digital exclusion and social exclusion: economic, cultural, social and personal. According to Helsper, such theoretical models could be applied in different environments because they are at the same time broad enough and capable of taking into account certain specific characteristics. Meanwhile, Novo-Corti and Barreiro-Gen (2014) have applied this theoretical model to the group consisting of the prison population in Spain, to study the possible development of public policies, which try to prevent digital exclusion. These authors have concluded that the lack of access to information technologies and communication in Spanish prisons acts as a factor of social exclusion in a crucial area for prisoners, because prisoners cannot improve their skills in ICTs due to security issues in prison. Therefore, it becomes necessary to find a way to reconcile security with ICTs training for the inmates.

The most common profile in prison corresponds to people with low education and low skills; this also leads to poor training preparation in digital environments and communication technologies. In addition, a significant percentage of the prison population is functionally illiterate, and many of them have not completed primary school. Moreover, there are many foreign prisoners, whose knowledge of the Spanish language is quite low. We can say that there is a high correlation between education and the ability to get a job (Barreiro et al. 2013). Considering the above, the situation of prisoners is not easy to solve. The academic literature suggests that the digital divide is caused by social factors, such as age, gender, education, social status, income and local infrastructure (Clayton and Macdonald 2013; Hindman 2000; Kingsley and Anderson 1998), so that digital isolation not only affects individuals but also groups with similar characteristics. These groups generally tend to be those who already suffer some form of exclusion, although digital exclusion can affect people who do not belong to any of the aforementioned groups.

Lack of access to Internet connections inside prisons creates a situation of deprivation of some of the tools of professional and technological training, commonly used by most of the population. Although it is true that many activities can be done offline, this solution would require some adjustments.

Digital exclusion usually sits in the same areas as social exclusion, namely: economic, cultural, social and personal (Helsper 2012). Thus, it affects different areas and affects them in different ways, among which are on-line purchasing, socialization via the Internet or online learning. In any case, people who are in custody are forbidden to link themselves with the outside world, including via digital connections, related to ICT. So, people who are already in social exclusion, run the risk of falling into the digital divide and finding themselves involved in a circuit that feeds back and which should be broken to achieve effective reintegration and therefore give these people a chance.

51.2 Methodology

The sample consisted of inmates from the five Galician prisons (Teixeiro, Bonxe, Monterroso, Pereiro de Aguiar y A Lama) and from the two Centres of Social Reintegration (CIS), in A Coruña and Vigo. The Centre of Vigo depends on the A Lama prison.

Quantitative and qualitative analysis have been carried out. 380 questionnaires, that represent 10.27 % of the Galician prison population, were collected in proportion to the number of inmates in each Centre or prison: Teixeiro (33.2 %), Bonxe (9.2 %), Monterroso (9.4 %), Pereiro de Aguiar (9.2 %), A Lama (35.5 %) and CIS of A Coruña (3.5 %). The sample was composed by 92.4 % of men and 7.6 % of women, in line with the total population. Moreover, 64.5 % was Spanish and 35.5 % was foreign. More than 40 % of the sample had not had any schooling or only had primary studies, while only 5 % had gone to the University.

A Factorial Confirmatory Analysis was carried out in order to deal with both observable and unobservable variables, also known as latent variables or constructs. To do this analysis, AMOS was used. A model has been estimated: ITC Skills was the dependent variable. Independent variables were "Social Skills", "General skills" and "Attitudes" (Fig. 51.1).

On the other hand, a qualitative analysis was carried out in order to complete the quantitative one. 30 in-depth interviews were conducted (9 in Teixeiro, 3 in Bonxe, 3 in Monterroso, 2 in Pereiro de Aguiar, 9 in A Lama and 4 in A Coruña CIS). They were coded by hand, using F4 program.

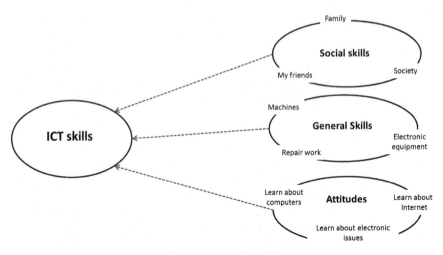

Fig. 51.1 Determinants of ICT skills. *Source* Own elaboration

51.3 Results

In the quantitative analysis, the estimation of the adjustment ($\chi^2 = 95.158$, $p < 0.01$) showed the fit was adequate in global terms. Moreover, other commonly-accepted indices, such as CFI (Comparative Fit Index) or RMSEA (Root Mean Square Error Adjusted) supported it (CFI = 0.915 and RMSA = 0.051).

All of the variables, not only the items formed the constructs, but also the constructs acting as explicative variables, had expected signs. However, only general skills have been shown to be relevant (Table 51.1).

Faced with these results, a qualitative analysis was considered necessary, with the aim of explaining why the "attitudes" variable was not shown to be relevant, in contrast to theoretical models.

Approximately 91 % of individuals have shown their willingness to learn, that is to say, most of them have shown a positive attitude towards ICTs and learning more in this area. Two factors may explain this fact: It has to be taken into account that respondents participated in this research voluntarily. This factor may introduce an impossible to avoid skewing of results, because the respondents showed their

Table 51.1 Results

	Estimator	Standardized estimator	Standardized error	t Student	P
Social Skills	0.010	0.036	0.019	0.509	0.611
Attitudes	0.062	0.029	0.166	0.372	0.71
General Skills	0.269	0.476	0.065	4.113	0.000

Source Own elaboration

interest in improving their situation through their participation in the research. Moreover, the questions made reference to "to learn", but it is possible that it was confused with "wanting to know". It is different wanting to know something and wanting to learn something. Learning is, according to Royal Spanish Academy, "acquiring knowledge of something through study or experience". In fact, learning implies effort (Novo-Corti et al. 2011).

While it is true that digital natives have shown more skills than the others, they need to improve their knowledge in new technologies, because, in general terms, their knowledge is low.

51.4   Discussion and Conclusions

The quantitative analysis, through Structural Equation Modelling, has shown that ICT skills can be explained using general skills as an independent variable. Therefore, future policies with the aim of avoiding the digital divide should promote all kinds of general skills, improving the inmates training.

General skills are directly related to vocational training. Therefore, it would be advisable to increase the inmates' knowledge in any of the spheres with activities that can be implemented in jail.

On the other hand, it is necessary to rethink the options available to balance security requirements in jail and the need to improve the ICT skills of the prison population. Solving problems by restricting access to the network could be a complementary line of action.

Attitude has not been shown to be relevant in the study of ICT skills. Respondents participated in this research voluntarily, that is to say, their attitude is more positive than the others. The possible reasons to explain this phenomenon have been achieved in qualitative analysis (the difference between "learning" and "wanting to know").

It is proposed, in line with Novo-Corti et al. (2011), to offer basic computing training, beginning with the elementary aspects of using a computer. In a second phase, the generation of interactive applications is proposed, whereby simulating real wed environments fosters participants' skills in ICTs.

References

Barreiro, M., Novo, I., & Ramil, M. (2013). Employment, education and social exclusion: Analyzing the situation of people at prison in Galicia. [Mercado de trabajo, formación y exclusión social: Análisis de la situación de la población reclusa de Galicia] *Revista Galega De Economía*, *22*(2), 225–244.

Clayton, J., & Macdonald, S. J. (2013). The limits of technology: Social class, occupation and digital inclusion in the city of Sunderland, England. *Information, Communication & Society*, *16*(6), 945–966. doi:10.1080/1369118X.2012.748817

European Commission. (2013). *Digital agenda for Europe*. Retrieved from http://ec.europa.eu/digital-agenda/

Gallie, D., Paugam, S., & Jacobs, S. (2003). Unemployment, poverty and social isolation—Is there a vicious circle of social exclusion? *European Societies, 5*(1), 1–32.

Helsper, E. J. (2012). A corresponding fields model for the links between social and digital exclusion. *Communication Theory, 22*(4), 403–426.

Hindman, D. B. (2000). The rural-urban digital divide. *Journalism & Mass Communication Quarterly, 77*(3), 549–560.

Kingsley, P., & Anderson, T. (1998). Facing life without the internet. *Internet Research-Electronic Networking Applications and Policy, 8*(4), 303–312. doi:10.1108/10662249810231041

Naraine, M. D., & Lindsay, P. H. (2011). Social inclusion of employees who are blind or low vision. *Disability & Society, 26*(4), 389–403.

Novo-Corti, I., & Barreiro-Gen, M. (2014). Barreras físicas y barreras virtuales: Delito y pena en la era digital. Nuevas políticas públicas para la reinserción. *RiHumSo. Revista de Investigación del Departamento de Humanidades y Ciencias Sociales, Universidad Nacional de La Matanza, 2*, 3–22.

Novo-Corti, I., & Barreiro-Gen, M. (2015). Walking from imprisonment towards true social integration: Getting a job as a key factor. *Journal of Offender Rehabilitation, 54*(6), 445–464. doi:10.1080/10509674.2015.1055036

Novo-Corti, I., Barreiro-Gen, M., & Varela-Candamio, L. (2011). Las TIC como instrumento de inclusión social a través de la formación académica y profesional en los centros penitenciarios: Análisis de las percepciones de la población reclusa en la región de Galicia, España. *Inclusão Social, 5*(1), 58–67.

Wong, Y. C., Fung, J. Y. C., Law, C. K., Lam, J. C. Y., & Lee, V. W. P. (2009). Tackling the digital divide. *British Journal of Social Work, 39*(4), 754–767.

Chapter 52
The Usefulness of Social Networks as Research Tools for the Media

Diana Lago Vázquez

Abstract The advent of social networks and the Web 2.0 has led to a revolution for journalists and researchers. However, its recent emergence implies a lack in methodologies, theories and analysis for supporting their work and avoiding ambiguity in results. The research explores the main methodologies—qualitative and quantitative—in the analysis of social networks within the information and communication field. For this purpose, it was collected a sample of academic papers published in the last five years on the analysis of the two most popular networks (Twitter and Facebook). The aim is to study the tools used in those studies, in order to assess and interpret the data provided by these digital networks. The modus operandi and trends of research on social networks is analysed to verify their degree of usefulness as research tools that report scientifically valuable conclusions.

Keywords Social networks · Research · Twitter · Facebook · Methodology · Media

52.1 Introduction

52.1.1 The Era of Convergence and the Boom of Social Networks

The media transformation is one of the effects of the emergence of the Web 2.0 and has been analysed by many researchers. The traditional system based on one-way communication has evolved into a convergent scenario in which the media has lost relevance in favour of the audience. Nevertheless, the fast inclusion of these changes has resulted in a context of uncertainty about how to face the challenges and take advantage of them. So there is still a need to determine the terms under which the traditional system and the new digital era are related (Jenkins 2008).

D. Lago Vázquez (✉)
University of Santiago de Compostela (USC), Santiago de Compostela, Spain
e-mail: dianalago20@gmail.com

© Springer International Publishing Switzerland 2017
F.C. Freire et al. (eds.), *Media and Metamedia Management*,
Advances in Intelligent Systems and Computing 503,
DOI 10.1007/978-3-319-46068-0_52

Today, new and old media converge and invent a media system of multiple dimensions and hybridizations between traditional media and services offered by the WWW. This process has not been free and has required an exercise of adaptation to new circumstances (Scolari 2013; Gillan 2010).

Some have even talked of this context as a "third industrial revolution" based on the information and knowledge society (Pérez and Acosta 2003) and the culture of participation (Jenkins 2010). The figure of the prosumer (Rincón 2008; Pérez and Acosta 2003) or consumer-producer of content, stems from this need for collaboration and triggers a redefinition of the relationship between the media and the public in order to create virtual communities (Tiscar 2005, 2008) or even fandoms (Jenkins 2010). At this point, social networks have revolutionized the already troubled context, presenting new challenges and opportunities for both public and media and creating phenomena like social TV (Gallego 2013; Lorente 2011). In terms of usage, it is unquestionable that they are acquiring power (Domínguez 2010). According to the latest Social Networking research made by organization IAB Spain (2015), more than 14 million users between 18 and 55 used social networks in 2015. Another similar study, the Digital Democracy Survey (2014) by consultancy Deloitte suggests that 47 % of consumers in Spain visit every day their profile on social networks. The most popular social networks—Facebook and Twitter—get a lot of audience around a product—information or fictional—and convert the already called social audience into a valuable tool to brands and advertisers (Quintas Froufe and Neira 2014; Aguado and Martínez 2012).

52.1.2   Social Networks in the Field of Research

Social networks make it possible to find news and get feedback from users, as well as analyse political strategies of the candidates for election or identify the so-called opinion leaders (Marwick and Boyd 2011). In the research area, social networks are very useful tools. However, the amount of data they generate every day is excessive. To process this volume of information, researchers require quantitative and qualitative techniques and methodologies for drawing conclusions with the greatest accuracy.

Sádaba (2012) notes that methodologies must evolve in parallel with changes in society and distinguishes four major approaches to social research. These could become the most successful methods in the coming years: virtual ethnography, audience analysis, network analysis, and social and online audiovisual analysis. The so-called social network analysis or SNA (Lozares 1996; Sanz 2003; Molina 2001) is focused on the relationship between groups of society, and is part of social sciences and psychology. However, recent years have seen a disciplinary hybridization (Gualda 2005), which allows the use of SNA in other areas such as communication and information. In this regard, the tools developed for this type of studies can be used for the extraction of data from Facebook and Twitter. After all, social networks also work as a network of interactions like SNA studies about society.

52.2   Objectives

The new scenario of media convergence requires knowing better than ever every element. Social networks are emerging as tools that provide data and useful information for understanding the new context. The problem is how to access these data. The present research aims at doing a literature review of studies published over the past five years on social networks in the field of communication and information. The purpose is to construct a map of the main methodologies and techniques used by the experts in recent years to obtain data from social networks and specifically Twitter and Facebook. The methodologies are classified and analysed to make a contribution to the field of research of social networking about the new trends. As a final note, an alternative methodology based on the SNA is proposed.

52.3   Methodology

The methodology used for the construction of this research is based on qualitative techniques of bibliometrics. First, three databases of professional scientific articles have been chosen—Scopus, WOS and Google Scholar. To maintain the relevance of the results, it was decided to analyse the articles published in the last five years (between 2010 and 2015) and the search has focused on two social networks—Twitter and Facebook. After obtaining all items resulting from the search process, they have been analysed through an Analysis tab in which the following information was classified: the country of publication, year, format, magazine or media on which has been published, type of methodology, approach, research technique, subject of the article and social networks investigated. The types of methods, approaches and research techniques are classified according to the studies of Hernández et al. (2003). A total of 121 articles of the three databases have been found, while only free-access articles (64) were analysed.

52.4   Results

The use of social networks as tools for the analysis of general and case studies is booming in all research fields. Areas of communication and information are taking advantage of these new features of digital networks. Facebook and Twitter have become waterfalls of continuous information. This information is valuable for understanding the current digital context and because of that, social networks are increasingly necessary in the field of research.

After classifying the articles of the sample, an increase in the volume of publications is observed. In 2010, 8 articles about the subject studied in this research were published, while 33 articles were published in 2015. The social network most used and most analysed in the articles has been Twitter (60 %) over Facebook (29 %) and other social networks (11 %). Because of its peculiarities, Twitter is a very useful tool for the field of communication. It provides real-time information and we can use it to analyse and track trends in a short period of time.

In terms of methodology, quantitative studies predominate. In 49 % of cases, the researchers extracted data from social networks and processed quantitatively, using statistics to develop their conclusions. However, 41 % also use a qualitative study. These investigations combine both methods to analyse the object of study from different angles. In the case of this sample, qualitative studies have only represented 10 % of the total. As regards the type of methodology, the prevailing trend is the exploratory study. That is, there is more research aimed to explore circumstances, events or facts which, by their present or not know the subject, cannot formulate precise hypotheses and cannot predict results. Descriptive methodologies, which represent 23 % of the total, are used to define the characteristics of specific cases and explanatory methodologies (3 %) are used to explain a situation through assumptions or cause-effect relationships (Chart 52.1).

The emergence of social networks as tools in research studies is recent and there was no time to develop a unified technique. Therefore, there are still doubts about which tools to use. Although there are many online tools to obtain basic data, we cannot verify its accuracy. For this reason, 52 % of the publications of the sample choose the method of manual observation. They collect data manually and directly from the account of Twitter or Facebook. In 36 % of the articles, one of the online tools was used, while 4 % of the publications contracted the services of software measurement companies.

There are increasing tools for analysing social audience on Twitter and Facebook, but there are no specific techniques and methods. At present, there is a trend within the field of social sciences where methodologies and programs of SNA (Social Networks Analysis) are used to analyse popular social networks as Twitter and Facebook. "The analysis of social networks (ARS), also known as structural analysis, has been developed as a tool for measuring and analysing social structures

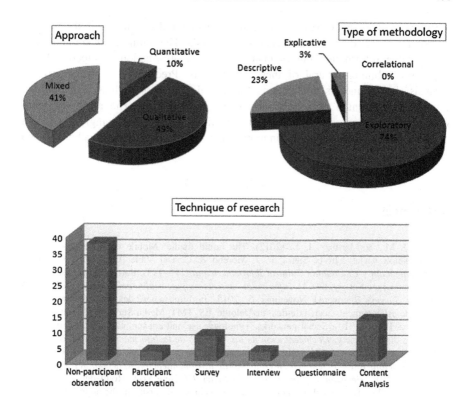

Chart 52.1 Types, approaches and research techniques. *Source* Prepared by the author

emerging from the relationships between various social actors (individuals, organizations, nations)" (Sanz 2003). In fact, there are already some tools and programs
to apply the method of SNA to Facebook and Twitter (Netvizz, Gephi, etc.).

52.5   Conclusions

In a context of continuous change of the media and communication, social networks
are becoming major global actors, as they are understood as virtual spaces of
interaction between the media and the audience. For researchers, the new scenario
of convergence and technology has been and is the focus of great interest. And at
this point, social networks are being used as tools for understanding the behaviour
of the audience. However, the most popular digital networks were created ten years
ago and although their success has risen rapidly, there is still a problem to extract
and handle the amount of data they provide. Methodologies evolve while researched objects are being developed, and this is precisely the point where is the
investigation of social networks. This research was performed in order to classify

the most used methodologies in the last five years and conclusion is the need to investigate more accurate and specialized methods. The online tools are helpful to get valuable information but in many cases this information is superficial and requires combining several of these instruments. Precisely, the social network analysis (SNA) and its applications for digital platforms such as Facebook or Twitter are presented as a good alternative methodology. The results of these studies help to understand the current situation and are increasingly needed in different areas, from political communication or advertising, to psychology and social sciences.

References

Aguado, J. M., & Martínez, I. J., (2012). The liquid media: Mobile communication in the information society. In F. Sierra, F. J. Moreno & C. Del Valle (Coords.), *Políticas de Comunicación y Ciudadanía Cultural Iberoamericana* (119–175). Barcelona: Gedisa.

Digital Democracy Survey. (2014). *Differential contents support traditional devices October 2014 Media Survey*. Recovered November 6, 2015, in http://www2.deloitte.com/content/dam/Deloitte/es/Documents/acerca-de-deloitte/Deloitte_ES_TMT_Media-Survey-2014.pdf

Domínguez, D. C. (2010). Social networks. Typology, use and consumption of 2.0 networks in today's digital society. *Documentación de las Ciencias de la Información, 33,* 45–68.

Gallego, F. (2013). Social TV analytics: New metrics for a new way of watching TV. *Index comunicación: Revista científica en el ámbito de la Comunicación Aplicada, 3*(1), 13–39.

Gillan, J. (2010). *Television and new media: Must-click TV.* New York: Routledge.

Gualda, E. (2005). Theoretical, methodological and technical diversity in the approach to social networks: Toward disciplinary hybridization. *Redes: Revista Hispana para el análisis de redes sociales, 9*(1).

Hernández, R., Fernández, C., & Baptista, P. (2003). *Investigation methodology*. México: McGraw Hill.

IAB Spain. (2015). *VI study social networks January 2015*. Recovered November 6, 2015, in http://www.iabspain.net/wp-content/uploads/downloads/2015/01/Estudio_Anual_Redes_Sociales_2015.pdf

Jenkins, H. (2008). *Convergence culture: The culture of the convergence of the media.* Barcelona: Paidós.

Jenkins, H. (2010). *Pirates of texts: Fans, television and participatory culture.* Barcelona: Paidós.

Lorente, M. (2011). Social TV in Spain: Concept, development and implications. *Cuadernos de Gestión de Información, 1*(1), 55–64.

Lozares, C. (1996). Social network theory. *Papers: revista de Sociología, 48,* 103–126.

Marwick, A., & Boyd, D. (2011). To see and to be seen: Celebrity practice on Twitter. *Convergence: The International Journal of Research into New Media Technologies, 17*(2), 139–158.

Molina, J. L. (2001). *The analysis of social networks: An introduction.* Barcelona: Bellaterra.

Pérez, A., & Acosta, H. (2003). Media convergence: A new stage for the management of information. *Acimed, 11*(5).

Quintas Froufe, N., & Neira, A. G. (2014). Active audience: Social audience participation on TV. *Comunicar: Revista Científica de Comunicación y Educación, 22*(43), 83–90.

Rincón, O. (2008). No more audiences, all we become producers. *Comunicar: Revista científica iberoamericana de comunicación y educación, 30,* 93–98.

Sádaba, I. (2012). Online Social Research. In M. Arroyo Menéndez & I. Sádaba (Coord.), *Metodología de la Investigación social: técnicas innovadoras y sus aplicaciones* (197–216). Madrid: Síntesis.

Sanz, L. (2003). Social Network Analysis: or how to represent the underlying social structures. *Apuntes de Ciencia y Tecnología, 7,* 21–29.

Scolari, C. (2013). *After the broadcasting TV: Hipertelevision, networks and new audiences.* Recovered November 6, 2015, in http://hipermediaciones.com/2013/06/15/la-tv-despues-del-broadcasting-hipertelevision-redes-y-nuevas-audiencias/

Tiscar, L. (2005). Towards a more participatory television. *Comunicar: Revista científica iberoamericana de comunicación y educación, 25*(44).

Tiscar, L. (2008). The new public sphere: the media as social networks. *Telos: Cuadernos de comunicación e innovación, 76,* 128–131.

Chapter 53
Academic Social Networks and Communication Researchers from Universities in the North of Portugal: An Analysis of Academia.edu and ResearchGate

María Isabel Míguez-González, Iván Puentes-Rivera and Alberto Dafonte-Gómez

Abstract Academic social networks emerge as an opportunity for researchers to improve their work and to potentially increase both the dissemination of their research and their collaboration with other scholars. However, there are still many gaps, uncertainties and reluctance surrounding academic social networks. This is not surprising, as they are a new phenomenon and are still not en a par with traditional media in terms of prestige and recognition. Considering such platforms as objects of study may contribute to their consolidation as valid tools for academic research. This study analyzes the presence of communication scholars and researchers from Universities in the North of Portugal in the two major academic social networks, Academia.edu and ResearchGate. Our work is in line with recent incipient studies in other geographical areas, such as Galicia, Spain. This article not only offers and overview, but also detailed insights into the presence or absence of researchers on these platforms, number of documents uploaded, followers, visits; analysed by gender, university or department.

Keywords Academic social networks · Communication · Portugal · ResearchGate · Academia.edu

M.I. Míguez-González (✉) · I. Puentes-Rivera · A. Dafonte-Gómez
University of Vigo (UVigo), Vigo, Spain
e-mail: mabelm@uvigo.es

I. Puentes-Rivera
e-mail: ivanpuentes@uvigo.es

A. Dafonte-Gómez
e-mail: albertodafonte@uvigo.es

© Springer International Publishing Switzerland 2017   405
F.C. Freire et al. (eds.), *Media and Metamedia Management*,
Advances in Intelligent Systems and Computing 503,
DOI 10.1007/978-3-319-46068-0_53

53.1 Theoretical Framework

The academic world has become one of the realms where the expansion of Web 2.0 tools has had a greater impact. This space offers new possibilities for the development of two of science's core elements: communication and collaboration (Codina 2009). Discussions around Science 2.0 have evolved in recent years. This is defined as "using social web technologies in the scientific process" (Merlo et al. 2010) as a way of understanding science in a less uptight fashion, imbued with the values of free access to information, free collaboration and the overcoming of physical distances as barriers to research. This model of "open science" (Merlo et al. 2010) is mainly understood in three ways: sharing research, resources and results.

The potential of Web 2.0 technologies is particularly explored in social networks. Thus, generalist social networks (Facebook, Twitter, Linkedin…), as professors Mercedes Caridad and Xosé López (Campos Freire et al. 2014) claim, can be used from an academic perspective, offering researchers the possibility to disseminate their studies fast to a wide audience. Despite the possibility of using generalist social networks, the need for more specific tools, with more reduced and specialised audiences, allowing for networking with researchers of similar backgrounds, led to the creation and development of so-called academic social networks.

Campos et al. (2014) have defined them as "ecosystems of software services, repositories and open on-line communication platforms" offering academics the possibility of easily accessing millions of scientific studies, contacting academics in their areas of interest, fostering cooperation and sharing knowledge, managing reputation or "scientific social capital" of researchers and institutions, or opening up many possibilities of working with metadata.

They are, for (Martorell Fernández and Canet Centellas 2013) a "meeting point for researchers from all over the world" with three working principles: "communication, cooperation and sharing knowledge in a virtual and democratic environment, perfectly suited for the dissemination of research, as long as participation and loyalty to academic rigour are observed".

The fact that these networks allow users to have easy and fast access to the contributions of other researchers in their areas of interest, multiplies the visibility of their research. This, in turn, increases the possibilities of a certain publication being quoted (a relevant indicator for scientific journals, as it contributes to increasing their impact index and therefore their interest for the scientific community).

Despite their constant growth, academic networks such as Academia.edu or ResearchGate still show some limitations (Martorell Fernández and Canet Centellas 2013; Cabezas-Clavijo et al. 2009). They are used mainly by younger researchers, rather than by those with established careers, and in many cases they are only used as repositories of knowledge previously published in conventional media or research uploaded without any quality filters to support the results published.

In any case, with their pros and cons, challenges and opportunities—such as for example ending redundancy and duplicity in research, as Ijad Madisch, founder of ResearchGate, claims (Becerra 2013), academic social networks are currently booming. It is the task of researchers to define this space, improve it and bring it into their daily activities, making the best of the opportunities it offers for scientific development.

With this background, this study analyses the use of academic social networks for a specific field, that of communication sciences, as this is an area where researchers are expected to be particularly aware of the need to disseminate their results, and they are also expected to be familiar with social networks (Túñez and Sixto 2012; Mendiguren et al. 2012; Subires and Olmedo 2013).

Thus, in the framework of the Galician project XESCOM, International Network for Research in Communication Management, and as a continuation of previous research on the way Galician communication scholars used Academia.edu and ResearchGate (Dafonte-Gómez et al. 2015), this study tries to understand how academic social networks are used by communication researchers in the North of Portugal.

53.2 Methodology

The networks selected for this research were Academia.edu and ReserachGate as they were both the oldest (founded in 2008) and those with the highest user figures: at the beginning of 2015 Academia.edu had over 17 million registered users and ResearchGate over 6 million. Besides, both networks ranked first and second in the benchmark of academic social networks developed by Martorell Fernández and Canet Centellas (2013); ResearchGate matched 84 % of their criteria for the ideal academic social network, while Academia.edu was around 75 %, the same value assigned to Mendeley.

In total, we analysed the presence of 78 scholars in the field of communication (42 women and 36 men) of three public universities in the North of Portugal: 18 from the University of Porto (UPorto), 30 from the University of Minho (UMinho) and 31 from the University of Trás-os-Montes e Alto Douro (UTAD), according to the data published by each of these universities' websites. The geographical focus responds to the strategic interest of the area for the European Cooperation Partnership Galicia-North of Portugal (GNP-AECT).

The record-card used registered the presence or absence of each researcher in the networks, the number of documents uploaded (papers, book chapters, etc.), the number of followers and following, the number of views and the use of the available documents. These indicators helped assess the usefulness of these networks for the researcher in terms of visibility and dissemination of their work. The data regarding the researcher's following and the documents uploaded offered some insights into the level of activity that the researcher showed on the network, while

views and followers showed their success level on the platform (against the activity
levels of other researchers).

The data were gathered during the first week of November 2015.

53.3 Results

59 % of communication researchers in the North of Portugal are present in at least
one of the two academic social networks. The most widely used network is
ResearchGate, 4 points ahead of Academia.edu, although half the scholars in the
sample have profiles on both networks. The percentage goes down to 29 % at
UTAD, which is the university with the highest percentage of researchers without a
profile in any academic network (55 %).

The percentage of women in some of the networks is far higher than that of men
(64 % vs. 53 %), with a clearly marked difference in the case of ResearchGate.
Likewise, the percentage of women having a profile in both networks is relatively
higher than that of men (56 % vs. 42 %). The data are consistent in the three
universities in the study, although at UTAD, the gap between women and men is
less significant (Table 53.1).

Regarding the documents uploaded by researchers onto the networks, the
average in ResearchGate is higher than that of Academia.edu (18 vs. 11 documents per person), although the data varies depending on the institution.
Researchers of UMinho share more documents than the rest, and they prefer to do
so through ResearchGate. After them comes the average of documents shared by
the lecturers of UMinho, who use both networks in a similar fashion. Researchers
at UTAD are those who disseminate their works less, and they do so mainly on
Academia.edu.

Likewise, seven of the 46 researchers (15 %) who have a profile on any of these
academic social networks (four of them have profiles on both) have never uploaded
any documents onto the network, which could initially mean that their presence is
merely symbolic and lacks any repercussion in the dissemination of their work. Out
of these seven researchers, five are women.

Table 53.1 Percentage of researchers with a profile on academic social networks (against the total
researchers of each university and the total of researchers in the sample)

	UPorto (%)			UMinho (%)			UTAD (%)			Total (%)		
	M	F	Tot	M	F	Tot	M	F	Tot	M	F	Tot
Profile in any other network	56	75	65	62	76	70	43	47	45	53	64	59
Profile in Academia.edu	44	50	47	54	53	53	29	29	29	42	43	42
Profile in ResearchGate	44	75	59	38	71	57	21	35	29	33	57	46
Profile in both networks	33	50	64	50	62	57	17	38	29	42	56	50

Source Prepared by the authors

In fact, a gender-wise analysis of the sample shows that the average upload of documents is higher by men than by women in both networks. Only in UMinho and in ResearchGate women have a higher upload average than men (Table 53.2).

Regarding the remaining indicators (Table 53.3), in the sample we see how the average views, followers and following are considerably lower for ResearchGate than for Academia.edu. Thus, although researchers are slightly more present and upload more documents onto ReserachGate than onto Academia.edu, their activity level in the former is lower than in the latter. These data help understand that the visibility of a researcher in academic networks depends more on the dynamics of the network than on the content introduced by the author.

The university-wise analysis of the data tends to be consistent with the general data. All universities perform better in terms of views, followers and following in Academia.edu than in ResearchGate. UTAD is the one with the lowest average views in ResearchGate and lower-than-average followers and following in both networks, despite the fact that, paradoxically, it is the university with the highest views on Academia.edu.

Despite the fact that women are more present in the networks, the three activity indicators show better results for men. The difference is especially significant in the average views in Academia.edu, with figures of 1656 for men and only 403 for women. There are no significant differences in the breakout of data per gender in each of the analysed universities.

Table 53.2 Average of documents uploaded by researchers on academic social networks (against the number of researchers present on social networks, by universities and total)

	UPorto			UMinho			UTAD			TOTAL		
	M	F	Tot	M	F	Tot	M	F	Tot	Male	Female	Tot
Academia	17	12	14	15	7	10	18	2	9	16	7	11
ResearchGate	29	6	15	22	29	27	4	3	3	20	16	18

Source Prepared by the authors

Table 53.3 Average views, followers and following by researchers (against the number of researchers present on social networks, by universities and total)

	Academia.edu			ResearchGate		
	Views	Followers	Following	Views	Followers	Following
UPorto	896	188	117	109	35	18
UMinho	941	188	116	94	52	44
UTAD	1.098	120	80	28	15	24
GLOBAL	973	169	106	82	38	31
Male	1656	289	168	111	54	33
Female	403	70	55	67	30	31

Source Prepared by the authors

53.4 Discussion and Conclusions

Almost 60 % of researchers in the field of communication in the North of Portugal are present in academic social networks and exactly in half of the cases, those who use these networks are present in both.

Despite the fact that Academia.edu has more users at world level, their penetration in the sample is slightly lower than that of ResearchGate. Besides, ResearchGate shows higher success in becoming the repository of publications, with a higher average of uploaded documents. However, Academia.edu is a more active network in terms of average views, followers and following.

A similar contradiction can be found in gender-wise data analysis. Women are more present in academic social networks, especially in ResearchGate, and the presence of women in both networks is also higher. Nevertheless, their activity rates are lower: they upload fewer documents than men and follow fewer researchers. Maybe as a consequence of this lower level of activity, their visibility levels are also lower, both in terms of the average views per profile as well as the average number of followers.

In fact, most of the users in the sample who have a profile on these networks, but have never uploaded a document, are women. These profiles have a low level of views and followers, therefore their usefulness in terms of increasing the visibility of research results or increasing citations is low. In any case, only one sixth of researchers have created profiles that are then filled-in with data.

Regarding the number of documents uploaded onto both networks, except for the University of Porto, the average uploads are diverging. This seems to show an unplanned on-line presence of researchers, or at least not an optimised use.

The online dissemination of academic works does not only depend on their quality, but also on accessibility and being positioned in a numerous and active peer community. Therefore, researchers have to become aware of the need to strategically plan the dissemination of their scientific work as part of their regular professional routines. In this sense, the differences revealed by this study in terms of the way communication researchers in the North of Portugal use Academia.edu and ResearchGate, as well as an analysis of these network's internal operations, leave the door open for future research into other geographical or knowledge areas.

Acknowledgments This article was developed within the International Research Network of Communication Management (R2014/026 XESCOM), a research project funded by an open call for proposals by the Galician Ministry for Culture, Education and Universities.

References

Becerra, V. (2013). *ResearchGate, una red social para la comunidad científica* (on-line). Retrieved from: https://artvisual.net/blog/researchgate-una-red-social-para-la-comunidad-cientifica/

Cabezas-Clavijo, Á., Torres-Salinas, D., & Delgado-López-Cózar, E. (2009). Ciencia 2.0: catálogo de herramientas e implicaciones para la actividad investigadora. *El profesional de la información, 18(1)*, 72–79.

Campos Freire, F., Rivera Rogel, D., & Rodríguez, C. (2014). La presencia e impacto de las universidades de los países andinos en las redes sociales digitales. *Revista Latina de Comunicación Social, 69*, 571–592. Retrieved from: http://www.revistalatinacs.org/069/paper/1025_USC/28es.html

Codina, L. (2009). *Ciencia 2.0: Redes Sociales y Aplicaciones en Línea para Académicos* (on-line). Hipertext.Net, 7. Universitat Pompeu Fabra. Retrieved from: Http://Www.Upf.Edu/Hipertextnet/Numero-7/Ciencia-2-0.Html

Dafonte-Gómez, A., Míguez-González M. I., & Puentes-Rivera, I. (2015). Redes sociales académicas: presencia y actividad en Academia.edu y ResearchGate de los investigadores en comunicación de las universidades gallegas. En A. Rocha, A. Martins, G. Paiva, L. Paulo y M. Pérez (Eds.), *Sistemas e Tecnologias de Informação. Atas da 10ª Conferência Ibérica de Sistemas e Tecnologias de Informação, 1*(1) 1233–1238. Águeda, Portugal: Asociación Ibérica de Sistemas y Tecnologías de Información—Universidade de Aveiro. Retrieved from: https://www.researchgate.net/publication/283765702_Academic_social_networks_Presence_and_activity_in_Academia.edu_and_ResearchGate_of_communication_researchers_of_the_Galician_universities

Martorell Fernández, S., & Canet Centellas, F. (2013). Investigar desde internet: Las redes sociales como abertura al cambio. *Historia y Comunicación Social, 18* (nov.), 663–675.

Mendiguren, T., Meso, K., & Pérez, J.A. (2012). El uso de las redes sociales como guía de autoaprendizaje en la Facultad de Comunicación de la UPV-EHV. *Comunicación Social, 6*, 107–122. Retrieved from: http://iesgtballester.juntaextremadura.net/web/profesores/tejuelo/vinculos/articulos/mon06/07.pdf

Merlo Vega, J. A. (Coord.), Ferreras Fernández, T., Galle León, J.P., Agosto Castro, A., Maestro Cano, J.A., & Ribes Llopes. (2010). *Ciencia 2.0: aplicación de la web social a la investigación* (on-line). Madrid: REBIUM. Retrieved from: https://www.Academia.edu/375082/Ciencia_2.0_aplicaci%C3%B3n_de_la_web_social_a_la_investigaci%C3%B3n

Subires, P., & Olmedo, S. (2013). Universidad, sociedad y networking: perspectivas ante el uso de las redes sociales de perfil académico profesional. *Estudios sobre el Mensaje Periodístico, 19*, 1037–1047. Retrieved from: http://revistas.ucm.es/index.php/ESMP/article/view/42188

Túñez, M., & Sixto, J. (2012). Las redes sociales como entorno docente: Análisis del uso de Facebook en la docencia universitaria. *Píxel-Bit, Revista de Medios y Educación, nº 41*, 77–92. Retrieved from: http://acdc.sav.us.es/pixelbit/images/stories/p41/06.pdf

Chapter 54
Visibility and Impact of the Microcredit and the Digital Social Media: A Case Study of Financial Institutions in Ecuador

Viviana Espinoza-Loaiza, Rosario Puertas Hidalgo, Valentín Alejandro Martínez Fernández, Aurora Samaniego-Namicela and Eulalia-Elizabeth Salas-Tenesaca

Abstract This study analyses the impact and visibility of financial institutions that promote the microcredit product through digital social media such as Facebook and Twitter in Ecuador. It analyses 70 Ecuadorian financial institutions that according to data from the Ecuadorian Superintendence of Banking and Insurance and the Ecuadorian Superintendence of Popular and Solidary Economy, until July of 2015, they offer the microcredit within their portfolio of products. The registered information in Facebook and Twitter was gathered by using: Netvizz, Followerwonk, Klear, Twitalyzer and Twopcharts; from which, it analyses the number of publications, type of publications, among other elements. It concludes that digital social media sites such as Facebook and Twitter have a great amount of users within the Ecuadorian financial system; however, the activity generated by the financial institutions is very limited. The visibility of micro-credits in these digital social media is very low. The existing information as well as the use of multimedia elements is minimum.

Keywords Financial institutions · Digital social media · Microcredit · Visibility

V. Espinoza-Loaiza (✉) · R. Puertas Hidalgo · A. Samaniego-Namicela
E.-E. Salas-Tenesaca
Technical Particular University of Loja (UTPL), Loja, Ecuador
e-mail: vdespinoza@utpl.edu.ec

R. Puertas Hidalgo
e-mail: rjpuertas@utpl.edu.ec

A. Samaniego-Namicela
e-mail: afsamaniego3@utpl.edu.ec

E.-E. Salas-Tenesaca
e-mail: eesalas@utpl.edu.ec

V.A. Martínez Fernández
University of a Coruña (UDC), Coruña, Spain
e-mail: valentin.martinez@udc.es

© Springer International Publishing Switzerland 2017
F.C. Freire et al. (eds.), *Media and Metamedia Management*,
Advances in Intelligent Systems and Computing 503,
DOI 10.1007/978-3-319-46068-0_54

413

54.1 Theoretical Framework

54.1.1 Microcredit

Nowadays, the microfinance sector has a great development in Ecuador. From 2012, there is a legal framework that regulates the actions and procedures of microfinance institutions, as well as popular and solidary economy (Martínez Castillo 2008). Microfinance institutions (MFIS) have become one of the favourite tools to fight poverty in developing countries. Micro-finance is responsible for providing financial service, such as providing micro-credit to poor families (Ivatury and Mas 2008; Charitonenko and Campion 2003; Rhyne 2001; Ledgerwood 1999; Cardoso 2011). Gutiérrez (2005) considers that microcredit is a search for the extension of loan services to poor people who have been excluded from formal financial services.

In positioning the microcredit Cano et al. (2014) considered that financial inclusion is an essential element in achieving an efficient transmission of monetary policy. The lack of access to micro-credit in the vulnerable sector, limits the development of entrepreneurship. This is a great challenge the country faces in the popular and solidary financial sector, as financial institutions have the great challenge of reaching the largest part of the population. These days, through the web, digital social media are a very interesting way to reach the market. The ease of access to digital social media has softened the issue of lack of knowledge of the products and financial services.

54.1.2 Visualization of Companies on Digital Social Media

Digital social media are a channel for companies to achieve identity and visibility in online media, allowing them to know the needs to develop new products, to meet the existing products, to receive feedback from customers, to segment customers, to position a brand. Digital social media are also informative, users make and answer questions of any kind, thanks to the reply time and the quantity and quality of responses (Ringel et al. 2010; Bolotaeva and Cata 2011; Valerio et al. 2014).

Digital social media have significant presence in the different service activities such as tourism, education, financial institutions, among others. In tourism as Martínez-Fernández et al. (2015) state, the perceived image of the users of some touristic complexes is assessed by users (Sánchez-Amboage et al. 2014), as well as the visibility through the promotion on Facebook. In the educational scope, according to the authors Valerio et al. (2014) who studied the publications of Mexican universities in Facebook, as well as likes, shares and comments, conclude that messages that generate more interaction with users of Facebook are short and longer messages use other types of resources to make them attractive to the reader.

For the financial institutions, the case of Bradesco in Brazil can be highlighted; it operates in the Brazilian market since 1943 with more than 56,000 points. Bradesco, through Facebook and AG2 Publicis Modem Agency is developing F. Banking, which allows customers to make regular financial operations such as balance inquiries, money transfers between Facebook friends, paying bills and shopping. The most important results of this strategy are that *more than 30 thousand clients access their accounts from Facebook; the number of monthly users has increased by 20 %, brand mentions increased from 6000 to 20,000 in one week after its launch* (Facebook 2015). In Argentina Creditos.com.ar has been created, a platform that speeds up money loan through an Internet webpage; its goal is to contact the credit petitioner with entities that can supply it. The process carried out lets potential customers of credit to fill the application online and creditos.com.ar sends their requests to financial institutions. Forty percent of customers come from Facebook users. Having a good communication strategy such as using original photographs and common vocabulary allowed *the increase of clicks by more than 100 %, the increase of daily visits by more than 50 % and an increase of 34 % of daily reach and almost 2.1 times more people carrying out actions on the platform* (Facebook 2015).

Digital social media in many countries have become an important link between the company and society. It is easy to measure <<I like>> on Facebook and the <<retweets>> on Twitter, but it would be more interesting to quantify how much of these reactions on digital social media are worth for organizations to place their products (De Rojas Giménez 2012). In his study, De Rojas Giménez (2012), mentions that every 5 s there are 56 new blogs in the network published and every 5 s, 40 people creates a new Facebook account, 250 tweets are published, 62,500 videos are viewed on YouTube, 187 images are seen in the Flick, 35 new people gain internet access, 125,000 searches are performed on Google and more than one million SMSs are sent to the world via the web.

54.2 Methodology

This study got the information of the websites of 70 financial institutions which offered in August 2015 microcredit products and were regulated by the Ecuadorian Superintendence of Banking and Insurance and the Ecuadorian Superintendence of Popular and Solidary Economy. Using a technical analysis data constructed by the authors, the Website, the Facebook Fan page and the Twitter Account of the account financial institutions (IFIs) were analysed for a month. Netwizz tool was used to get information from Facebook, and Followerwonk, Klear, Twitalyzer and Twopcharts from Twitter; through these, the number of publications, type of publications, among other elements of utility for the purpose of this study could be determined. Facebook and Twitter had a greater amount of users within the Ecuadorian financial system. However, the activity generated by the financial institutions was limited. The visibility of micro-credits in these digital social media is very low, the existing information as well as the use of multimedia elements are also limited.

54.3   Results

Information obtained from the companies selected, allow us to present the following results as the main reason of the study. According to the classification of the 70 analyzed institutions (IFIs), 22 are private banks, 40 are cooperatives, 6 are financing companies and one is a Mutualist, where 57 % of institutions that offer micro-credit products are cooperatives, followed by private banks with 31 %. While the 70 IFIs have a Website, only 62 % includes links to official digital social media. The main social media that have been used by users are Facebook, Twitter, Google, YouTube, Flickr, and also Instagram, and Video and Streaming. Although 56 (80 %) of the 70 IFIs have a profile on Facebook, only 39 links it to the Website. On Twitter occurs the same, 45 have an account but 37 (64 %) are linked between the Website and the social media account.

The Chart 54.1, compares the IFIs accounts and their number of followers of the two social media.

Pacific Bank stood as the institution with the highest number of followers and the best community management performance on Facebook and Twitter account. Even if the number of monthly publications were still low on Twitter, Guayaquil Bank tops the list with 228 posts, the competitors did not do exceeded it, as for example: Pacific Bank did 117 posts and the rest of IFIs did not do more than 100 monthly posts. On Facebook, JEP Cooperative even with only 52 publications and 135,604 followers in the month of September, reached the highest value in engagement, publications, likes and comments. Pichincha Bank with 54 publications reached 23,117 in engagements, 21,009 I like publications and an extremely high amount 23,117 times shares. It is interesting to compare the cases also with Guayaquil Bank, who published 228 posts, but it reached fewer values than their competitors who posted less, but reflected a greater engagement for competitors.

On Twitter, Pacific and Guayaquil Bank are the community manager leaders, for each 100 published tweets, in response; the Pacific Bank got 154 replies, and a

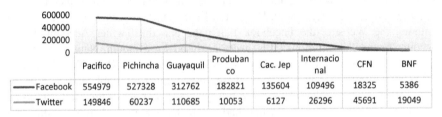

	Pacifico	Pichincha	Guayaquil	Produban co	Cac. Jep	Internacio nal	CFN	BNF
Facebook	554979	527328	312762	182821	135604	109496	18325	5386
Twitter	149846	60237	110685	10053	6127	26296	45691	19049

Chart 54.1 IFI's account of Facebook and Twitter and followers number—September 2015. *Source* Authors, September (2015)

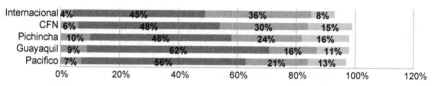

Internacional	4%	45%	36%	8%
CFN	6%	48%	30%	15%
Pichincha	10%	48%	24%	16%
Guayaquil	9%	62%	16%	11%
Pacifico	7%	56%	21%	13%

■ 12-17 years old ■ 18-24 years old ■ 25-34 years old ▪ 34-49 years old ■ 50-64 years old

Chart 54.2 Age of digital social media users of Ecuadorian IFI's, 2015. *Source* Authors, September (2015)

popularity of 190 RTs; and Guayaquil Bank got 166 replies and 194 RTs in popularity.

The average user is between the ages of 18 and 24 years old, followed closely by those of 25–34 years and the ages of 34–49; what provides evidence that the use of these channels is more likely to be higher for people who are within the range of the economically active population of the country (Chart 54.2).

The use of the studied digital social media as a channel of products portfolio communication and in particular of micro-credit, is still inconclusive; so within the observed period, there were 41 publications related to these lending products, of which 38 correspond to multimedia such as photos and three videos; as well as also 22 links that allow the user to get more information. In terms of publications that are related to credit and microcredit, these reached 1279 likes, 112 shares, 111 comments in total. The hashtags application is limited, in total of six hashtags associated with microcredit in Facebook and 16 were recorded on Twitter.

54.4 Discussion and Conclusions

While the total number of financial institutions analysed in this study have a website, 80 % of them have an account on Facebook and 64 % on Twitter; this provides evidence that, on one hand, there is more inclination to use Facebook as an alternative to a permanent communication with their customers. However, on the other hand, it got our attention the low level of publications and interactivity that it has been generated within this network, as for example, the financial institution with the highest number of daily posts did around of seven.

The results reflect that financial institutions that offer the product of microcredit in Ecuador, have not considered the potential that digital social media offer as a marketing tool to access a greater number of users of their products and/or services, achieving it in both urban and rural areas throughout the country. Financial institutions should be consistent in the publications of advertisements and advertising campaigns, especially when positioning and promoting microcredit campaigns, the advertising strategies should be expanded in a daily basis and at any time, since digital social media allows, through mobile phone service, to users to remain informed at any time and in any place.

Financial institutions should guide the digital social media objectives: to attract new customers; to retain old customers; to enhance micro-credit portfolio; to launch micro-credit campaign in a more creative manner; to place the financial products and services in the consumer's mind; to make use of the benefits of digital social media of turning viral and have visibility.

References

Bolotaeva, V., & Cata, T. (2011). Marketing opportunities with social networks. *Journal of Internet Social Networking and Virtual Communities, 8.*

Cano, C. G., Esguerra, M. D., Garcia, N., Rueda, J. L., & Velasco, A. M. (2014, May 2). Retrieved from http://www.banrep.gov.co/sites/default/files/eventos/archivos/sem_357.pdf (Accessed August 5, 2015).

Cardoso, G. (2011). *De la Industria del Micro-crédito a las Finanzas Populares. Finanzas Populares como parte de un nuevo sistema económico, social y solidario.* Quito, Ecuador.

Charitonenko, S., & Campion, A. (2003). Expanding commercial microfinance in rural areas: Constraints and opportunities.

De Rojas Giménez, A. (2012). Cuando los públicos hablan: estrategias de publicidad y comunicación en el entorno 2.0. En U. Cuesta (Ed.), *Planificación estrategica y creatividad* (p. 257). Madrid: ESIC EDITORIAL.

Facebook. (2015). *Facebook business.* Retrieved from https://www.facebook.com/business/ (Accessed October 21, 2015).

Gutiérrez, B. (2005). Antecedentes del microcrédito. Lecciones del pasado para las experiencias actuales. *Revista de Economía Pública, Social y Cooperativa, 25–50.* A este no lo encuentro.

Ivatury, G., & Mas, I. (2008). The early experience with branchless banking. *CGAP, 16.*

Ledgerwood, J. (1999). *Manual de microfinanzas Una perspectiva institucional y financiera.* Washington D.C.: Banco Mundial.

Martínez Castillo, A. (2008). El microcrédito como instrumento para el alivio de la pobreza: Ventajas y limitaciones. *Cuadernos de desarrollo rural, 61.*

Martínez-Fernández, V. A., Sánchez-Amboage, E., Mahauad-Burneo, M. D., & Altamirano-Benítez, V. (2015). La gestión de los medios sociales en la dinamización de destinos. *Hologramática, 15.* Retrieved from www.hologramatica.com.ar (Accessed October 21, 2015).

Rhyne, E. (2001). *Mainstreaming microfinance: How lending to the poor began, grew, and came of age in Bolivia.* A Kumarian Press Book.

Ringel, M., Teevan, J., & Panovich, K. (2010). What do people ask their social networks, and why? A survey study of status message q&a. *CHI '10 Proceedings of the SIGCHI Conference on Human Factors in Computing Systems* (pp. 1739–1748). New York, USA: ACM.

Sánchez-Amboage, E., Rodríguez-Fernández, M., & Martínez-Fernández, V. A. (2014). La Promoción de los Establecimientos Termales de la Región Norte de Portugal en Facebook. In F. J. Herrero Gutiérrez, S. Toledano Buendía & A. Ardévol Avreu (Eds.), *La democracia no es un editorial. Patrones neoliberales en los medios de comunicación* (pp. 1–27). La Laguna, Tenerife: Sociedad Latina de Comunicación Social. Retrieved from http://www.revistalatinacs. org/14SLCS/2014_actas/096_Sanchez.pdf (Accessed April 3, 2016).

Valerio, G., Herrera, N., Herrera, D., & Rodriguez, M. (2014). En facebook el tamañao sí importa. Engagement y el impacto de la longitud delmensaje en las fanpages de las universidades mexicanas. *Revista digital universitaria, 11.*

Chapter 55
Evolution of the Semantic Web Towards the Intelligent Web: From Conceptualization to Personalization of Contents

Blanca Piñeiro Torres and Aurora García González

Abstract While Web 2.0 meant the raising of social media, the semantic dimension of Web 3.0 targets artificial intelligence, operating a method based on ontology to classify webpages, which allows users to find and understand information. Search engines and computers control the process, collecting knowledge from profiles and interactivity of users.

Keywords Artificial intelligence · Ontology · Conceptualization · Personalization · Semantic web · Syntactic web

55.1 Theoretical Framework

55.1.1 Static Web 1.0

The initial version of the website was the static and unidirectional Web 1.0, which was created to provide information about corporate organizations, news or specific content of diverse subjects. Due to the high costs of web publishing, updating was performed infrequently, so the aim of Web 1.0 consisted of a presential strategy on the Internet.

In the new concept of web 2.0 as an interactive online community (Cebrián Herreros 2008), tools like content syndication, messaging services, forums, weblogs, wikis and social networks permit open communication to users.

B. Piñeiro Torres (✉) · A. García González
University of Vigo (UVigo), Vigo, Spain
e-mail: blancapineiro@hotmail.com

A. García González
e-mail: auroragg@uvigo.es

© Springer International Publishing Switzerland 2017
F.C. Freire et al. (eds.), *Media and Metamedia Management*,
Advances in Intelligent Systems and Computing 503,
DOI 10.1007/978-3-319-46068-0_55

419

55.1.2   Syntactic Web 2.0

The origin of the social web 2.0 is a collaborative work on Internet written by Levine et al. (1999) to examine the impact of the Internet, in both organizations and consumers. Their communication achieved the support of more than a thousand cosigners, including businessmen and Internet users. To define how companies speak with customers through the Internet, these authors published the Cluetrain Manifesto in 2000, consisting of 95 theses.[1]

In 2001, the crisis of digital companies also contributes to the emergence of Web 2.0. However, it was not until 2005 when Tim O'Reilly, founder and CEO of O'Reilly Media, provides theoretical foundation to define the concept in his article "What is Web 2.0" (O'Reilly 2005), which explains the evolution happening in internet: web pages become bidirectional platforms, where the receiver mediates in the network of communication, with the ability to share content, comment and collaborate (Fig. 55.1).

O'Reilly and Battle reviewed the basics of web 2.0 and after five years they redefined collective intelligence, reinforcing the value of communities in the construction of the online media. The Internet collective intelligence depends on the management, understanding and responding to a massive amount of user-generated data in real time.

From a conceptual revision, O'Reilly and Battle announce a transition to the semantic, social, mobile web and virtual reality:

> Is it the semantic web? The sentient web? Is it the social web? The mobile web? Is it some form of virtual reality?

> It is all of those, and more (O'Reilly and Battle 2010).

55.1.3   Web 3.0

From a basis on syntaxis and connection between users, Web 2.0 evolves into the semantic web or web 3.0, where a deep understanding is allowed and the user experience is developed by obtaining more accurate results in content search, introducing new applications, such as geolocation or biometrics, and developing artificial intelligence technologies. Berners-Lee (1997) predicted that machines

---

[1]These conclusions were organized in 8 sections: markets as conversations (thesis 1–6), the hyperlink as subversion of the hierarchy (thesis 7), the connection between new markets and companies (thesis 8–13), the entry in the market organizations (thesis 14–25), the marketing and response organizations (thesis 26–40), the impact of the intranet in the control and structure of organizations (thesis 41–52), the connection between the market for Internet and corporate intranets (thesis 53–71) and new market expectations (thesis 72–95).

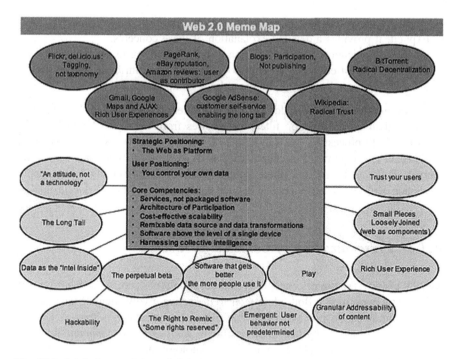

Fig. 55.1 Original map of the web 2.0. *Source* http://www.oreilly.com/

would be able to read web pages as easily as humans in his book 'Weaving the Web'.

Web 3.0 is based on artificial intelligence techniques using natural language in their searches, based on data mining, machine learning and attendance software agents to convert information into collective knowledge. O'Reilly and Battle (2010) explain that the inferential web 2.0 learning method is based on the explicit meaning of the processed data.

Thus, the semantic web uses language to find, share and integrate information. Among the advances in Web 3.0 are cross data or big data analysis, the use of 3D technology, based on the three-dimensional design, augmented reality, facial recognition in images, telematic applications for health monitoring or geolocation systems and triangulation used in mobile telephony.

Semantic web site keeps syntactic principles: decentralization, sharing, compatibility, accessibility and content contribution. But in addition, the semantic web ontology releases the notion of artificial intelligence.

According to Gruber (1993), the definition of ontology used in artificial intelligence application is "a formal, explicit specification of a shared conceptualization" (Gruber 1993: 199). In Gruber (1995), the author specifies the principles of ontology oriented towards knowledge sharing. In the conceptualization of ontology, the following aspects can be identified to understand it:

- Specification: description of characteristics and performance of concepts.
- Explicit: it provides a detailed explanation of these concepts.
- Formal: organization of terms and connections according to normalization.
- Concept: a set of concepts, described by their entities, attributes and relations.
- Sharing: common knowledge accepted by a group of users.

As Urrego-Giraldo and Giraldo Gómez (2005) explained, the ontology used to classify and index information filters out queries from users and facilitates their interaction with processors that can infer relevant information about their needs.

Metadata are required to organize and classify the information contained in the Internet. These texts are structured alphanumeric data representing bibliographic description of electronic resources and provide structured descriptions to locate objects.

55.1.4 Personalization of Contents for Users

Web 2.0 inspires the use of concepts as *emerecs* (emitters—receivers) or *prosumers* (producers—consumers). Both ideas refer to the active involvement of the receiver in the communication process, while that of *emerec* (*emetteur—être—recepteur*) was originally proposed by Jean Cloutier[2] in his '*Petit Traité de communication*' in 1973 and revised to analyze the effects of technology on the processes of human communication (Cloutier 2001).

Fernández Castrillo (2014) recognizes the increasing involvement of users, who called prosumers in content creation UGC or user generated content and the combined use of internet media, which contextualized storytelling in the transmedia, reasoning on the research conducted by Jenkins (2009) at the Massachusetts Institute of Technology, as a process in which the elements of fiction are systematically disseminated through multiple distribution channels in order to create a unique entertainment experience.

55.1.5 The Concept of Cibermedia

The proposal of *Cibermedia* is based on the research group Novos Medios (López García et al. 2003: 40) and refers to the online media using the journalistic

---

[2]Cloutier's own work has a hypertext structure, composed of six modules (Guylaine Martel 2014), which deal with: (1) the universal communication, (2) communication functions—to inform, educate, encourage and entertain-, (3) *emerec* description (its characteristics, influences, media, resources and technical applications used), (4) the media, languages and messages in the new technologies of information and communication (5) the size of the cyberworld (state, virtuality and interactivity) and (6) the law of the three thirds, favoring the balance of communication, due to a tandem between the *emerec*, the other participants and their status.

techniques, the multimedia language, interactivity and hypertext to update information on the Internet.

From this concept, several perspectives arise:

- The first point of view focusing on the media, with their own editorial, narrative and discursive structures.
- The perspective of the subject who broadcasts informative, commercial and entertainment communication or infomediation.
- The public, which provides a personalized and individualized communication in which "the user leaves the passive role that often plays in the mass media, becoming active (and interactive) subject" (Díaz Noci and Salaverría 2003, p. 41).
- The professional criteria, structure, writing and ethics relating to journalism.
- The hypertext, multimedia technology and interactivity.
- The update.

This definition of online media has recently been reviewed and concretized by Díaz Noci and others, in order to provide a methodology for studying digital media or online media: "A medium that uses a digital interactive online platform in the form of web site or as a mobile web app" (Codina et al. 2014).

Díaz Arias (2009) distinguishes between digital media (which refers to newspapers and magazines) and online media, since the source of the content is directly published on the Internet, the diffusion and the interaction occur in the context of Cyberspace (Díaz Noci et al. 2014).

While characterizing the online media, López García et al. (2003) detected the degree of dynamism of cibermedia, which determines four dimesions: hypertextuality, multimedia, interactivity and refresh rate. With the new version of semantics, this academic proposal adds the personalization of content, which considerates the user profile and preferences.

In content writing for online media, Pavlick (2005) explains that adaptation to the Internet has led to the emergence of contextualized journalism, which meets five conditions: communication modes; hipermediality; increasing audience participation; dynamic content; and customization.

According to Mayoral Sánchez and Edo Bolós (2014), terms such as hypertext, interactive or multimedia have been employed often to characterize cyberjournalism.

55.2 Methodology

The method applied for this research is qualitative content analysis, using as sources academic publications in electronic scientific journals, academic literature and Internet fonts.

55.3 Results

The purpose of the present research is to update the evolution of online media during the period between 2010 and 2015, which corresponds with the arrival and evolution of the semantic web.

Díaz Arias (2009) finds the distinction in the historical development of the online media, for which the author classifies three stages: during the first five years newspapers were the first generation of network information, with minimal input, a second phase in which newspapers, magazines, radio and television create their own Internet publication for professionals who use technical language and multimedia resources, and a third phase that gives prominence to the audiovisual and interactive content, media convergence and online media. The collaborative, dialogic and interactive nature of Web 2.0 opens the way to a new vision of journalism, as Díaz Arias recognizes: the right to freedom of expression and information, namely the right to seek, receive and disseminate ideas, opinions and information. But in the media society, the individual is a consumer of the information. Therefore, access to relevant public information allows him to become a citizen (Díaz Arias 2009: 6).

The web 2.0 users can be editors, customizing their access to information, and community broadcasters, assessing, discussing and sharing content published by consulting sources. In addition, they can create their own content on social networks and blogs.

Many theorists see citizen journalism as an opportunity on global integration, rather than a threat. According to Rheingold (2002), this collective intelligence finds resources in social networks and aims to produce user-generated content (UGC).

The growing influence of social networks has been undertaken by public and private agencies. The user-generated content have recently been integrated by institutions and considered by the political parties in their campaigns, since they consider social participation through networks such as Facebook, Twitter or Youtube to prepare their agendas and electoral programs.[3]

At this level, it is required to introduce the category of digital native cibermedia classified by Salaverría and Negredo (2013).

If the advent of computers and internet represented a paradigm shift with the transition from the mainstream media to online media, the expansion of mobile terminals and tablets also involves changes in the design, editing and digital content consumption, which adapt to portability and multiple screens, according to Aguado and Castellet (2013).

In the same way that the online media had evolved from adaptation of content to the creation of multimedia, hypertext and interactive online creation, the evolution

---

[3]The main political parties that run for the general elections on 20 December 2015 enhanced the participation of their followers in social networks for making electoral programs and electoral monitoring their activity.

of mobile devices and tablets has experienced two stages: at first, they only offered text messaging services (SMS) or multimedia (MMS), email and web browsing, but similar content to those offered in the web; the launch of the iPhone in 2007 and the advent of digital tablets led to the development of specific content more dynamic and adaptive design for ubiquitous touch devices (responsive design).

Meso Ayerdi et al. (2014) speak of mobile as the fourth screen and conclude that this new channel of communication poses new challenges for journalism to increase competitiveness: the development of narrative forms and adapted genres and effective formulas for participation of active and prosumer audiences. However, mobile applications favor the consolidation of brands, a new form of advertising revenue and user loyalty. Thus, they affirm that the future of online media goes through a strategic approach to multiplatform content.

55.4 Discussion and Conclusions

55.4.1 From Semantic to Intelligent Web

Technological advances in infrastructure and applications allow us to mention a new version of the web, known as intelligent web. The potential uses and applications of this new website are in an experimental phase.

55.4.2 Customization in Cibermedia

The development of applications of artificial intelligence and advances in mobile media and tablets promote the diversity of formats and customization of content in online media or cibermedia, which can improve usability, considering the user-generated content.

55.4.3 Coexistence of Diverse Web Generations

In the current Internet ecosystem, there are four coexistent generations of web (static, syntactic, semantic and intelligent), whose phases of development, characteristics and applications are specified in the Table 55.1.

Table 55.1 Evolution of the web to the artificial intelligence (1990–2015)

Version	Web type	Characteristics	Applications
Web 1.0 1990– 2000	Static	Static: The websites are unidirectional, with static elements and little update	Search engines Websites Databases Email Cibermedia: Digital edition
Web 2.0 2000– 2010	Syntactic	Dynamic: web pages create dynamic content extracting information from one or more databases Social: users contribute to content development. Changes in editing and publishing information procedures Example: Wikipedia, Youtube or Facebook	RSS Weblog Wikis Newsletters Newsgroups Forum SMS/MMS Social networks Cibermedia: Digital publications
Web 3.0 2005– 2020	Semantic	Based on the representation of meaning and connection of knowledge Add to semantic web metadata information by ontologies Collaborative creation Extends the interoperability of systems	Onthology Thesaur and taxonomy Semantic Search Semantic Web Semantic Weblog
Web 4.0 2015– 2030	Intelligent	Development of artificial intelligence Algorithms for natural language processing Semantic development communities Web applications can identify resources and manage them in the context of temporary or permanent sessions	Decentralized communities Artificial intelligence Intellectual property

Source Prepared by the authors

References

Aguado, J. M., & Castellet, A. (2013). Periodismo móvil e información ubicua. In J. M. Aguado, C. Feijóo & I. J. Martínez (Coords.), *La comunicación móvil. Hacia un nuevo ecosistema digital* (pp. 187–218). Barcelona: Gedisa.

Berners-Lee, T. (1997). *Realising the full potential of the web.* Retrieved October 28, 2015 from http://www.w3.org/1998/02/Potential.html

Cebrián Herreros, M. (2008). La Web 2.0 como red social de comunicación e información. *Estudios sobre el mensaje periodístico, 14,* 345–361.

Cloutier, J. (2001). *Petit traité de communication. EMEREC à l'heure des technologies numériques»*, *Communication* (online), Vol. 22/1|2003, published 8 of January of 2014. Retrieved October 28, 2015 from http://communication.revues.org/4816

Codina, L., Pedraza, R., Díaz Noci, J., Rodríguez-Martínez, R., Pérez-Montoro, M., & Cavaller-Reyes, V. (2014). Sistema Articulado de Análisis de Cibermedios (SAAC): Una propuesta sobre el qué y el cómo para estudiar medios de comunicación digitales. *Hipertext.net* (online), 2014, no 12. Retrieved October 28, 2015 from http://raco.cat/index.php/Hipertext/article/view/275560/364530.

Díaz Arias, R. (2009). *Avances y desafíos en los cibermedios.* In I Congreso Modelos Emergentes de Comunicación: de lo análogo a lo digital. Universidad Autónoma del Caribe. Barranquilla (Colombia) (11/5/2009).

Díaz Noci, J., & Salaverría, R. (Coord.) (2003). *Manual de redacción ciberperiodística.* Barcelona: Ariel.

Fernández Castrillo, C. (2014). Prácticas transmedia en la era del prosumidor: Hacia una definición del Contenido Generado por el Usuario (CGU). *Cuadernos de Información y Comunicación, 19,* 53–67.

Gruber, T. R. (1993). A translation approach to portable Ontologies. *Knowledge Acquisition, 5*(2), 199–220.

Gruber, T. R. (1995). Toward principles for the design of ontologies used for knowledge sharing. *International Journal of Human-Computer Studies, 43*(4–5, November 1995), 907–928.

Jenkins (2009). *Confronting the Challenges of Participatory Culture Media Education for the 21st Century.*The MIT Press Cambridge, Massachusetts.

Levine, R., Locke, C., Searls, D., & Weinberger, D. (1999). *Manifiesto cluetrain.* Retrieved September 13, 2015 from http://personal.us.es/mbmarquez/textos/cluetrain.pdf

López García, X., Limia Fernández, M., Isasi Varela, A., Pereira Fariña, X., Gago Mariño, M. & Calvo Diéguez, R. (2003). Tipología de los cibermedios. In R. Salaverría (Coord.), *Cibermedios. El impacto de internet en los medios de comunicación en España* (pp. 39–81). Sevilla: Comunicación Social Ediciones y Publicaciones.

Meso Ayerdi, K., Larrondo Ureta, A., Peña Fernández, S., & Rivero Santamarina, D. (2014). Audiencias activas en el ecosistema móvil. Análisis de las opciones de interacción de los usuarios en los cibermedios españoles a través de la web, los teléfonos móviles y las tabletas. *Hipertext.net [online], 2014. Núm. 12.* http://raco.cat/index.php/Hipertext/article/view/274309/364489

Mayoral Sánchez, J., & Edo Bolós, C. (2014). Evolución de la producción audiovisual en cinco cibermedios españoles. Fonseca, *Journal of Communication, 9*(July-December of 2014), 233–262. ISSN: 2172-9077.

O'Reilly, T. (2005). *What is web 2.0. Design patterns and business models for the next generation of software.* Retrieved October 13, 2015 from http://www.oreilly.com/pub/a/web2/archive/what-is-web-20.html

O'Reilly, T., & Battle, J. (2010). *Web squared: Web 2.0 five years on.* Retrieved October 13, 2015 from http://www.web2summit.com/web2009/public/schedule/detail/10194

Pavlick, J. V. (2005). *El periodismo y los nuevos medios de comunicación.* Barcelona: Paidós Comunicación.

Rheingold, H. (2002). *Multitudes inteligentes: la próxima revolución social.* Barcelona: Gedisa.

Urrego-Giraldo, G., & Giraldo Gómez, G. (2005). La estructura de servicios y de objetos del dominio: Una aproximación al concepto de ontología. *Revista Tecno Lógicas*, 46–67. December of 2005.

Salaverría, R., & Negredo, S. (2013) Caracterización de los cibermedios nativos digitales. In M. A. Cabrera (Coord.), *Evolución de los cibermedios. De la convergencia digital a la distribución multiplataforma* (pp. 175–180). Madrid: Fragua.

Chapter 56
Social Networks in 20 Minutos, the One Survivor of Free Distribution Press in Spain

Ana Bellón Rodríguez and José Sixto García

Abstract Free press and social networks have become two important phenomena in the last two decades in the media landscape. Since 1995, when Metro started to be distributed, the public has the possibility of receiving a free information product entirely financed by advertising. With the rise of social networks and their incorporation to media, communication flows become fully bidirectional. Taking *20 Minutos* as a case study—the one survivor of the free press in Spain—it is analyzed how to influence citizens' contributions in the configuration of news. For this, we explored the paper and the web, and the presence and use in 2015 of social networks. Apart from providing a snapshot of the presence of the mark on social media, we summarized the effects that this new media landscape poses to media and audiences.

Keywords Social networks · Free press · Citizen journalism · Social interaction · Democracy

56.1 Theoretical Framework

56.1.1 The Social Networks

We can already assert that social networks are an essential tool of communication for all kinds of organizations. We are in a time in which communication flows have become totally bidirectional and the user has the chance of confirming the discourse

A. Bellón Rodríguez (✉)
University of Santiago de Compostela (USC), Santiago de Compostela, Spain
e-mail: ana.bellon@usc.es; ana.bellon@csic.es

A. Bellón Rodríguez
Spanish National Research Council (CSIC), Santiago de Compostela, Spain

J. Sixto García
Instituto de Medios Sociales, Santiago de Compostela, Spain
e-mail: direccion@institutomedios.com

© Springer International Publishing Switzerland 2017   429
F.C. Freire et al. (eds.), *Media and Metamedia Management*,
Advances in Intelligent Systems and Computing 503,
DOI 10.1007/978-3-319-46068-0_56

reception and understanding. The purely journalistic piece of news completes the perspective of citizen journalism and the receivers' opinions. The contents which the public provides hasn't got journalistic characteristics, but they add value. How citizen contributions influence journalistic schedule configuration?

Social networks have become a communication platform that enables users to strengthen their relations. A good online corporative reputation always has to face social rumours, so that products and brands use to wake up every morning with online information which can hurt them so much (Rodríguez 2011) if they do not know or are not capable of managing it. The impact of social media about informative enterprises is dual and it acquires particular hints regarding other kind of organizations: (1) Informative enterprises do not stop being an organization which offers a product to a public and it is logic that they apply the same marketing mechanisms as other kind of enterprises (Sixto 2012). The informative enterprises should use social networks as a store window of their informative product and as an interaction channel; and (2) the product which is offered by informative enterprises is not static, so within it they are participating the media, the sources, the advertising actions and the stakeholders.

This process of elaboration and distribution of information by citizens instead of being distributed by communication enterprises is known as citizen journalism, one which thanks to the new social media, gains a relevance which has never been achieved until now. Marhuenda and Nicolás (2012) assert that brands presence in social networks has meant a revolutionary change in marketing. In the following Decalogue we summarize the main effects that this new media setting means for media and public: (1) The new social environment allows the public to comment and interact about the news at the same time in which they are produced. The public has become social; (2) different screens and devices are used in order to receive the same content; (3) receivers become content distributors and diffusers and the basic contents are nourished with various individual or group contributions that enrich them; (4) the most collaborative receivers create contents too. The media is over all than ever, although only journalists continue to maintain the legal authority to inform rigorously; (5) media is more democratic than ever. There has been an addition to news search methods that refers to social contributions through which users contribute to enrich the media and its news; (6) neither everything is worth, nor everything is worthless. The journalistic elaboration techniques must be more thorough than ever in order to discern between rumour and information; (7) mass media has more and more the need to stimulate user's participation. Prestige falls if there is not social interaction, in such a way that audience measuring parameters are modified; (8) information is extremely contemporary. All of us have the capacity of influencing contents at the same instant in which we consider that something has changed; (9) crisis can be generated in a matter of seconds and with a capacity of destruction over the brand and its image which can be brutal; and (10) social networks are also communication media. Thanks to the option of sharing contents of traditional means they are the access point of many users to the official mean. Their use is increasing the traffic web and the SEO.

56.1.2 The Free Press Phenomenon and the 20 Minutos Case

The free distribution transformed in the last decades press market in the whole world, so this publication model broke with the established dynamics in the sector: the reader did not have to pay for the copy and nor it did not have to move to a place in order to get it. Everything began with *Metro* launch in Stockholm in 1995 and it continued with the circulation entrance of other heads with similar characteristics (Santos 2008). Their implementation and success were favoured by the rising of advertising inversion and the appearance of low-cost printing systems (Nieto and Iglesias 1993). Free press is the result product of the publishing activity of an enterprise which delivers it without economical compensation by the receiver side and whose only and mainly income comes from the advertisements (Nieto 1984). It is part of the regular publication sector and it offers informative products to the general public (Rojo 2008). These characteristics explain that the reader profile in the press model is also different concerning age and gender criterion. The age average of the paying reader is around 48 years old in front of the 27 years old of the free ones, publication in which more than the half of its public are women (Fernández 2003).

In Spain we had to wait until the year 2000 for starting the distribution of free diaries of general information in the main Spanish cities promoted by both great groups of specialized communication in this type of products and by groups linked with paying press. During the golden years (2000–2008), *20 Minutos, Metro, Qué!* and *ADN* came to spread 3.6 millions of copies. Another illustrative data from this period of splendour for the sector are that in 2004, according to EGM, in Spain three of each four persons read free press. Nevertheless, in November of 2015 from the four ones only one remains in circulation: *20 Minutos*. The sector, therefore, had lived in a little more than a decade both an exponential growth and its restructuring (Bellón 2013).

56.2 Methodology

This paper's aim is to determine, taking *20 Minutos* as the object of study, what use are free paper diaries making of the new social environments and how they influence the ultimate content configuration. Beforehand it seems an apparatus in which everything can be free, but is it actually like this? What effect these circumstances have in the media scene?

The methodology used is qualitative, so we have resorted to the content analysis in order to going into detail about the use which *20 Minutos* make of its spaces in social networks. We have also carried out the corresponding bibliographic revision about social networks and free press. We have carried out a tracking and a monitoring of the free ride diary's content in Europe from 2008 to 2014, although for the measuring of social virality we have focused the study on the months of September, October and the first week of November of 2015. It is about an exploratory-descriptive study in which attention is paid, on one hand, on the structure and bet for the citizen press and the interactivity in *20 Minutos* paper diary and its online edition and, on the other hand, on the *20 Minutos* brand presence within social networks. In order to do this, we have carried out a search of *20 Minutos* official accounts and, once they have been identified, we have analysed the cover message in the same and we have gathered the audience data.

56.3 Results

The *20 Minutos* paper edition is presented as "the social media" and it consists of the following sections: locality of edition; current issues; sports; Area 20. When dealing with citizen press issue in this diary it convenes to highlight Area 20, a space where voice is given to the readers. The online edition, 20Minutos.es, is divided in seven main sections: cover; national; international; economy; your city; sports; technology; arts. It is important to highlight that *20 Minutos* measures the social effect of the news through ECO. The value is calculated though the internal parameters, typical of *20Minutos.es* community activity around a piece of news, and of other external ones, such as their effect upon social networks. The participation data are obtained in a periodic manner and they are calculated while the piece of news is active, while the measuring ranks are modified according to time zones, in a way that none of the values are never static elements. From the upper part of a piece of news it can be observed the maximum point of social activity of this information and also to value it, to twitter it, click on "I like it" on Facebook, sharing the information, sending it by e-mail, etc.

20 Minutos concedes very much importance to blogs. If you want to participate with your texts, photos, videos, sound recordings, sketches, you do not have more than writing to zona20@20 minutos.es. Moreover, since 2006 the group convenes annually The Awards 20 Blogs with three categories: (1) Best Blog, (2) Best Blog for each category, (3) Best blog for voting. Inside this 2.0 journalistic scene where the user is situated as the center of all communicative actions, the bet of *20 Minutos* brand for social networks is decisive.

The group is present is seven horizontal or general-interest social networks, to which we can access through its respective buttons inside the *20 Minutos* main page and which has the following monitoring rates in the 8th of November of 2015: (a) Official page on Twitter (@20 m), where it is presented as "the social and citizen media. Information, analysis and personal touch with *20 Minutos* readers the 24 h of the day". In this social network it has 971,872 followers and it follows 49,031 people, therefore its popularity rate is of 0.05 and, so, it is an influential profile (on Twitter it is considered as relevant or influential those profiles with a popularity rate of < 0.5, result of dividing the number of persons that it follows—following— between those who are following this profile -followers-. Since its start-up, in April

2009, 127,350 tweets have been published. (b) Page on Facebook, where it is presented as "the social media. Follow at every minute the current issues in www. 20minutos.es". It was launched in 2008 and in November 2015 it has 623,179 followers. It is assessed with four points over five within a total of 2318 qualifications. (c) Corporative page on Google+, with 560,788 followers, 131,449,173 visits. In this account *20 Minutos* is presented in this way: "Our aim is to inform, to entertain and… to take a risk. We were first in counting on our readers for almost everything, and the touch with them inspires us. In addition we are innovative and each certain time we like to reinvent ourselves". The strategy of giving relevance to this Google page favours positively its SEO positioning. (d) Account on Instagram (@20m), merely face-to-face, with just only 28 publications. Even so, it has 2974 followers and it follows 122 users. (e) Enterprise page on LinkedIn with 1955 followers. (f) YouTube channel, where they are uploaded, between others, videos about the digital encounters. This space was created on August 2009, and it has 4281 subscribers and a total of almost 2,400,000 displays. (g) Pinterest, where it makes itself known as "we are a communication media and our aim is to inform, to entertain and… to take a risk. The touch with the readers inspires us". It has 2600 followers to which they have presented 31 panels about current issues of great impact. They only follow 84 users, something that, the same as it occurs on Twitter's case, increases its popularity rate. (h) Storify, where it has 658 followers and it is presented as "the social and citizen media. Information, analysis and contact with the readers during the 24 h of the day". Again, the group bets for the strategy of people following it without being followed and it decreases its monitoring rate to just only 40 users. As it occurs in the other networks, the other social spaces of the group are connected or advertised, in such a way that it is expected that users interact with the brand through its more habitual network of use, that is to say, it is not the reader the one who has to do the effort of following the media, otherwise the media appears as one more in the habitual spaces of users' social interaction. (i) *20 Minutos* was present on Tuenti, where it came to have more than 2000 followers and where it did use of this social network in order to reach a younger public section. Now the page appears as empty.

All these bets fostered from the *20 Minutos* group so that paper readers and web users could interact are joined together with the facilities to access to the information where and whenever they want. The group is presented as a multiplatform and from 20minutos.es the application for iPhone, iPad, Android and Nokia can be downloaded. In this sense, the group summarize its philosophy of action in three pillars: (1) *20 Minutos* in your day-to-day, (2) You make us improve, (3) Thought to all. We are in front of a communication proposal which at every moment bets for citizen collaboration as an element that provides added value to the offered contents and that, in effect, it offers a large number of social accessibility mechanisms with the aim of favouring this information flow (all with all) which facilitate information without economic cost.

56.4 Debate and Conclusions

The change which has meant the emergence of social media in the enterprise relationships with its public is one of the most relevant phenomena from the communicative point of view of the last decade. Enterprises which elaborate informative products are not alienated from this phenomenon. We have to take into account that they must apply the same marketing strategies than any other business organization. But to this condition it is also added the distinctive feature about the fact that the offered product is clearly influenced from a social point of view. However, the new social media have provided these public receivers of a great ability of manoeuvre over these products which themselves end up in consuming them.

The diagram of fast display contents fosters users´ interactivity and it ends up in generating a great reliance in relation to the creation of contents by users. Citizen journalism takes on great relevance in this type of publications, so a part of published news get access through social mechanisms. It is, precisely, *20 Minutos* the one which concedes more importance to social participation. This is one of the unquestionable motives that makes it the only survival of the initiatives that have been arising from 1995. Social media represent one of the power sources for this kind of publications.

References

Asociación para la Investigación de Medios de Comunicación (2000–2015). *Estudio General de Medios*.

Bellón, A. (2013). *20 minutos: tres productos y un modelo de redacción*. Faculty of Communication Sciences. University of Santiago de Compostela.

Fernández, T. (2003). Diarios gratuitos: Nuevos diarios que salen del túnel y buscan un lugar en el sol. *Innovaciones en periódicos. Informe Mundial*, 54–63.

Marhuenda, C., & Nicolás, M. A. (2012). *Herramientas para la medición de los social media*. In M. A. Nicolás & M. M. Grandío (Coords.), *Estrategias de Comunicación en Redes Sociales. Usuarios, aplicaciones y contenidos* (pp. 31–50).

Nieto, A. (1984). *La prensa gratuita*. Navarra: EUNSA.

Nieto, A. & Iglesias, F. (1993). *Empresa informativa*. Barceona: Ariel Comunicación.

Rodríguez, O. (2011). *Community manager*. Madrid: Anaya.

Rojo, P. A. (2008). *Modelos de negocio y consumo de prensa en el contexto digital*. Murcia: University of Murcia Servicio de Publicaciones.

Santos, M. T. (2008). *El auge de la prensa gratuita en España*. Bilbao: Editorial Service of the Universidad of El País Vasco.

Sixto, J. (2012). *Las redes sociales como estrategia de marketing*. Lisboa: Media XXI.

Chapter 57
Spanish TV Series on Twitter: What Social Media Audiences Say

Verónica Crespo-Pereira and Óscar Juanatey-Boga

Abstract Online social networking users reflect opinions and attitudes towards the topic discussed. This study aims to analyze Spanish social audiences' habits and their perceptions and attitudes towards national TV series in Twitter. The results show the influence of American and British audiovisual products in social audiences' speech. Evidence suggest that national audiovisual products are compared to foreign standards due to their good reputation and acceptance. In this context, traditional Spanish TV fiction is criticised by some parts of the social audience because of its distance from international fictional productions' standards and some narrative practices considered negatively. In contrast, another part of the social audience reflects their acceptance of a new wave of producing national TV series.

Keywords Social media audience · Spanish TV series · Audiovisual consumption · Twitter

57.1 Theory

The use of social networking sites and second screens to talk about the TV content watched are transforming TV audiences' habits. This combination has given rise to a new way for audiences to interact and exchange views, not only within the circle of friends but also with any potential viewers, in real time. These audiences, known as social audience (Deltell 2014), have the power to affect the social repercussion of a TV production (González-Neira and Quintas-Froufe 2014).

The introduction of second screens while watching TV is highly widespread. In Spain, one in two people uses them while viewing television. This practice is

V. Crespo-Pereira (✉)
University of Vigo (UVigo), Vigo, Spain
e-mail: veronicacrespopereira@gmail.com

Ó. Juanatey-Boga
University of a Coruña (UDC), Corunna, Spain
e-mail: oscarjb@udc.es

© Springer International Publishing Switzerland 2017
F.C. Freire et al. (eds.), *Media and Metamedia Management*,
Advances in Intelligent Systems and Computing 503,
DOI 10.1007/978-3-319-46068-0_57

435

particularly high among young people. In fact, 86 % of people from 14 to 24 and 76 % of young adults from 25 to 34 years employ a second device (Zenith 2015). Those who use them access social networking sites as the second most common activity (Zenith 2015).

In this context, the social networking site Twitter has rapidly become social audiences main SNS to publish opinions and generate debate about TV content in real time. The data supports this idea. In 2013, 32 % of tweets published during prime time were about a TV program and 1.5 million people tweeted about television content. One year earlier there were only 600.000 people (Tuitele 2013). The use of this microblogging social network is also growing in importance among TV channels since they screen their official hashtag, a textual marker to facilitate the tracking of a conversation about a topic (Highfield et al. 2013), as a strategy to gain attention (Neira 2015).

Although it is necessary to point out that social audience does not match the full spectrum of TV viewers, yet Twitter may become a useful tool for channels given its capability of encouraging public discussions and monitoring audiences' engagement, feelings (Tuitele 2013) trends and habits (Nielsen 2015) and thus facilitating alternative audience measures, in terms of quantitative and qualitative data (González-Neira and Quintas-Froufe 2015).

57.2 Methodology

Given the possibility to monitor audiences' attitudes and behavior on audiovisual productions in social media, it is our goal to analyze Twitter social dialogue. Two Spanish science fiction TV series have been monitored manually: *El Ministerio del Tiempo* (TVE) and *Refugiados* (Atresmedia Group). The TV series analyzed were chosen for being an uncommon genre (science fiction) in Spanish TV series and their potential attractiveness for young audiences, main users of social networking sites. In order to establish the results, this work has been inspired by two previous studies about the TV series mentioned above (Crespo-Pereira and García-Soidán 2015; Crespo-Pereira et al. 2015). In the first part of the study, our aim is to highlight audiences' behavior and attitude towards the mentioned TV fictional contents. More than 2800 tweets have been analyzed the day of their respective premières under the hashtag #MdT1 (2015, February, 24) and #Refugiados (2015, May, 7) (for quantitative analysis go to mentioned papers). Hashtags were employed in the study for providing a space to share opinions among users (Highfield et al. 2013). Tweets were classified in three groups: before, during and after the broadcast and in three feelings: positive, negative and neutral. Positive and negative messages represent social audiences' interests when seeing and analyzing TV series. Monitoring the tweets allowed us to know how speeches evolve and what engages the audience the most. Thus each feeling category has its own classification as a result of social audiences' treated topics. Audiences' talks were about all kind of aspects of a TV series: professionals involved (showrunners, actors), script, characters, direction, direction of photography, production design, visual effects, music, sound effects, costume and makeup. In the second part of the study, authors will discuss audiences' patterns and opinions when talking about Spanish TV productions in general.

57.3 Results

It's common for both audiences (*El Ministerio del Tiempo* and *Refugiados*) to make positive general judgments based on "the good impression" before the content is aired. A new release on TV, especially if it is a prime time showing, is carefully treated by the media that produces it. The massive publicity reached on news and different programs attracts people's interest in seeing the première as the analyses confirm. The expectancy and the desire to see the TV series are the main positive topics treated previously to a première. Publicity influences audiences' idea about the quality of a content in a positive manner. However, the study shows social audiences' caution when talking about their expectations because good publicity is not a guarantee of a good content as they mention.

Negative messages in this period, low in number, are not related to the TV series themselves but linked to TV channels' prime time practices such as the amount of ads broadcasted, the starting and finishing hour of prime time slot, the scheduling clashes and web site emission technical problems among others.

Before the broadcast, consumers' preferences are revealed. Audiences manifest alternative choices to traditional live TV such as online web emission. Audiences who mention their preference when seeing the content in a specific language do it in favor of the original version whenever possible. That is the case with *Refugiados* since it was shot in English as a result of being a coproduction between the BBC and the Atresmedia Group.

During the broadcast of both TV series, social audiences give their opinion on all kind of aspects of a TV series as mentioned before. The topics analyzed are the product of the elements treated during social dialogue on Twitter. While on *El Ministerio del Tiempo* positive tweets are predominant ones, on *Refugiados* negative messages prevail over positive ones. The messages analyzed are those with the biggest representation on each TV series. Although the initial controversy at comparing *El Ministerio del Tiempo* 's positive tweets and *Regufiados*' negative messages, there are interesting behavior patterns to underline.

El Ministerio del Tiempo 's script almost gets the majority of positive comments (42 %). The "I like it" is the second topic discussed (24.95 %), followed by the liking of some of the characters (15.8 %). These characters seem to have a special connection with social audience possibly because of a combination of their link to Spanish pop and traditional culture and the humor employed.

On *Refugiados*, the first episode broadcast the "I don't like it" or/and its low rhythm category attracts the first position of negative tweets (28.3 %). The second one is the script (17.2 %), followed by the dubbing (12.5 %). On the second episode aired right after the first one,[1] the "I don't like it" or/and its slow rhythm category has a 35.2 % of negative comments. The bad review about the script has another 35.2 % of negative messages. The wrongful treatment of Spanish customs and traditions on the TV series is placed in third position (12.6 %).

Whether the predominant part of the messages are positive or negative, the script and the statement about the liking or not of the content are the most treated topics in both TV series. In the third position of the *El Ministerio del Tiempo* and on *Refugiados'* second episode, Spanish culture seems to play certain role in the social discussion.

After the broadcast, both TV series reach a higher number of positive messages over negative ones, even despite the number of *Refugiados'* bad reviews. In this case, the harsh speech found on *Refugiados* during the broascast is clearly softened. Social audiences make a general valuation of the TV series and some also congratulate the professionals for their work. Again, the script (whether in a positive way or negative one) and "I like it" or "I don't like it" are the main categories in this period.

57.4 Discussion and Conclusions

The opinions of social audiences enlighten how TV products are thought and thus consumed. In our study we have found some underlying patterns on both Twitter TV series' audiences concerning the topic: Spanish TV series.

Part of the social audience is proud of the products made in Spain and argue that Spanish TV series are enjoying a Golden Age. To base such opinion people give examples of some recent Atresmedia Group's TV series such as *Vis a Vis*, *Allí abajo*, *Sin identidad*, and TVE's *Isabel*, *Cuéntame* or *Águila Roja*. Some products are mentioned by both social audiences: *Bajo sospecha* (Atresmedia Group) and *El Ministerio del Tiempo* (TVE). Only TVE (public broadcaster) and Atresmedia Group's TV series are mentioned for its perceived good quality. Mediaset España, the other biggest private broadcaster in Spain, would not be on that list as far as we are concerned.

In contrast to this position, prejudices against national TV fiction productions can be found in tweets analyzed in both TV series. The label "Spanish TV series" have a pejorative meaning. Evidence suggests that our TV series are considered to be of a poor quality according to some parts of social audience, thus any element that improves their opinion takes audience by surprise as some tweeters publically manifest. On the other hand, national fictional productions seem to be linked to old-fashioned audiovisual practices. The appearance of females topless in both

---

[1] *Refugiados'* première is characterized for broadcasting two episodes in a row. Thus our analysis incorporates the data of the first and the second episode.

productions is highly criticized by both audiences since it is believed to be a embarrassing symbol of national productions. Some of the critics even make the connection with the Spanish "destape"[2] period.

International products and foreign channels are alluded to by social audience when talking about the TV series analyzed. The results manifest that international TV series standards are used as the measure to compare our productions. American and British TV content is the only foreign content mentioned in both studies. The examples given refer to their most popular TV series worldwide. Their successful products are the content which national projects are compared to. The comparisons highlight some aspects of our audiovisual products the public would like to be different. The incorporation of some foreign standards such as the length of the content or the treatment digital visual effects or photography would be considered positively by social audience who are used to seeing international products.

On the other hand, narrative similarities between foreign fictional content and national content is judged differently. The idea that Spanish products incorporates elements based on international films and TV series' successful ones is publically manifested and does not enjoy a good reputation. Narrative similarities (in plots for example) are mainly considered negatively and even a "copy" (a term used by some tweeters). In addition, visual similarities pointed out by the audience (such as in the opening credits) may be either thought of as a lack of originality in our productions or as the introduction of the innovations required to produce modern and attractive TV series.

Acknowledgments This study has been possible thanks to Xunta de Galicia's predoctoral program Plan I2C (2011-2015). Authors appreciate the contribution to this work done by Valentín A. Martínez Fernández (UDC) and Pilar García Soidán (Uvigo).

References

Crespo-Pereira, V., & García-Soidán, P. (2015). Análisis de la audiencia social de Twitter. Caso de estudio: Refugiados. In Ó. Díaz-Fouces & P. García-Soidán (Coords.), *Redes y retos. Estudios sobre comunicación en la era digital* (pp. 45–66). Barcelona: Ediciones Octaedro.

Crespo-Pereira, V., Martínez-Fernández, V.-A., & Juanatey-Boga, Ó. (2015). Análisis de la audiencia social de El Ministerio del Tiempo. In J. Rúas-Araujo, A. Silva-Rodríguez, & I. Puentes-Rivera (Eds.), *Dos medios aos metamedios de comunicación [CD]* (pp. 945–953). Pontevedra: Xescom.

Deltell, L. (2014). Audiencia social versus audiencia creativa: caso de estudio Twitter. *Estudios sobre el Mensaje Periodístico, 20*(1), 33–47.

González-Neira, A., & Quintas-Froufe, N. (2014). *Twitter, la televisión y la audiencia social. ¿Por qué triunfa un espacio en la audiencia social?* Retrieved January 21, 2016, from http://ruc.udc.es/handle/2183/15401

---

[2]The "destape" period is the name given to a cinematographic phenomenon in which naked bodies, mainly women, started to appear in Spanish films in a context of a political and cultural change during Franco's dictatorship.

González-Neira, A., & Quintas-Froufe, N. (2015). Revisión del concepto de televisión social y sus audiencias. In N. Quintas Froufe & A. González Neira (Coords.), *La participación de la audiencia en la televisión: de la audiencia activa a la social* (pp. 13–26). Madrid: AIMC. Retrieved January 04, 2016, from http://www.aimc.es/-Participacion-Audiencia-en-TV-.html

Highfield, T., Harrington, S., & Bruns, A. (2013). Twitter as a technology for audiencing and fandom. *Information, Communication & Society, 13*(3), 315–339.

Neira, E. (2015). Audiencia social: ¿consiguen las redes sociales que veamos más televisión?. In N. Quintas Froufe & A. González Neira (Coords.), *La participación de la audiencia en la televisión: de la audiencia activa a la social* (pp. 47–59). Madrid: AIMC. Retrieved January 04, 2016, from http://www.aimc.es/-Participacion-Audiencia-en-TV-.html

Nielsen. (2015). *Nielsen social*. Retrieved January 21, 2016, from http://www.nielsensocial.com/

Tuitele. (2013). *Un año de Televisión Social en España. Septiembre 2013-agosto 2014*. Retrieved September 17, 2015, from informes.tuitele.tv/emailing/Tuitele_1_año_tv_social_en_España.pdf

Zenith, E. (2015). *Estudio Multipantalla Zenith: 'Del punto de cruz al multipantalla: ¿tejes o enriqueces?*. Retrieved September 16, 2015, from http://es.slideshare.net/ZenithES/estudio-multipantalla-zenith

Chapter 58
Twitter as a Communication Tool for Local Administrations: The Cases of São Paulo and Madrid Municipalities

Flávia Gomes-Franco e Silva

Abstract The development of interactive digital tools has prompted a raft of changes in the traditional model of communication, replacing vertical and unidirectional discourse for a shared content construction on the Internet. Nowadays, networks and social media related to the web 2.0 promote closeness between all parties involved in the communication process. Local administrations are also affected by the new interactive channels, having to adapt themselves to the demands of citizen-users. The present research mainly aims to compare, using an approximate analysis, the use of Twitter by the City Halls of two emblematic cities: São Paulo (Brazil) and Madrid (Spain). A web content analysis on a sample of tweets posted on the official Twitter account of the two City Halls has been conducted. The study reveals an under-use of communication and interactive resources at this microblogging platform, with accounts transformed in mere shop windows especially destined to generate traffic to the websites in each municipality.

Keywords Internet · Interactivity · Social media · Twitter · Municipalities

58.1 Introduction

The development of information and communications technology (ICT) has provoked structural changes in the traditional communication model. According to Internet World Stats, the Internet penetration rate increases year after year, having registered in 2015 73.5 % of Internet users in Europe, 87.9 % in North America and 53.9 % in Latin America and the Caribbean. In the specific case of Brazil and Spain a penetration rate of 54.2 and 74.8 %, respectively, is observed. At this point, people become users, exercising citizenship both in an offline and online way.

F. Gomes-Franco e Silva (✉)
Rey Juan Carlos University (URJC), Móstoles, Spain
e-mail: flavia.gomes@urjc.es

© Springer International Publishing Switzerland 2017
F.C. Freire et al. (eds.), *Media and Metamedia Management*,
Advances in Intelligent Systems and Computing 503,
DOI 10.1007/978-3-319-46068-0_58

441

With the arrival of social media, users assume protagonism, promoting a multi directional discourse. In turn, the classical discourse of local administrations (vertical and unidirectional) is affected by the dialogic context 2.0. The contemporary communication ecosystem invites professional actors in the field to rethink relations with the recipients of messages.

This study proposes an approach to communication processes performed by the municipalities of São Paulo and Madrid from their official accounts in Twitter. These two cities have great importance in the economic, tourism and cultural landscapes of their countries, being Madrid the capital of Spain and São Paulo the capital of the Brazilian Federal State named after the city and the most populous settlement in South America, hosting around 11,967,825 inhabitants (IBGE 2015). Madrid, in terms of population, is the home to 3,198,645 inhabitants according to the last Census of Population and Housing (INE 2011).

Both in Brazil and Spain local administration consists of three significant levels: the administration under the central government, the administration under the regional government (Federal states in Brazil and autonomous communities in Spain) and local level authorities (municipalities). The latter is often perceived by citizens as the closest and therefore it's the one more likely to deal with complaints, suggestions and demands of citizens.

Given the advance of ICT, social media have become indispensable communication tools when establishing a connection between citizen-users and local administrations. Being a microblogging service, Twitter has the ideal characteristics for effective communication with a significant scope and potential impact, creating an appropriate environment to inform and discuss.

58.1.1   Social Media as a Channel of Communication Between Citizens and Local Administrations

In the field of political communication, Ballester (2013) draws attention to the power that social media have to inform citizens and to interact with them. Despite intrinsic challenges in incorporating ICT in classical and institutionalized communication processes, Pardo Baldeón (2014) considers these tools play a fundamental role and that they can be used to know the opinion of citizens and their satisfaction or dissatisfaction degree towards government policies.

E-government and parliamentary cyber-democracy consolidation is associated with the construction of a communication context that includes interactive platforms, where society can express opinions and political thoughts. If on one hand the importance of having a presence in social media is acknowledged, on the other, there is a tendency on the part of local administrations to confuse presence with participation.

Regarding the use of Twitter by local administrations, authors such as Balcells et al. (2013: 66) claim that this tool: "becomes a clear channel of interaction when responding to public consultations directly addressed to it."[1]

However, several studies point out to a largely unidirectional use of this technology by political actors, leaders and local administrations (Ellison and Hardey 2013; Bonsón et al. 2012; Waters and Williams 2011).

58.2 Objectives, Hypothesis and Methodology

This study, as a main objective, aims to conduct a comparative analysis of the use of Twitter by the municipalities of São Paulo and Madrid. In a more specific goal, the research intends to understand the dynamics of postings in both official accounts (@prefsp and @MADRID) and the types of contents in their timelines, taking into consideration the premise that there is a low discussion rate between local administrations and citizen-users.

The analysis has been conducted through a ten-day period comprising two non-consecutive weeks (Monday to Friday): from 22nd to 26th June and from 27th to 31st July, 2015. This data gathering procedure facilitates to obtain a sample of tweets collected at two different moments, allowing a look at the development of communication processes performed by the municipalities on the Web.

Two independent analysis protocols has been prepared, based on the web content analysis method (Herring 2010):

(a) Protocol 1: contains variables regarding the Twitter accounts of the two selected municipalities (ID and URL, bio content, the date of the account opening, followers, followings, among others).

(b) Protocol 2: contains variables regarding the contents from the timelines (tweets categories, date of posting, total amount of "likes", etc.)

In order to know the types of contents released by the official accounts of São Paulo and Madrid municipalities, the tweets comprising the sample (n = 149) have been framed into two categories: information tweets and discussion tweets. Most of the data directly came from the accounts and timelines analyzed. To complement the analysis, we also used the tools Twopcharts and MetricSpot.

---

[1]In Spanish in the original: "se convierte en un canal claro de interacción cuando responde a consultas ciudadanas que se le dirigen directamente".

58.3 Results

58.3.1 Bio, Followers and Followings

Data on general information about the municipalities have been collected on July 21st, 2015, date on which the following contents were observed in their biographies:

(a) São Paulo (@prefsp): "Official Twitter of the Municipality of São Paulo. Here you will find news and information on the city we love".[2]
(b) Madrid (@MADRID): "Official Account of Madrid City Hall".[3]

Despite being accounts verified by Twitter, municipalities prefer to use the term "official" in the bio. Besides, they facilitate the location and email address of their websites. The City Hall of São Paulo joined Twitter on May 22nd, 2014, while the City Hall of Madrid is on this social media platform since March 6th, 2007.

As of July 21st, the City Hall of São Paulo had 25,806 followers and followed 53 Twitter users. On the other hand, the City Hall of Madrid had 129,509 followers and followed 75 tweeters. The ratio of followers/following in the municipalities of São Paulo and Madrid is around 486.9 and 1726.8, respectively.

58.3.2 Timelines Updates

During the first period of analysis (P1: from 22nd to 26th June, 2015), the municipalities of São Paulo and Madrid published a total of 28 and 59 tweets, respectively. In the second period (P2: from 27th to 31st July), São Paulo and Madrid published in this order, 36 and 26 messages.

The posts released by the City Hall of São Paulo represent the 42.95 % of the total sample, while the number of tweets issued by the City Hall of Madrid stand for the 57.05 %. On July 29th, the date in which more tweets have been published by the City Hall of São Paulo (12 in total), a full coverage on the signing ceremony for a cooperation agreement between the municipality and the Public Ministry to end violence against youngsters was conducted on the platform. In the case of Madrid City Hall, both the 25th and the 26th June, 15 tweets were posted in the timeline, being this the record of publications of the two municipalities in a single day. Specifically, on June 25th, the Madrid City Board meeting took place, an event whose coverage was reflected in Twitter.

---

[2]In Portuguese in the original: "Twitter oficial da Prefeitura de São Paulo. Aqui você encontra notícias e informações sobre a cidade que a gente ama".

[3]In Spanish in the original: "Perfil oficial del Ayuntamiento de Madrid".

Table 58.1 Total of retweets made by each City Hall

City Hall	P1: June	P2: July	Totals
São Paulo	12 (35.3 %)	22 (64.7 %)	34 (100 %)
Madrid	32 (86.49 %)	5 (13.51 %)	37 (100 %)

Source Author's own elaboration

Regarding shared contents, a total of 71 retweets have been accounted, distributed in periods of analysis in Table 58.1, both in absolute values and percentages. The last column contains all retweets made by each municipality.

Retweeted messages by the municipalities of São Paulo and Madrid link to posts made by similar institutional accounts, such as their mayors and local public agencies.

58.3.3   Analyzing the Tweets

During the analysis, we registered a significant percentage of purely information/diffusion posts by both the City Hall of São Paulo, with 81.25 % of information tweets, as the City Hall of Madrid, with 74.11 %.

As an example of the content we analyzed here, the post released on June 26th by the account of the São Paulo City Hall dealt with the launching of the first mobile unit of LGBT (Lesbians, Gays, Bisexuals and Transgender) citizens with a message that has no textual element whatsoever to encourage dialogue. Next we show a tweet posted by the City Hall of Madrid about RPA (Residential Priority Area) in which we detected a message with an inclusive effect, by treating the user as an equal to the institution:

(a) São Paulo (@prefsp, June 26th): "The City Hall launches the first mobile unit of LGBT citizens to assist in the cases of homophobia\o/: goo.gl/ZWW20Y".[4]

(b) Madrid (@MADRID, July 31st): "From August 1st, Opera is RPA: traffic priority for residents. Register your vehicle at: bit.ly/1lb2y1Y".[5]

We found a wide variety of topics in the composition of timelines. The most frequent posts respond to the diffusion of events promoted by the municipalities (23.5 % of the sample) and information on public events the mayors attend to (16.8 %).

---

[4]In Portuguese in the original: "Prefeitura lança primeira Unidade Móvel de Cidadania LGTB p/atender casos de homofobia\o/: goo.gl/ZWW20Y".

[5]In Spanish in the original: "Desde el 1 de agosto, Ópera es APR (Área de Prioridad Residencial): tráfico prioritario para residentes. Registra tu vehículo: bit.ly/1lb2y1Y".

58.3.4 The Use of URLs, Images, Mentions and Hashtags

We came across with a recurring use of URLs in tweets (56.4 % of the total sample includes an email address) almost always intended to lead users to the City Hall official website.

41 % of the tweets sample has at least one picture. The images depict topics such as events promoted by the municipalities or public actions involving the mayors. We didn't register the presence of any video in the timelines.

Mentions have been used in a 38.2 % of the sample, with a maximum of two entries per tweet. The mentioned users reinforced affinities: @smcsp (after the Local Secretary of Culture) or @Haddad_Fernando (after the current mayor) in the case of São Paulo and @EMTmadrid (after the Local Transport Company) or @ManuelaCarmena (after the current mayor) in the case of Madrid.

It has also been found that the use of hashtags only affects a minority of posts. Merely 16.8 % of the sample has a tag, indicating an under-use of one of Twitter's main resources.

58.3.5 "Likes", Retweets and Replies

By checking the data provided in Table 58.2, a higher level of influence of Madrid City Hall account is observed. Data represent, in absolute values and percentages, the total of "likes" received by the City Hall posts, considering both periods of analysis and the totality of "likes" in each account studied. Similarly, in Table 58.3 the total amount of retweets in both timelines is presented.

From data it's clear that the content published by the City Hall of Madrid have a wider scope and a greater impact (real and potential) that the posts of São Paulo City Hall. It should be noted that there has been no reply by the municipalities during the period of analysis.

Table 58.2 Total of "likes" received

City Hall	P1: June	P2: July	Totals
São Paulo	291 (45.33 %)	351 (54.67 %)	642 (100 %)
Madrid	2.072 (72.52 %)	785 (27.48 %)	2.857 (100 %)

Source Author's own elaboration

Table 58.3 Total of retweets produced

City Hall	P1: June	P2: July	Totals
São Paulo	349 (45.92 %)	411 (54.08 %)	760 (15.03 %)
Madrid	2.930 (68.19 %)	1.367 (31.81 %)	4.297 (84.97 %)

Source Author's own elaboration

58.4 Discussion and Conclusions

From the study conducted, the premise that there is a low discussion rate between local administrations and citizen-users is confirmed as for the great prevalence of posts with mere information purposes and the absence of public responses or replies in the City Hall accounts under analysis. Tweets intended only to diffusion, sending the user to the municipality official website, generate a hierarchical, one-way redundant discourse (due to duplication of content in different platforms, on the institution, which is the type of praxis closer to classic organizational realities than to the distinct interactivity of social media.

We didn't find any regularity in the frequency updating at the accounts observed, which may reflect the absence of a clear strategy in this regard. However, this dynamic can be justified by the fact that more tweets are issued on dates related to specific events, when Twitter is used for live coverage at the same time of the information provided by mass media. The real-time coverage on the events in which mayors and City Halls themselves have strong presence gives users content backed up by the credibility of the local administrations.

The study shows that users and topics supporting the authorities in power determine the use of communication resources and even the selection of contents to share in this microblogging platform. Apart from this finding, we came across with the fact that there is any reciprocity at all between followers and followings so it's clear the lack of willingness to dialogue with citizens by municipalities in this case.

To overlook the interactive and conversational nature of social media in general is a risk because you may be ignoring the views of citizens, critical point of views and claims that are not only present in the offline dimension, but that are increasingly more evident on digital platforms. It is necessary, therefore, to ensure a constant monitoring on social tools not only to listen to citizens but also to provide answers through dialogue. The proximity provides confidence and acts positively on the image and reputation of municipalities and local governments.

References

Balcells, J., Padró-Solanet, A., & Serrano, I. (2013). Twitter en los ayuntamientos catalanes. Una evaluación empírica de usos y percepciones. In J. I. Criado & F. Rojas Martín (Eds.), *Las redes sociales digitales en la gestión y las políticas públicas. Avances y desafíos para un gobierno abierto* (pp. 62–81). Barcelona: Escola d'Administració Pública de Catalunya.

Ballester, A. (2013). Análisis de la política de comunicación en Twitter de las Administraciones Públicas en la Comunidad Valenciana. In R. Bañón & R. Tamboleo (Dirs.), *Gestión de la Escasez, participación, Territorios y Estado del Bienestar. Experiencias de Democracia y Participación* (pp. 209–219). Madrid: GOGEP Complutense.

Bonsón, E., et al. (2012). Local e-government 2.0: Social media and corporate transparency in municipalities. *Government Information Quarterly, 29*(2), 123–132.

Ellison, N., & Hardey, M. (2013). Developing Political Conversations? Social media and English local authorities. *Information, Communication & Society, 16*(6), 878–898.

Herring, S. C. (2010). Web content analysis: Expanding the paradigm. In J. Hunsinger, L. Klastrup, & M. Allen (Eds.), *International handbook of internet research* (pp. 233–249). Netherlands: Springer.

Instituto Brasileiro de Geografia e Estatística (IBGE). (2015). *O Brasil município por município. São Paulo.* At http://www.cidades.ibge.gov.br/xtras/perfil.php?lang=&codmun=355030&search=sao-paulo|sao-paulo. October 25, 2015.

Instituto Nacional de Estadística (INE). (2011). *Censos de Población y Viviendas 2011.* At http://www.ine.es/censos2011_datos/cen11_datos_inicio.htm. October 25, 2015.

Internet World Stats. (2015). *World internet usage and population statistics.* At http://www.internetworldstats.com/stats.htm. October 25, 2015.

Pardo Baldeón, R. S. (2014). Análisis sobre el uso de Twitter en las administraciones locales de la provincia de Castellón. *Miguel Hernández Communication Journal, 5,* 361–379.

Waters, R. D., & Williams, J. M. (2011). Squawking, tweeting, cooing, and hooting: analyzing the communication patterns of government agencies on Twitter. *Journal of Public Affairs, 11*(1), 353–363.

Chapter 59
The Environment of Web 2.0 as a Relational Factor in the Use of Loyalty-Raising Parameters Within the University Environment

José Rodríguez Terceño, Juan Enrique Gonzálvez Vallés and David Caldevilla Domínguez

Abstract The new narratives that have emerged from the social media profiles have shown that public demand closeness of those institutions to which they belong one way or another. The case of universities is not alien to this phenomenon and they have adapted the form and substance of communications through social networks with the various stakeholders with which they interact. From this initial premise, we must seek successful cases that serve as the basis for a good analysis of the content of such profiles. The typology of content and frequency of publication should be two conducting axes of our research as they provide the tone one wants to give to this type of communication. We also believe that, unlike other types of descriptive research that base their depth on the long run, we must limit our observational period to a short range that also delimits our analysis process. This way, we can reach specific conclusions that are relevant, innovative and effective.

Keywords Web 2.0 · University · Social networks · Content

J. Rodríguez Terceño (✉) · D. Caldevilla Domínguez
Complutense University of Madrid (UCM, Madrid, Spain
e-mail: josechavalet@gmail.com

D. Caldevilla Domínguez
e-mail: davidcaldevilla@ccinf.ucm.es

J.E. Gonzálvez Vallés
CEU San Pablo University (USP CEU, Madrid, Spain
e-mail: juanenrique.gonzalvezvalles@ceu.es

© Springer International Publishing Switzerland 2017
F.C. Freire et al. (eds.), *Media and Metamedia Management*,
Advances in Intelligent Systems and Computing 503,
DOI 10.1007/978-3-319-46068-0_59

59.1   Theoretical Framework

When formulating a sufficiently broad framework to accommodate the heterogeneous aspects of our research but necessarily specific to its principles to fit our
method, we should refer first to a proper definition of statistics, since our object of
study is the analysis of statistics-provided data.

So, Murray R. Spiegel defines statistics as the science that "studies scientific
methods to collect, organize, summarize and analyze data and to draw valid conclusions and make reasonable decisions based on such an analysis" (Spiegel: 2011,
20) but we can supplement this view with that of David Ruiz Muñoz, who speaks of
"science that aims to apply the laws of the amount to social events to measure their
intensity, deduce the laws governing them and make their next prediction" (Muñoz
2004, 3). It is obvious that the social and prospective vision contributed by Muñoz
reinforces the rational view of Spiegel, reaching a holistic nature in this regard.

Having already addressed the first point of our object of study, then we will
frame the significance of social networks, as the core of our analysis. When
referring to Web 2.0, we have to focus on what its appearance has meant from the
point of view of users and how it has begun to relate them to the Internet. From the
technological and programming point of view, it has not been a substantial change
but now they have set new relational frameworks with the subject who is on the
other side of the screen.

As Santiago and Navaridas (2012: 19) points out, Web 2.0 "has been a further
level of development of the 'traditional' web that has meant, in fact, to achieve a
high degree of collaboration-cooperation and interaction among internet users,
which tends to be massive for many applications." The degree of development of
Web 2.0 is certainly already high at all levels and many authors point to the
imminent addition of Web 3.0 into our lives.

59.2   Methodology

The demonstrational and exhibition method is most suitable for the present case,
since it serves to obtain relevant conclusions. Knowing all these points is relevant
because only then we can assess the sustainability of the project and, if so, scrutinize the possible future actions that can be carried out.

We sought, ultimately, to obtain positive results that also can be applicable to
other projects both within the world of sport and in what relates to areas of marketing that affect all aspects of our life.

The main objective of this paper is to find the keys to the successful use of social
networks by the faculties of communication and/or information within the
Community of Madrid, through the analysis of one of the tools provided by Web 20
itself. Teachers, students, management staff and services and even governmental
institutions of all levels of government have brought about an unleashed singular

and unique phenomenon in the social networks, especially on the basis that there are few things in life there as important as passing through the university world.

59.3   Results

59.3.1   Social Networks in the Faculty of Humanities and Communication Sciences of San Pablo CEU University

In late July 2015, a change in the deanship team of the Faculty of Humanities and Communication Sciences (CCHF) of San Pablo CEU University took place, José María Legorburu Hortelano being appointed Dean of the Faculty. Moreover, in a clear commitment to give visibility to the life of this site and all the activities it develops, Marilé pretzel was appointed Assistant Dean. Both, along with the rest of the deanship team, drive a new form of communication where social networks are going to be protagonists.

After the deserved month off in August, a new communication model hovers in the social media profiles of this faculty, standing out for making students protagonists but without forgetting corporate obligations and the fact that the faculty belongs to a larger entity. Therefore, CCHF endeavored to carry out this work on social networks, focusing its efforts on Facebook and Twitter.

59.3.1.1   Facebook

Although San Pablo CEU University already had a corporate fan page (http://www.facebook.com/USPCEU) with more than 9,000 followers, a page for this faculty was created (http://www.facebook.com/USPCEUhumanidades) that accumulated followers since it was put into operation.

Then, as we also do with the Twitter profile, we will present the data collected during the two-month follow-up at the beginning of the academic year 2015–2016. First, the history of followers on Facebook Profile created by CCHF, which has been a landmark since September 1, 2015 (Fig. 59.1).

The analysis of this figure leads us to use the word awesome first. In less than two months, 'USPCEU humanities' had gained almost 100 followers more within a small audience, something available to few profiles of such entities. Out of the 295 'I like it' accumulated by the profile on September 1, 2015, it increased to 387 on October 31, 2015, i.e., only two months later.

As we say, this event took place mainly between the months of September and October 2015, certainly due to an effect because of the quantity and quality of content provided in this profile. We do not go into a detailed analysis of them since that will be the subject of a subsequent piece of research, but we aim that the vast

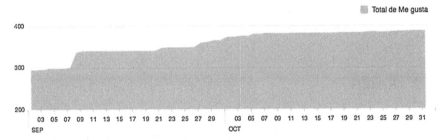

Fig. 59.1 Historical fans Facebook. *Source* Made by the author

quantity and quality of the photos, videos and communications provided there were a great claim to create a viral effect which contributed followers to this profile.

The second strong point of this historical analysis comes at a period that can be considered as a 'valley', since after the first week of beginning of the course, there is no extraordinary activity in the academic calendar. However, continuous updating of the profile with new content, combined with the presence of teachers and the deanship team in various events that included the presence of media, complementary strategy certainly very effective, managed to continue increasing the number of followers in an spectacular way.

The third significant increase in followers resulted from the creation of different sections within this profile, which also moved to the Twitter profile. Fans recognized this effort, making the current ones loyal and increasing virality which produced a greater number of new additions. It is logical that, in offering again high quality content and, above all, content being almost 100 % different from other universities except the Twitter profile of Nebrija University, this increase of nearly 400 followers occurred when the month of October 2015 was over.

From that moment, additions are still occurring in the number of followers of this profile on Facebook. We still have to know, maybe at a later analysis, if the digits have already reached their peak and is difficult to raise them or there is still room to increase them further.

59.3.1.2 Twitter

Facebook's success also moved to Twitter, since none of the two social networks was set aside and, at all times, the necessary efforts were made to feed both (Fig. 59.2).

As was the case with Facebook, San Pablo CEU University already had a corporate account on Twitter (USPCEU) but FHCCse decided to create an official profile (HumanidadesCEU) to tell all the news about the activities of the faculty. Contents followed those that were published on Facebook and most of them were referenced to this network.

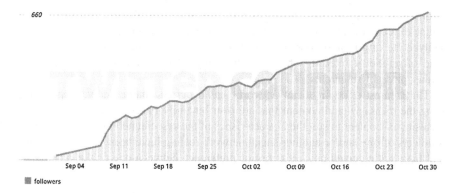

Fig. 59.2 Historical followers Twitter. *Source* Twitter counter

Still, the increase was sustained during the first two months, highlighting a sharp increase in the first week of the academic year 2015–2016, more than 20 new followers. The numbers are just a few, but the profile on Twitter already has more than 650 followers and in any week there has been a decline in followers, so it is expected that, within the 2015–2016 course, significant numbers can be reached.

59.4   Discussion and Conclusions

The results of this piece of research show that the communication strategy of niches of the Faculty of Humanities and Communication Sciences of San Pablo CEU University in social networks has been a wise move and it has been successful in all respects. We have seen how the university entity held various general corporate profiles on various social networks, which did not imply that it communicated effectively and efficiently with students of a certain faculty.

A result of this was the creation of said profiles of CCHF, in order to transport the positive values entailed by the activities undertaken within the Faculty to its specific niche. In this regard, the commitment of the new deanship team by raising the visibility of the Faculty through social networks was an excellent testbed for future strategies in this area.

The social profiles of CCHF have proved to be not only a winning bet but they have also gone on to become a symbol for the faculty, which even now is looking forward to the arrival of certain days of the week to see what is new in the new sections. It is not that this initiative has enhanced the image that the stakeholders of the faculty had but the interest shown by the high number of followers within a limited target audience entails that there is a perfect promotional space to transmit the values that the university entity wants for itself.

With nearly 1000 followers, uniting the followers of the Facebook and Twitter profiles, CCHF has become a reference for other faculties of communication and/or information of private universities of the Community of Madrid, well above any initial expectation and overcoming the competition. At this point, it remains to analyze the origin of supporters, and it remains to establish who belong to every kind of public of interest, which will be analyzed in an article resulting from a further piece of research.

The greatest effort in social networks has occurred, undoubtedly, in personalizing the messages and content that are distributed by Facebook and Twitter. The inclusion of content to suit the characteristics of these social networks has led to the followers of one and/or another social network feeling have been cared for enough and have not received any contempt or preference between one or the other.

References

Muñoz, D. (2004). *Manual de Estadística*. Sevilla: Eumed.
Santiago Campión, R., & Navaridas Nalda, F. (2012). La web 2.0 en escena. *Pixel-Bil: Revista de medios y educación, 41*, 19–30.
Spiegel, M. (2011). *Schaum's Outline of Stastics*. Londres: Copyright Materials.

Part VIII
New Media and Metamedia

Chapter 60
Trends in Journalism for Metamedia of Connectivity and Mobility

Ana Isabel Rodríguez Vázquez and Xosé Soengas Pérez

Abstract The evolution of the networked society is defining journalistic practices and trends. While the elements of journalism remain, there is no doubt that the possibilities offered by existing tools feed experiencing renewed dimensions, from which there arise formats and pieces that help professionals to develop more quality information, according to the characteristics of the present communication environment. As this new ecosystem is formed, journalism is defining its elements in the age of metamedia. The work explores some of the ongoing news practices of recent years, such as data journalism, immersive journalism, slow journalism, real-time journalism and drone journalism. It also includes results on new media trends, contributions on metamedia and news added value.

Keywords Data journalism · Immersive journalism · Slow journalism · Drone journalism · Metamedia

60.1 Theoretical Framework

Journalism, which is undergoing a period of far-reaching restructuring (Casero 2012), is seeking its future every day. The renewed professional practices, encouraged by labs belonging to some of the main media, and by scientific research, serve us as reference to understand the evolution of this communication technique, which has already very much proved its worth in democratic societies.

The result of the development of present information technologies is the setting of a complex communication environment, which is highly dynamic and feedback from a dense network of interconnected nodes that update contents instantly (Cruz Álvarez and Suárez Villegas 2012).

A.I. Rodríguez Vázquez (✉) · X. Soengas Pérez
University of Santiago de Compostela (USC), Santiago, Spain
e-mail: anaisabel.rodriguez.vazquez@usc.es

X. Soengas Pérez
e-mail: jose.soengas@usc.es

© Springer International Publishing Switzerland 2017
F.C. Freire et al. (eds.), *Media and Metamedia Management*,
Advances in Intelligent Systems and Computing 503,
DOI 10.1007/978-3-319-46068-0_60

457

The digital setting has opened up new opportunities for journalism, which has the tools to work in the technological environment. Multimedia (Deuze 2004); Interactivity (Scolari 2008); Participation (Masip et al. 2015; Customization (Thurman 2011); Memory/Documentation (Guallar 2011); Mobility and the use of new platforms (Westlund 2014), are some of the main dimensions that new media should consider.

There is a widespread agreement that the future of journalism is telling good stories (Boynton 2015), using present technologies, and targeting at connected citizens, who need to be given the best possible information for making everyday decisions. Good non-fiction stories remain the central core of journalism, but the challenge today is to produce clear and efficient pieces of news with all available resources and techniques. The use of some of the current technologies, formats, platforms, disseminating channels, and reception devices, has to be at service of that objective, since the essence of journalism relies on the story of what is happening in the world.

In the age of metamedia, understood as new media with new properties (Manovich 2013), journalism has diversified and expanded trends, which show the opportunities that connected and mobile society is opening up. Journalism, rather than being worried about its death or disappearance, is concerned about its future. Also, journalism is making efforts to show its potential as information techniques, which not only contributes to the proper functioning of democratic societies, but also help citizens to be involved in news and to be well informed.

60.2   Methodology

Framed within the present context, the work is carried out from an exploratory and descriptive perspective. It analyses data journalism, immersive journalism, slow journalism, real-time journalism, and drone journalism, are some of the emergent news practices. The work is part of a broader research and includes results of the above-mentioned project on news trends and movements for new media ("Innovation and development of Spanish cybermedia. Architecture of news interactivity on multiple devices: news formats, conversation and services", reference: CSO2012-38467-C03-03); analysing a broad sample of cybermedia in the last five years. In the case of data journalism, there was carried out a specific analysis of The Guardian (UK) and El Mundo (Spain), from January to July 2015. After a three-step research process—design, empirical and analytical stage—(Igartua 2006), specific contributions for metamedia have been presented, emphasising the added value of news pieces, produced with updated techniques. The case study has revealed some of the keys of the present news practices in various successful cybermedia that are presented at the end of this paper.

60.3 Results

60.3.1 Data Journalism

Data, which are part of the basis of journalism, are now the main players of one of the most innovative journalism trend, with stories based on the use of online information and a good visualization. The relevance of data in the present-day society—open data and big data-, has given unprecedented impetus to a technique that seeks to contribute to news pieces with added value, taking advantage of databases and reports information—available online-, and making it useful for citizens.

Data journalism is rooted in precision journalism, which on his day was a change in journalism techniques, combining informatics and social and statistical sciences (Meyer 1993). Data journalism makes use of statistical and visualization tools, for the purpose of telling old stories in an innovative and more clear way, and discovering others hidden until now (Flores 2012). Also, it goes beyond simply using traditional genres; it employs multimedia and infographics resources to make data clear (Chaparro 2014). Its excellent adaptation to the current environment gives it special power to inform in the networked society.

The choice of some media companies to invest in data journalism has been their key to success. In Europe, the case of The Guardian (Rogers 2011); in the EE.UU, there are also good examples, from The New York Times to the Chicago Tribune and Los Ángeles Times (Guallar 2013). Some of the major media The New York Times, The Guardian, BBC, National Geographic, and The Economist) have found in this trend an opportunity to turn it into a distinctive quality factor to maintain their privileged status (Salaverría 2015). Their reports using data journalism techniques revealed interesting research results.

The combination of journalism research techniques and available online data as a source for telling stories has been successful. The public impact of a large number of works published in mentioned newspapers—research on tax havens and tax evasions, among others-, has been instrumental in creating the brand's image of data journalism, in its race to occupy a place as a quality journalism technique.

Supporters of this trend say that the future of journalism lies in knowing how to search, analyse and visualize data (Bradshaw 2010). The analysis of The Guardian and El Mundo (2015) prove this point, even though there are differing nuances. The Guardian pays close attention to image and visualizations, while El Mundo is more focused on the text and contextualization and extended explanations (Tuñas 2015). But trends indicate an ongoing care when visualizing results of data analysis.

Related to data journalism, there has emerged the fact checking. Although major successful initiatives are closely related to political information, it has also been incorporated by the main media as a technique for offering news services and for strengthening the credibility of information products. This trend, which has succeeded in the EE.UU since the beginning of the present decade (Guallar 2013), has arrived to other countries such as Spain.

60.3.2 Immersive Journalism

Immersive journalism, based on interfaces and action (Domínguez 2013), has emerged as a trend that allows first person experiences. Using virtual-reality goggles and other technological devices, users can move around stages and experience virtual situations. The aim is to bring the user to the scene, so that it becomes a witness of facts. Virtual reality is the chosen allied for this renewed technique in journalism, which has already many examples.

The first known work based on immersive journalism is "The Des Moines Register" (http://www.desmoinesregister.com/), which is part of Gannett Digital. The project, called "Harvest of change", is a five-part immersive and interactive work that tells the story of a farm family. Using the Oculus rift, users can walk through an environment in which there appear information, videos, and real pictures that bring real but also innovative context.[1] The initiative, promoted to demonstrate the possibilities of virtual reality, has attracted attention among many cybermedia, which are working out strategies for the foreseeable future.

Nonny de la Peña—former correspondent for Newsweek—is one of the pioneers in experimenting with this new way for journalism. Her immersive project may be accessed at: http://www.immersivejournalism.com/. Her news pieces are reports on real facts—most of them on human rights-, in which she introduces virtual animations and gives them realism, without forgetting reality. Her 3D immersive experiences are followed by the most innovative cybermedia, whose aim is to explore this new way of telling stories.

In Spain, works by Eva Domínguez (http://www.evadominguez.com/), both for her dissertation and follow-up of these initiatives (http://blogs.lavanguardia.com/elcuartobit), make her a guide in the field of immersive journalism, which is expected to advance as technological companies (Sony, Google, Apple, Microsoft...) present new virtual reality devices.

60.3.3 The Challenge: Continuous Renewal

While data and immersive journalism are two established trends, there are other initiatives emerging with strength, such as the drone journalism—very popular because of the different news coverage—the slow journalism versus real time journalism; and multimedia journalism, which has become very popular, building

---

[1]Immersive Journalism is expected to revolutionize the way of telling non-fiction stories in terms of journalism. Some of the forecasts appear in this article by Miriam García Martín (Media-tics): http://www.media-tics.com/noticia/5173/medios-de-comunicacion/el-periodismo-inmersivo-revol ucionara-la-forma-de-contar-historias.html.

on the success of Snow Fall—http://www.nytimes.com/projects/2012/snow-fall/#/?
part=tunnel-creek—and los *webdocs*. There are initiatives to renew narratives and
techniques, which sometimes turn into trends and, in other cases, journalistic
movements.

The technological renovation also entails challenges. For example, the *screen-less* technology requires the production of messages with very high levels of
interactivity. Success of new devices (goggles and watches), will be critical to the
success of small-format narratives, who should gets the maximum efficiency and
integration of the present framework, which allows interactivity of the connected
and mobile society. The way is still in its initial phase, but there have emerged the
first products seeking to settle formats that ensure efficient communication.

The combination of explanatory narratives with good data management and
visualization are the objective o much of the innovative experiences in online
journalism. Teams made up of experts in narratives, design, and informatics—
computer engineers-, are one of the proposals for addressing present challenges in
journalism. At the same time we are witnessing a greater presence of machines in
communication processes—news written by robots and pictures captured by
drones-, while journalists continue to maintain a relevant role when designing
strategies and producing news.

60.4   Discussion and Conclusions

Journalism does not reject the idea of exploring new territories that configure the
networked and mobile society. While the essential elements remain, there are new
possibilities for exercising quality journalism, providing accurate and useful
information to citizens. Emergent trends, such as data and immersive journalism,
offer ways to cultivate another possible journalism.

Although data and immersive journalism are the two main trends for metamedia
of connectivity and mobility, there are many proposals, such as the Hi-tech journalism, slow journalism, real time journalism, and *glocal* journalism. These are
ways of responding to the ongoing change and of taking advantage of the new
techniques, which can be combined to develop news products. The proliferation of
designations coincides with renewal proposals that, in some cases, become trends
and movements in journalism that offer a response to some of the challenges facing
journalism today.

Acknowledgments   The research is part of the results obtained in a research Project, promoted by
the Ministry of Economy and Competitiveness. The title is "Innovation and development of
Spanish cybermedia. Architecture of news interactivity on multiple devices: news formats, conversation and services". Reference: CSO2012-38467-C03-03.

References

Boynton, R. S. (2015). *El nuevo periodismo. Conversaciones sobre el oficio con los mejores escritores estadounidenses de no ficción.* Barcelona: Ediciones de la Universidad de Barcelona.

Bradshaw, P. (2010). El futuro del periodismo pasa por saber buscar, analizar y visualizar los datos. Retrieved from http://www.lavanguardia.com/tecnologia/20101027/54060875060/paul-bradshaw-el-futuro-del-periodismo-pasa-por-saber-buscar-analizar-y-visualizar-los-datos.html. Accessed August 10, 2015.

Casero Ripollés, A. (2012). Contenidos periodísticos y nuevos modelos de negocio: Evaluación de servicios digitales. *El Profesional de la Información, 21*(4), 341–346.

Cruz Álvarez, J., & Suárez Villegas, J. C. (2012). Ética de la participación ciudadana em los discursos periodísticos digitales. *El Profesional de la Información, 21*(4), 375–380.

Chaparro Escudero, M. A. (2014). Nuevas formas informativas: el Periodismo de datos y su enseñanza en el contexto universitario. *Historia y Comunicación Social, 19*, 43–54. Retrieved from http://revistas.ucm.es/index.php/HICS/article/view/45009. Accessed August 10, 2015.

Deuze, M. (2004). What is multimedia journalism? *Journalism Studies, 5*(2), 139–152. Retrieved from http://jclass.umd.edu/classes/jour698m/deuzemultimediajs.pdf. Accessed August 10, 2015.

Domínguez, E. (2013). Periodismo inmersito: fundamentos para una forma periodística basada en la interfaz y la acción. Retrieved from https://es.scribd.com/doc/133994321/Periodismo-inmersivo-Fundamentos-para-una-forma-periodistica-basada-en-la-interfaz-y-en-la-accion. Accessed August 9, 2015.

Flores Vivar, J. (2012). Sinergias en la construcción del Nuevo Periodismo derivadas del Data Journalism y el Transmedia Journalism. En *Actas III Congreso Internacional Comunicación 3.0* (pp. 476–487). Salamanca: Universidad de Salamanca.

Guallar, J. (2011). La documentación en la prensa digital. Nuevas tendencias y perspectivas. Retrieved from http://eprints.rclis.org/16326/1/ciberpebi2011_guallar_documentacion%20prensa%20digital.pdf. Accessed August 09, 2015.

Guallar, J. (2013). Prensa digital en 2011–2012. In T. Baiget & J. Guallar (Eds.), *Anuario ThinkEPI 2013. Análisis de tendencias en Información y Comunicación*. Barcelona: EPI SCP.

Igartua, J. J. (2006). *Métodos cuantitativos de investigación en comunicación.* Barcelona: Bosch.

Manovich, L. (2013). *Software takes comand.* Nueva York: Bloomsbury Academic.

Masip, P., Guallar, J., Suau, J., Ruiz Caballero, C., & Peralta, M. (2015). News and social networks audience bahavior. *El Profesional de la Información, 24*(4), 363–370. Retrieved from http://www.elprofesionaldelainformacion.com/contenidos/2015/jul/02_esp.pdf.   Accessed August 09, 2015.

Meyer, P. (1993). *Periodismo de precisión.* Barcelona: Bosch.

Rogers, S. (2011). *Data journalism at the guardian: What is it and how do we do it?* Retrieved from http://www.theguardian.com/news/datablog/2011/jul/28/data-journalism. Accessed August 10, 2015.

Salaverría, R. (2015). Periodismo en 2014: Balance y tendencias. *Cuadernos de Periodistas, 29*. Retrieved from http://www.cuadernosdeperiodistas.com/periodismo-en-2014-balance-y-tendencias/. Accessed August 10, 2015.

Scolari, C. (2008). *Hipermediaciones. Elementos para una teoría de la Comunicación Digital Interactiva.* Barcelona: Gedisa.

Thurman, N. (2011). Making 'The Daily Me': Technology, economics and habit in the mainstream assimilation of personalized news. *Journalism: Theory, Practice & Criticism, 12*(4), 395–415. Retrieved from http://openaccess.city.ac.uk/58/2/neil-thurman-making_the_daily_me.pdf. Accessed August 10, 2015.

Tuñas, E. (2015). *El periodismo de datos en 2015: Análisis de El Mundo y The Guardian. Trabajo Fin de Grado.* Santiago de Compostela: Facultade de Ciencias da Comunicación.

Westlund, O. (2014). The production and consumption of mobile news. In G. Goggin & L. Hjorth (Eds.), *The mobile media companion.* New York: Routledge. Retrieved from https://pure.itu.dk/ws/files/44277439/Westlund_Mobile_News_Mobile_Media_Companion_2013_PREPRINT.pdf. Accessed August 10, 2015.

Chapter 61
The Future of Video-Journalism: Mobiles

Martín Vaz Álvarez

Abstract This communication aims to present and study the trends in mobile video-journalism in the production of news and television/internet programs, as well as its possible repercussions in the industry, the improvements in the quality of information and the reduction of production times. The latest technical improvements in mobile devices (smartphones and tablets), together with the proliferation of apps and gadgets adapted to these systems open up a new spectrum of possibilities for journalists, being already possible to film and edit audiovisual pieces simultaneously with broadcast quality and a much lower budget than what a standard television equipment needs. This turns the journalist into an active agent in the edition of the news and puts him in full control of the story. This communication will also study the cases of newschannels using mobile devices for program and news-making. Since the 29th of June 2015, local genoese channel Léman Bleu produces its newscast using strictly iPhones, BBC produced earlier this year its show Click in the Mobile World Congress of Barcelona exclusively with smartphones and tablets, Al Jazeera also has a section dedicated exclusively to video-journalism, and corporations such as CNN, Channel News Asia or Singapore NewsMedia are investing in formative courses for their reporters.

Keywords Video-journalism · Mobile communication · Audiovisual · Mobile devices · Mobile video-journalism

61.1 Theoretical Framework

In order to understand the field in video-journalism which is analysed in this communication, we first need to define what is 'mobile video-journalism' in the context of information. Mobile video-journalism is a composed word that refers both to the mobile version of journalism and an audiovisual platform. This might

M. Vaz Álvarez (✉)
University of Santiago de Compostela (USC), Santiago de Compostela, Spain
e-mail: martin.vaz.alvarez@gmail.com

© Springer International Publishing Switzerland 2017
F.C. Freire et al. (eds.), *Media and Metamedia Management*,
Advances in Intelligent Systems and Computing 503,
DOI 10.1007/978-3-319-46068-0_61

seem as an obvious remark, but it is key to define each concept separately in order to understand the actual moment in this branch of journalism where the terminology and it's meaning changes as technology moves forward. Journalism has always been linked to the scenarios of information, where stories are born, and that's why some professionals state that mobility is inextricably linked to journalism. In this article, however, we understand mobile journalism (MoJo) as all journalistic practices that involve the use of mobile devices (such as tablets or smartphones). Regarding audiovisuals, when we put together the terms 'mobile' and 'video-journalism' we understand that every audiovisual piece produced with mobile devices that respects the basic rules of journalistic production belongs to this category. This means that mobile video-journalism does not only consist in filming with phones and tablets, but it also involves an editing process. It is precisely this element, editing, and its transition to mobile, which fills up with meaning the term 'mobile video-journalism'.

We will start the technical analysis in the year 2012, when Vericorder was already offering a full featured mobile editing system and also a content manager for storing and publishing the clips produced within the application. We will go all the way to the present state of mobile video-journalism, analysing the most recent changes in mobile software and the characteristics of the new devices.

61.2 Methodology

The methodology used to elaborate this article consists in the revision of the existing literature in mobile video-journalism (articles, books and technical manuals), as well as the author's personal experience in the use of these technologies and contents extracted from the contact with other professionals in the audiovisual sector.

61.3 Evolution or Revolution?

It's very difficult to determine the precise moment when mobile video-journalism entered the scene, given the fact that the technologies used for this purpose have been around for a long time now, and it was through the constant actualization of these, more in the qualitative aspects than functional ones. This means that mobile video-journalism didn't enter the scene at the time when it was possible to film and edit in mobile devices (from 2010 with Apple's iMovie software) but it really made a difference when the camera lenses and the editing softwares were powerful enough to produce audiovisuals pieces in a short time and with enough quality to compete with the traditional forms of video-producing. This also included the agility factor, key in the production of information.

61.3.1 Technical Improvements

One of the highlights of mobile videos was the improvement of camera lenses in mobile devices. In June 2009 Apple released the first iPhone capable of recording video: the iPhone 3Gs, which filmed in 480p at 30fps with 3.15 megapixel camera. A little less than 6 years after that, in September 2015, the last generation of iPhones hit the market. These new devices are now able to record in 4K at 30fps or 1080p at 60fps, amount other possibilities, with a 12 megapixel camera. Regarding editing softwares, the major improvements have all been focused in adapting to the increasingly larger video formats, although there have also been important functional progresses. Stephen Quinn wrote in his book 'MOJO: Mobile Journalism in the Asian Region' (Quinn 2012) about the software Vericorder Technologies, which allowed to edit and manage a great range of multimedia contents, from radiophonic reports, slideshows and video packages for TV. But what really made a difference was that this software also included the possibility to use a dedicated platform to send all the generated content back to the newsroom. This was the first software to offer an effective solution for one of the biggest challenges in mobile production: data transmission.

In the same book, the author also admits that given the speed with which all these technologies develop, new applications might take over the space in the next years. That's precisely what happened with Apple's last iMovie version, which has become the standard for mobile editing in the world of video-journalism.

61.3.2 Apple Takes Advantage

All that is said about mobile technologies must be constantly revised, as the agility in terms of development of both software and hardware make it almost impossible to state which one is at the top in each moment. Nevertheless, nowadays we can all agree that Apple has a certain advantage over the rest of the operating systems, for various reasons. The first reason is that Apple produces both the software and the hardware of its products, which means that their native applications have a level of precision and fluency that has been so far unmatched. Furthermore, it has one of the most powerful cameras in the market, and the video it produces is perfectly optimized to be edited in their native iMovie software. This software offers a simple and complete user experience, allowing the audio and the video to be edited separately, the inclusion of pictures and even subtitling and voice recording. The function of the dedicated server that Vericorder included has now been substituted by the channel's own internal servers or CMS (Content Management Systems) like iSite 2 or iBroadcast in the BBC (Elson 2013).

Another aspect in which iMovie has facilitated the transmission of contents from the device to the newsroom was the upgrade in the compression of videos, as well as their render times, although the size of the clips is still an issue to be solved.

A one and a half minute video edited in iMovie in 1080p, with pictures, voiceover and music has an approximate size of between 150 and 200 MB. With a standard mobile transmission set composed by data cards and stable environmental conditions in a 4G net, it would take one minute or less to transmit the video.

61.4 Field Trials in Mobile Video-Journalism

Mobile video-journalism is a very open style of journalism, offering different sets of possibilities regarding of both the purpose of the journalist and the technology in use. This also allows different degrees of immersion, which translates into the actual situation, with some media groups integrating mobile technologies completely (Léman Bleu), others including them in special channels as part of their mobile strategies (AJ+) and some others using these technologies in a practical way, using mobiles as support devices in their traditional filming setups. According to Judd Slivka, Assistant Professor in the Missouri Journalism School, the present panorama can be divided into 3 parts:

> Mobile gathering: the most practiced by organizations when they say they are doing 'mobile journalism'. It basically consists in gathering all audiovisual material through mobile phones or tablets to then come back to the newsroom an do the editing in professional non-mobile systems. It's very typical in established media that have made big investments in editing systems and therefore possess powerful equipment which allows mobile footage to be treated the same way as traditionally gathered footage.

> Uniplatform production: consists in the filming and editing of all the footage in the same device. It substitutes the traditional form of uploading and importing every single piece to the table editor, legating all the process on the mobile device.

> No gatekeeper journalism: all the footage filmed in extreme circumstances (wars, demonstrations, riots), which therefore does not follow the quality standards of an audiovisual clip for broadcast. (Slivka 2014: 1).

61.4.1 BBC Click in the 2015 Mobile World Congress of Barcelona

In March 2015, BBC Click's filmed and edited in 5 days its program about the Mobile World Congress of Barcelona using a wide range of mobile devices, softwares and gadgets, in order to test the real possibilities of mobile productions. The interesting part of this experiment was the diversity of the devices they used, which make the conclusions drawn from the experience much more precise.

The softwares used for the filming were Filmicpro and Cinema-FV5, for iPhone and Android respectively. Both softwares allow the user to control practically every parameter from focus points (for the depth of field) to the compression rate. The conclusion drawn by the Click team in this aspect was that the software Filmicpro had a great advantage over Android's software, as it only had to adapt its features to one particular system (iPhone), whereas Cinema-FV5 had to work on multiple devices made by multiple manufacturers, which made some features unavailable in certain devices (Kelly 2015). Image stability was also a problem, so they decided to use a Lanparte Three-Axis gimbal, a mobile stabilizer which costs around 350 €. Sound, on the other hand, was one of the most delicate elements during the shooting. The original idea was to use special Rode microphones, but these couldn't deliver enough quality so they had to use a Pro-Jive which allowed them to use their professional microphones linked to the mobile devices. Out of all the mobiles and tablets they used, the iPhone was the only one that allowed to monitorize sound during the filming.

For the editing they used Pinnacle on the iPad, which works in a simple and intuitive way, but only with material filmed in iPhone or iPads, which only allowed them to edit a small portion of the footage. They then decided to use Premiere Pro in a Microsoft Surface Pro 3, which seemed like the only software capable of editing 23 min of program with footage encoded in 9 or 10 different formats. Despite the constant freezing and lagging of the software, they managed to deliver the program on time, which made Click's World Mobile Congress of Barcelona the very first example of a solely mobile produced program to be broadcasted in professional television.

61.4.2 Léman Bleu and Their Mobile Newscast

From the 29th of June 2015 and until the end of that summer, local genovese channel Léman Bleu decided to produce their daily newscast using exclusively iPhones. The purpose to reduce costs and equipment weight lead them to try out this new format, which resulted in an inspiring experience in the way it forced them to explore new types of audiovisual language and offered them the chance to raise a question about how television should be made (Favre 2015). Lauren Keller, Léman Bleu's director once said in an interview to Le Temps, that he considers erratic thinking that the iPhone delivers a worse image quality than a standard camera.

Editing was also made on mobile devices, which made the whole experience 100 % mobile. Comparing this experiment with what Click did in the MWC of Barcelona we notice that they both had to use special equipment for the sound. This is the main weak point of mobile production, it is still not able to compete with the quality of professional microphones.

61.4.3 AJ+ (al Jazeera Media Network)

AJ+ was born on the 15th of September 2014, an exclusively digital platform, part of the the Al Jazeera network, but independent of this last in terms of style and content. There are no anchors in AJ+, no schedule, no television. The idea of AJ+ is to eliminate the parameters that define a typical television in terms of information aesthetics and content consumption. It's not about attracting the audience, but going where the audience is. That's the reason why their contents are spread through social media: Facebook, Instagram and Twitter, always at hand for the user, and specially focusing in creating content with and for mobile devices.

Between November and December 2014, AJ+ filmed Ferguson's unrest using an iPhone 5, a Monfrotto monopod, a Rode VideoMic Pro, and an Audio Technica clip microphone for the interviews, an iKan led light for night shootings and a wide angle lens adaptor. The total worth of the equipment was of about 600 dollars without the iPhone, and it allowed them to be the first newsmedia to broadcast the incidents on the spot, without sacrificing any quality (Weiss 2014). When AJ+'s reporter Shadi Rahimi was asked about the future of mobile video-journalism in the channel, she responded that their goal was to create a new generation of journalists: the mobile army of AJ+.

61.5 Discussion and Conclusions

Limitations in mobile video-journalism have a fundamentally physical basis: video is helplessly dependent on internet connection, specially if we use these technologies for their agility and not as much for its versatility. This means that without a good internet connection all the journalist's work outside the newsroom might be fatally compromised. This is an issue specially in countries with underdeveloped connections, but there are reasons to believe that this will probably improve 5 years from now, when 5G is expected to arrive. In this fifth generation of mobile network we will no longer speak about data transmission in megabytes per second, but in gigabytes. In the year 2014 Samsung tested the 5G network in a simulated real environment in order to check, as precise as possible, how the real user experience is going to be with this new network. They obtained a transference rate of 7.5 Gb/s in a car that circulated at 4 km/h, simulating the speed of a walking person; and they obtained 1.2 Gb/s when the car was driving at 110 km/h. That's about 25 times faster than the optimal speed we can expect from a 4G net. On a different note, the University of Surrey, United Kingdom, managed to obtain up to 1 Tb/s of data transfer in an optimized circuit, in non-natural conditions. This is the equivalent of an optic fiber cable's performance, but without the cables. This new network might be the solution mobile video-journalism has been waiting for.

The final conclusions drawn in regards of the present situation in mobile video-journalism are several. The first and most important is that it is already possible to create high-end broadcasting material using only mobile devices. New technologies open a wide range of possibilities to create new types of journalism, less intrusive (more discreet and light equipment, for interviews), faster and more connected to a society that is increasingly growing towards the audiovisual. There are also things to be improved, like connectivity, and on the technical aspects, the specialization of editing softwares to cover the most specific needs of journalists.

References

Elson, S. (2013). Content management systems at the BBC. *BBC Academy.* Retrieved from http://www.bbc.co.uk/academy/technology/article/art20130716172702924. Accessed at October 18, 2015.

Favre, A. (2015). Léman Bleu lance le téléjournal <<100 % iPhone>>. *Le Temps.* Retrieved from: http://www.letemps.ch/suisse/2015/06/24/leman-bleu-lance-telejournal-100-iphone. Accessed at October 25, 2015.

Kelly, S. (2015). Technology TV programme click made on mobiles. *BBC News.* Retrieved from: http://www.bbc.com/news/technology-31747911. Accessed at October 10, 2015.

Quinn, S. (2012). *MOJO—Mobile journalism in the Asian region.* Singapore: Konrad Adenauer Stiftung.

Slivka, J. (2014). We should probably define mobile journalism first. *American Journalism Review.* Retrieved from http://ajr.org/2014/03/13/probably-define-mobile-journalism-first/. Accessed at October 1, 2015.

Weiss, J. (2014). How AJ+ covered Ferguson with just a mobile device. *International Journalist Network.* Retrieved from https://ijnet.org/en/blog/how-aj-covered-ferguson-just-mobile-device. Accessed at November 3, 2015.

Chapter 62
Key Features of Digital Media Consumption: Implications of Users' Emotional Dimension

Javier Serrano-Puche

Abstract This chapter contains two objectives: (1) To identify general factors that are distinctive of digital consumption in the current media ecosystem and (2) to describe the connotations that said consumption has on human beings' emotional dimension. It further attempts to present key elements of this consumption, namely (a) the overabundance of information available to the user, (b) the acceleration of time as it is experienced, (c) the emergence of attention as currency, (d) the multiplicity of screens and devices, and (e) the socialization of consumption. The chapter finally concludes that the development of a critical and conscious media consumption, which is associated with emotional management, is one of the main issues in the analysis of digital technology and its everyday use.

Keywords Media consumption · Digital communication · Multiscreen · Digital technology · Emotions · Users

62.1 Theoretical Framework

In recent decades, research on emotions has made a lot of progress, not only in disciplines where one might expect interest in the topic (such as psychology, medicine and neurology), but also generally in humanities and social sciences. Moreover, emotions are not just an object of study, but indeed look toward a new epistemological turn (Clough and Halley 2007). This growing academic interest is closely linked to the rise of people's emotional dimension in social life, where major changes in the expression of emotions, both in the private and public spheres, are readily identifiable.

The growth of the affective dimension in the academic world and in society over the last two decades has coincided with the social implementation of information and communications technology (ICT). Digital devices have been widely adopted

J. Serrano-Puche (✉)
Institute for Culture and Society, University of Navarra (UNAV), Pamplona, Spain
e-mail: jserrano@unav.es

© Springer International Publishing Switzerland 2017   471
F.C. Freire et al. (eds.), *Media and Metamedia Management*,
Advances in Intelligent Systems and Computing 503,
DOI 10.1007/978-3-319-46068-0_62

among the population and their omnipresence and ubiquity is accompanied by the possibility of permanent connection to these devices. We are living, therefore, in an era of hyper-connectivity, where technology is fully integrated into our daily lives, to the point that it is no longer possible to separate it from most daily activities.

Thus, the affective dimension in social life is increasingly important, on the one hand, and, on the other, digital technologies have acquired a significant role in everyday interactions, leading to the emergence of a rich and complex field of study at the crossroads of both realities.

62.2 Methodology

To achieve the research objectives (indicated in the abstract), this article consists of a literature review combined with ethnography, both in the physical world itself and in the online environment (digital ethnographic observation). This technique lends support to the literature review, helping to illustrate the types of phenomena on which the article theorizes.

62.3 Results

When referring to media consumption, we refer to the use and habits associated with different media, which, given the process of media convergence, at present largely live under the umbrella of the Internet. The information and entertainment that the user acquires is determined by factors relating to the status and situation of each person, and are reflected in specific forms of consumption relative to the type of media and content, frequency, timing, patterns of consumption behavior, etc. However, it should be noted that in the contemporary media ecosystem there are some general elements that frame online consumption and affect users' emotional dimension. The key factors include: (a) the overabundance of information available to the user, (b) the speed with which interactions occur in the digital realm, (c) the emergence of attention as a currency, (d) the multiplicity of screens, and (e) the socialization of consumption.

62.3.1 Flooded With Information

Unlike previous eras in history, information is no longer a scarce good. An extraordinary wealth of information is an advantage, with more content available to people that is often free and instantaneous, but it also brings new challenges and can lead to some cognitive and psychological problems. In 1970, Toffler warned of information overload as a specific harm derived from this new social context;

therein, a person lacks the tools or skills to properly absorb an excessive amount of information.

Although information overload as a phenomenon precedes the popularization of the internet, this new digital ecosystem makes the problems and challenges that it brings more visible and complex; access to information, which the internet has principally reinforced, is constant and ubiquitous thanks to the popularity of mobile devices like tablets or smartphones. Moreover, this proliferation of screens, with which increased information consumption is associated, has shifted and reduced the importance book printing has traditionally had in knowledge acquisition and generation. In the digital age, knowledge is no longer treated as a finite series of precise and reliable content stored in repositories, but rather as a flow, that is, as a network of unlimited discussions and arguments (Weinberger 2012). However, we should not forget that "knowledge is not synonymous with unlimited access and a greater flow of information, but rather with interpretation, critical understanding and even recreation of this information within a spatial, temporal and cultural context" (Barranquero-Carretero 2013: 429).

Therefore, one of the biggest challenges of digital literacy is to offer users guidelines and effective strategies with which they can face information overload. Otherwise, it can cause various pathologies and can affect the emotional dimension associated with the use of technology, including "technostress" (Brod 1984), boredom (Klapp 1986) or "information anxiety" (Wurman 1989), among others.

62.3.2 Hyper-velocity and the Inflation of the Present

One of the most important transformations related to the consolidation of digital technologies refers to the coordinates of space and time as they are lived. However, this is a phenomenon that dates back to late last century and that, in its origin, predates the popularization of digital sphere as a place for social relationships. Harvey (1989) stressed "space-time understanding" as a hallmark of our time, where there is "an acceleration in the pace of life, with so many spatial barriers overcome that the world sometimes seems to collapse on us" (Harvey 1989: 240).

In the process, digital technologies have historically helped to shape the "culture of speed" proper to this era; and, at the same time, they are this culture's best example. The enormous speed at which time is experienced (Wajcman 2015) has led to changes in production and consumption processes, the organization of work, lifestyles, and in the way the brain processes information. We are living in an inflation of "now," a "presentism" (Rushkoff 2013) that also influences media consumption, since, as Díaz-Nosty (2013: 137) notes, "the overflow of information could lead, in some cases, to a kind of media bulimia, to compulsive consumption in real time, that exhausts itself and does not solve the needs of cognitive sedimentation".

Faced with a tyranny of immediacy and information in real time, voices emerge advocating for a broader and deeper sense of temporality that offers cultural durability against the increasingly rapid obsolescence that the market produces. From the perspective of the temporal dimension, which in turn should be reflected in patterns of media consumption, we must aim to find the right speed, that is, to act quickly when it makes sense to do so and to slow down when a slower pace is appropriate.

62.3.3   The Attention Economy

Another social coordinate that helps us to understand media consumption's present configuration involves the realization that, in developed societies, people's attention is a scarce commodity. We live in the "attention economy" (Davenport and Beck 2001), where content producers compete to capture people's interest and occupy their spare time. Thus, a wide range of new media chases potential users at any time and in any place. People are clearly bombarded with messages and cognitive stimuli, especially in the digital environment. Since technology decreases the amount of time it takes to send information and there are ever more agents willing to share with potential recipients, the "bandwidth" of information that the user receives grows continuously. However, in parallel, the amount of time one can devote to information is ever decreasing.

From the point of view of those that edit media content, the challenge is to achieve an effective balance between the information flow and users' actual absorption capacity. If we look at this from the point of view of consumer experience, it is obviously advantageous to develop strategies and habits that properly manage attention spans. The ability to cultivate concentration, and to navigate the information stream following what Lucchetti (2010) calls the "principle of relevance," is a crucial skill in conquering the flood of information and even in achieving a full life (Goleman 2013).

62.3.4   One User, Multiple Screens

The proliferation of technological devices has diversified consumption across multiple screens, each of which offer features that affect digital media habits and that are reflected in everything from the time and frequency of use, level of interaction with others, and the type of content that is accessed through them. At the same time, the use of digital technology brings out emotions in people that are more or less intense and more or less positive or negative. The digital emotional regime is primarily a system of emotional intensities in which the amount of emotion matters (measurable in the number of likes, retweets, times viewed, etc.). Each technological device, application or communication channel is coupled with an "affective bandwidth," this is, it allows for a certain amount of emotional information to pass (Lasén 2010). Moreover, the internet encompasses different socio-technical environments that allow emotions to surface in varying degrees such that the affective dimension is not revealed equally in all interactions and communicative situations that take place in the digital realm.

Meanwhile, the proliferation of digital devices has resulted in some users employing an increasingly common practice of consulting multiple terminals simultaneously during an activity. One of the most common practices in this regard is the use of phones or tablets while watching TV, which has been dubbed as "second screening" with mobile devices playing a secondary and support role for the big screen at home. However, multi-screen use is becoming ever more complex and diverse. For example, Smith and Boyles (2012) indicate that using mobile devices is not always associated with TV viewing, but rather has its origins in television content that does not engage viewers (during a commercial break, for example) and thus they turn to a mobile or tablet as a distraction, without totally leaving the former aside. Thus, often the "second screen" becomes first one in reality because it captures the user's attention, whether for interacting with others, searching for information, checking email or social networks, etc.

62.3.5 The Socialization of Consumption

From the perspective of media consumption, mobility represents a disruptive factor in that it has taken media use out of the domestic environment, replacing it with users that have their own digital devices, thus making access highly individual. But paradoxically, consumption has, at the same time, become more socialized, given the prominence that interconnection across networks and interpersonal contacts acquires in the online environment. The process of media consumption socialization is reflected in various aspects. On the one hand, information overload has led people to rely more on their social networking contacts as a filter that gives meaning to the overwhelming amount of information. However, this response may have an adverse side effect, i.e., the user only surrounds himself with his own ideological group and the internet turns into a juxtaposition of closed spaces, not into an open space for the exchange of ideas. On the other hand, the increasing socialization of media consumption is found in the motivations and common practice behind multi-screen usage. Often the use of phones and tablets during a TV program boils down to the fact that people are interested in commenting on the content and/or reading related comments that other users post (Giglietto and Selva 2014), either through social networks, especially Twitter, or through chat applications like WhatsApp.

More broadly, the atmosphere of emotionality found online buoys the social dimension of digital consumption. With regard to the monitoring of current affairs, it is easy to see that emotions are at the heart of the act of sharing content and news in the digital environment. On networks like Twitter, the timeline around certain events of a political or social character is an amalgam of information, opinion, interpretation and emotions, repeated and amplified by the network itself, giving rise to what Papacharissi (2014) calls "affective news streams."

62.4   Discussion and Conclusions

As noted throughout these pages, there are some general factors that shape contemporary media consumption. At the same time, given the emotional charge that the use of technological devices carries with it, digital habits for acquiring information and keeping in touch with other users always have an emotional component. The first of those factors is the extraordinary amount of information available to the user, which prompts him to implement an effective system for filtering information so as not to succumb to the information avalanche. Second, the acceleration of time as it is experienced leads to short-range, instant information being overvalued. Proper management of the cognitive and attention component emerges, moreover, as a necessary skill in the process of media education. This is a challenge considering that one of the distinctive elements of media consumption is its realization through multiple devices that often converge simultaneously, converting digital practice into a multiscreen experience. Finally, digital consumption has taken on an increasingly important social dimension, giving greater weight to platforms such as Facebook and Twitter and to user recommendations at the expense of traditional media sources.

References

Barranquero-Carretero, A. (2013). Slow media. Comunicación, cambio social y sostenibilidad en la era del torrente mediático. *Palabra Clave 16*(2), 419–448.

Brod, C. (1984). *Technostress: The human cost of the computer revolution*. Reading: Addison Wesley.

Clough, P. T., & Halley, J. (2007). *The affective turn: Theorizing the social*. Durham: Duke University Press.

Davenport, T. H., & Beck, J. C. (2001). *The attention economy: Understanding the new currency of business*. Cambridge, MA: Harvard Business School Press.

Díaz-Nosty, B. (2013). *La prensa en el nuevo ecosistema informativo. ¡Que paren las rotativas!*. Barcelona: Ariel/Fundación Telefónica.

Giglietto, F., & Selva, D. (2014). Second screen and participation: A content analysis on a full season dataset of tweets. *Journal of Communication, 64*, 260–277.

Goleman, D. (2013). *Focus: The hidden driver of excellence*. New York: Harper Collins.

Harvey, D. (1989). *The Condition of Postmodernity. An Enquiry into the Origins of Cultural Change*. Malden, Oxford, UK: Blackwell.

Klapp, O. E. (1986). *Overload and boredom: Essays on the quality of life in the information society*. New York: Greenwood Press.

Lasén, A. (2010). Mobile media and affectivity: Some thoughts about the notion of affective bandwidth. In J. R. Höflich, et al. (Eds.), *Mobile media and the change of everyday life* (pp. 131–154). Frankfurt am Main: Peter Lang.

Lucchetti, S. (2010). *The principle of relevance. The essential strategy to navigate through the information age*. Hong Kong: RT Publishing.

Papacharissi, Z. (2014). Toward new journalism(s). Affective news, hybridity, and liminal spaces. *Journalism Studies*, 27–40.

Rushkoff, D. (2013). *Present shock. When everything happens now*. New York: The Penguin Group.

Smith, A., & Boyles, J. L. (2012). *The rise of the 'connected viewer'*. Pew Internet & American Life Project, Pew Research Center. Retrieved September 24, 2015 from http://www.pewinternet.org/2012/07/17/the-rise-of-the-connected-viewer/

Toffler, A. (1970). *Future shock*. New York: Random House.

Wajcman, J. (2015). *Pressed for time. The acceleration of life in digital capitalism*. Chicago: The University of Chicago Press.

Weinberger, D. (2012). *Too big to know: Rethinking knowledge now that the facts aren't the facts, experts are everywhere, and the smartest person in the room is the room*. New York: Basic Books.

Wurman, R. (1989). *Information anxiety*. New York: Doubleday.

Chapter 63
Reports in and from Smartphones: A New Way of Doing Journalism

Alba Silva Rodríguez and Francisco Campos Freire

Abstract In the current ubiquitous society mobile devices have a predominant role. Over time they have permeated the global culture and have become the main authors of a journalistic change which affects numerous levels: emission, transmission and reception of information. While it is true that, in its conception, the information designed to the "fourth screen" had to meet the patterns of brevity, clarity, concision and accuracy, mobile editorial platforms such as The Atavist or The Atlantic have demonstrated that the "new journalism" have also a place in handheld devices. This communication meddles in the presence of the journalistic report in the web and mobile app versions. The aim is to determine whether the contents addressed in depth have also a place in the new platforms and to discover whether we are faced with the appearance of a new genre with particular characteristics. To that effect, the mobile apps of the four most read digital media of Spain, according Comscore (2015) to are taken as a reference. The method is based on an analysis of content which deals with issues related to the quantification of this genre, its structure, design and the presence of the multimedia language, interactive and hypertextual. The study is complementary with a reflection about the features which should be followed in order to produce specific contents from mobile phones. The results show that the report has a secondary role in the mobile apps of the analysed newspapers, besides lacking self-identity.

Keywords Mobile journalism · Mobile report · Smartphones · Apps · Ubiquity

A. Silva Rodríguez (✉)
Pontifical Catholic University of Ecuador—Ibarra (PUCESI), Ibarra, Ecuador
e-mail: albasilvarodriguez@gmail.com

F. Campos Freire
University of Santiago de Compostela (USC), Santiago de Compostela, Spain
e-mail: francisco.campos.freire@gmail.com

© Springer International Publishing Switzerland 2017
F.C. Freire et al. (eds.), *Media and Metamedia Management*,
Advances in Intelligent Systems and Computing 503,
DOI 10.1007/978-3-319-46068-0_63

63.1 Theoretical Framework

63.1.1 The Mobile Phone Revolutionizes the Information Consumption

"Smartphones for headlines and tablets for longer articles". This is one of the conclusions drawn by a study on the use of mobile devices among North American Adults conducted by Pew Research Center. It is highly plausible that in-depth reading is less compatible with on-screen reading, since social networks, emails and online advertising compete for the reader's attention, making it more difficult for him to focus. However, it is clear that the emergence of mobile phones and then smartphones, has led media companies to an important challenge: producing contents that can be spread within a new context, in which information is consumed in mobility (Silva 2013).

Both mobility and digital connectivity increase day by day. Consumption in smartphones grows on a daily basis and, in the majority of the cases, it has a multiplatform nature. According to the data published in the 2014 Mobile Behaviour Report the use of multiplatform devices is prevalent. Two thirds of the interviewed population use their mobile device while watching TV at least once a day, and 41 % of tablet owners use their smartphones and tablets at the same time at least once a day (Salesforce 2014: 13).

The barometer of reading and buying habits in Spain (FGE 2012) shows that mobile users prefer to read in their e-reader. Even if these data refer to reading in general, including magazines, blogs, books and other formats, the difference between the number of people who read via smartphone and that do it via e-reader is not as big as it could seem. At the time of the study, 4.6 of interviewees read books in their smartphone and 6.3 in the e-reader. It seems logical that these figures may change over the years.

Although mobile devices (tablets, smartphones or phablets) were originally considered entertainment support systems in which consumption patterns were focused on social networks and games, they started to gain prominence in the field of reception of informative contents. So much so that nowadays smartphones have surpassed e-readers as the most used reading support for digital books. Since the introduction of the first e-readers in the market in the mid-90s, the digital reading revolution has significantly changed the world. With the arrival of new screens, there are more and more users who choose to read using new mobile devices, either tablets or smartphones.

In Spain, the last report on book-reading habits using smartphones in 2014 points out that 28.6 % of people read in e-readers, tablets and smartphones, followed by 16.4 % of people who read via smartphones, i.e. an average of 7.67 million people. Of these, there are 18.5 % of people who read books (2.47 million people) (Conecta 2014). This same research indicated that over the last three years, the number of readers who use mobile devices has tripled. Moreover, press reading has also increased. In December 2010 2.5 % of population consumed digital press, whereas in January 2014 this figure increased to 18.5 % (Conecta 2014: 10).

Besides, the population's consumption patterns have changed. The growing number of screens in the market has resulted into a diversification of informative consumption. Users channel their consumption through multiple devices at the same time. A study on TV consumption in mobile devices conducted by the Association for the Research of Media Communication (AIMC) performed in 2015 states that Internet users are increasingly connected to the network and watching TV at the same time. Indeed, the number of people who combine both activities increased from 66.3 to 72 % over the last years. When performing both activities, attention generally tends to focus on the Internet, since 43.1 % of people pay more attention to their Internet activity, 15.3 % are focus on TV and use the Internet in punctual moments, and 41.3 % pays attention to both to the same degree.

Mobile phones, tablets and TV have experienced an increase in the participation rate regarding Internet access as compared to traditional devices. 9 out of 10 surfers use the mobile to connect to the Internet, ahead of laptops/netbooks and desktop computers (AIMC 2015). It is therefore not surprising that Mark Zuckerberg, the creator of the successful network Facebook, felt the need to launch a new product called *Instant Articles*, aimed to allow editors to create quick, interactive articles on Facebook. Mark knew how to take advantage of the particularities of the mobile device and managed to substantially improve the user's experience by offering interactive maps, videos with autoplay, the option to manipulate pictures from the phone, and the possibility to listen to the sound of headlines, among others. Furthermore, he ensured that these pages take an average of only eight seconds to load (Facebook 2015). This success aroused controversy among those who thought that those media joining the social network would lose followers on their websites and thus benefits and those who considered that it was a good opportunity to reach a massive audience that would end up redirecting to the desktop website. Some of them decided to join this initiative, e.g. *The New York Times, BuzzFeed, NBC News, The Guardian, BBC, Spiegel Online, National Geographic* and *The Atlantic*.

The possibilities offered by the digital support could lead to think that there is enough room for stories of all kind, for different types of genres. Nevertheless, the trend of online journalism has not followed this path due to the difficulty of reading long stories, particularly on some devices. That given, journalist Evan Ratliff from *Wired*, thought on a way of facilitating the reading of long texts and, at the same time, take advantage of the web format. This is how *The Atavist* (Image 63.1) was born. This platform publishes non-fiction stories for reading mobile devices such as iPad, Kindle and Nook. Each text has videos and links that enrich the reading process. Similarly, it allows the user to hear the story while, for example, driving. Besides, these stories have other peculiarity: they can be read, heard and commented. Behind *The Atavist* there is a team of experienced journalists, who have worked in well-known media such as *Wired* or *the New Yorker*.

However, mainly within the Spanish context news sites are contenting themselves with adapting their design to allow access via mobile phones. Nevertheless, their strategy for implementing new business models should be broader, taking new

Image 63.1 Screenshot from *The Atavist*

technologies such as geolocation and mobile payment service into consideration. When combined, both technologies can revolutionize the advertising market for local businesses. Advertising could lead users to the closest shops and the anonymous information obtained from mobile payments could allow to draw precise conclusions on the effectiveness of advertisement.

63.1.2 New Supports, New Genres

Journalistic genres are discursive forms used by digital journalism to properly communicate current news with the aim of serving society (López 2005). This involves meeting the requirements implied in the practice of journalism since its traditional conception as well as the essential qualities required by cyberjournalism: interactivity, hypertextuality, multimodality, and update frequency. Initially, a simplified typology was imposed. It divided texts into two broad modalities: (i) reportage of events—whose aim was to explain and describe the facts—and (ii) comment—opinion texts whose objective was to expose and argue ideas (Larrondo 2008: 178). However, over the years, it was proposed to classify the genres into three comprehensive groups: (i) informative, (ii) interpretative, and (iii) opinion genres. Larrondo points out that this is one of the most extended systematisations and that it coincides with the historical development of these genres in the three main phases of journalism: opinionated journalism, informative journalism and interpretative journalism (Larrondo 2008: 195). The traditional categorization of journalism genres (informative, interpretative and opinionated) has undergone a transformation since the arrival of cyberjournalism, which favoured the emergence of new expression modalities. Journalism genres find in this new context a set of expressive resources that can modify, enrich or even transform their predecessors.

Right now, we are facing a phase of reinvention. There are some who think that were are observing the emergence of new genre typologies, whereas others believe that the renewal of the classic categorizations is the key to the future (Cantalapiedra 2009: 1). Nevertheless, it is clear that mobile devices have reopen the debate on genres, as it happened in the last decades with the emergence of websites. Traditional genres remain and are preserved in the net, and new narrative models such as "hypermedia and multimedia reports, interactive interviews between a character and the readers, chats, forums and multimedia infographics" have been added to those (Larrondo 2008: 206). Salaverría (2003: 3) is aware that cyberjournalism is undergoing significant changes, which are based on the reconfiguration of genres. The author mentions the presence of traditional genres and formats in the digital support, such as news, interviews, chronicles or reports but for him the difference lies in the mutation all these have suffered at a formal and structural level. Among them, the following should be highlighted: digital interviews, surveys and polls or comment areas. Professor Canavilhas believes that cyberjournalism differs from traditional journalism in the introduction of new specific genres. For the author, interactive multimedia infographics, panoramic photos (360° pictures that allow the user to browse an image) or *serious games* (interactive games) are examples of exclusive journalism genres that add value to online publications as a final product (Canavilhas 2010: 56).

Many authors have theorized about the development and the transformations genres have suffered over time. Salaverría highlights four major categories that explain these changes: repetition, enrichment, renewal and innovation.

1. Repetition: this model is determined by the literal reproduction of genres and textual formats in the cybermedia, these being taken from previous media, generally in print form.
2. Enrichment: it is reached when the genre adds textual, multimedia or interactive possibilities by taking advantage of the communicative features of cyberspace.
3. Renewal: it involves the full reconfiguration of a previous genre by using the communicative possibilities of cyberspace.
4. Innovation: this category is reached once new cyberjournalistic genres for cybermedia have been created without using prior referents from the print and audiovisual media. Salaverría exemplifies this case with *weblogs*, which he considers to be "one of the first innovative genres created by cybermedia while reporting on daily issues" (Salaverría 2005: 149).

Table 63.1 Cyberjournatistic genres

Informative genres	– News item
Interpretative genres	– Reportage (reportage of current events, special reports, documentary dossier) – Chronicle
Dialogical genres	– Interview – Forum or debate – Chats (Online interview, interaction with personalities, interaction amongst users) – Survey
Opinion genres	– Traditional: editorial, comment, criticism, letters to the editor, article, column, cartoons, etc. – Networked debates: forums, chat
Infographics	– Individual and collective infographics

Source Prepared by the authors on the basis of the data from Díaz Noci and Salaverría (2003)

Nowadays, it would be convenient to review the traditional categorization of genres and how it was adapted to those features of cybermedia that allow us to establish a new typology. For this purpose, we follow the proposal of professors Díaz Noci and Salaverría (2003: 40–41), shown in Table 63.1.

It should be highlighted that in cybermedia "the news item is the hegemonic genre" (Edo 2009: 57). However, in new narrative independent minimum units of content are a relevant feature and have to be taken into account (Gago 2011). Hypertextual construction transforms the structure of information in such a way that they cause the emergence of new formats that arouse the curiosity of users. Microformats or microblogging are some of the examples that explain this phenomenon. Texts are shorter, simpler and divided into parts. As some authors as Salaverría and Cores point out hypertextuality has marked a turning point in the development of genres (Salaverría and Cores 2005: 147).

The development of a mobile environment in journalistic production (Fidalgo and Canavilhas 2009) is progressively leading to a type of journalism based on shorter messages, with an instantaneous nature that has to be continuously updated. In this context, the mobile environment of production helps to reconfigure the field of journalism in terms of the generation and dissemination of information.

63.2 Methodology

The method applied in this study was based on the analysis of the content of reportages published on the front pages of mobile websites and apps from the four benchmark mainstream media, with a print format, in Spain (El País, El Mundo, La Vanguardia & ABC) during the months of September and October 2015. Those media were chosen taking into account their audience ratings, which make them

| Audiencia en España de diarios online | TODOS LOS DISPOSITIVOS | | VISITANTES EXCLUSIVOS DISPOSITIVOS MÓVILES |
| | Abr. 2015 | % variación mes anterior | Abr. 2015 |

		TODOS LOS DISPOSITIVOS Abr. 2015	% variación mes anterior			VISITANTES EXCLUSIVOS DISPOSITIVOS MÓVILES Abr. 2015
1	EL Pais Sites	13,643	-10%	1	EL Pais Sites	6,145
2	Elmundo.es Sites	12,768	-9%	2	20MINUTOS Sites	5,870
3	20MINUTOS Sites	11,212	-5%	3	ABC.ES Sites	5,742
4	ABC.ES Sites	11,070	-9%	4	Elmundo.es Sites	5,005
5	La Vanguardia	6,639	-21%	5	La Vanguardia	3,100
6	ElPeriodico	5,232	-2%	6	ElPeriodico	2.886
7	La Voz De Galicia	3,397	5%	7	La Voz de Galicia	1,998
8	La Razon - Spain	2,986	-11%	8	Ideal	1,582
9	Las Provincias	2,742	-16%	9	Las Provincias	1,578
10	Ideal	2,435	-17%	10	La Razon - Spain	1,233

Fuente: comScore Multiplataforma, Abril 2015 / millones de visitantes únicos. Categoría Newspapers

Fig. 63.1 Audience ratings of digital newspapers in Spain. *Source* Comscore Multiplataforma, February 2015

"benchmark". Since there are practically no data on the measure of mobile audiences in the chosen countries, we worked on the basis of the criteria of print run and dissemination of the different journals. In this case, the audience ratings of the traditional supports (Web) and the multiplatform devices provided by the consultancy firm Comscore match (Fig. 63.1).

All of them are print media and they are the newspapers with the highest digital audience ratings in their countries. They are considered "benchmark" journals. This terminology comes from Merrill (1968), who identifies a "quality newspaper" with "benchmark newspaper" or "elite newspaper". This synonymy is established as a guarantee of homologation for the different media, taking the prestige of their print version as a guideline, and being aware, at the same time, that their quality as benchmark media does not extend to their level of dissemination. Empirical observation was completed with theoretical contributions in order to establish an in-depth debate about the future of information in mobile devices.

63.3 Results

63.3.1 Analysis of the Journalistic Genres Used in Websites and Mobile Apps

First of all, we observed which journalistic genre was predominant on the front pages of the mobile apps from the different journals. In this section every piece was classified according to its genre. The starting point was the theory of genres from the Spanish journalistic tradition (Casasús and Ladeveze 1991), according to which they are classified into the main classic groups (informative, interpretative and opinion, to which the authors add argumentative and instrumental). Its adaptation to cybermedia was also considered (Díaz Noci and Salaverría 2003). The studies from Ainara Larrondo on cyberjournalistic genres and the influence of hypertextuality on the configuration of new genres (Larrondo 2009) were also taken as a reference. As noted by López (2007: 98), hypertext meant "changes in the conception of the close structure of information and causes modifications in the traditional elements of press informative genres: news item, chronicle, short reportage, etc."

On the basis that mobile screen was conceived for the consumption of short, simple, clear and direct texts, it seemed interesting to research about the presence of interpretative genres, particularly reportage, in the different versions of mobile websites and apps of the selected journals. Regarding this first category, the predominant genre in all of them is informative news. However, even if news item continues to be the genre par excellence, as shown in Table 63.2, it is noteworthy that there are some cybermedia that highlight the reportage. This is the case with *El País*, in which this genre has a clear predominance within the "national" section. If we compare the data from the selected national media with those of the European and international newspapers (Chart 63.1), we observe that the trend is the same, being *The New York Times* the digital newspaper which includes the highest number of reportages on its mobile front page.

Another interesting aspect is that chronicle is the genre that, after news item and reportage, takes on greater relevance on the main websites of mobile apps. This is probably related with the fact that this genre is normally used in international news, one of the most relevant sections on the front pages of mobile apps (Chart 63.2).

Table 63.2 Predominant genres on the front pages of mobile apps

Genres	EL MUNDO	EL PAÍS	LA VANGUARDIA	PUBLICO	ABC
News item	17	18	18	8	24
Reportage	2	4	0	1	0
Chronicle	3	0	0	0	0
Interview	0	1	0	0	0
Opinion	1	1	0	0	1
Editorial	0	0	0	0	0

Source Prepared by the authors

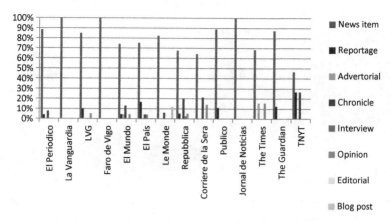

Chart 63.1 Journalistic genre used on the front pages of mobile apps. *Source* Prepared by the authors

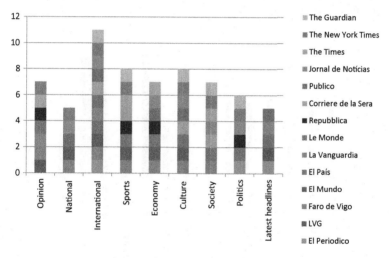

Chart 63.2 Highlighted sections on front pages classified by subject. *Source* Prepared by the authors

63.3.2 The Picture: The Focus of Attention in Reportages on Mobile Apps

Image 63.2 shows how a reportage is structured on the mobile app. If we compare the screenshot from the mobile phone with that of the desktop, we can determine that there are differences between them. Although these differences are not significant, they allow us to draw valuable conclusions in this regard. As shown in the screenshots, the interpretative genre in the native app from *El País* gives a clear

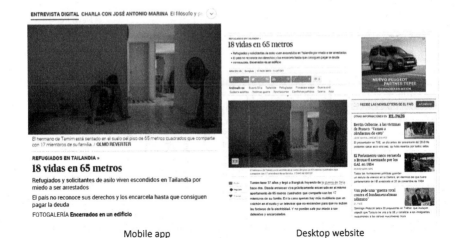

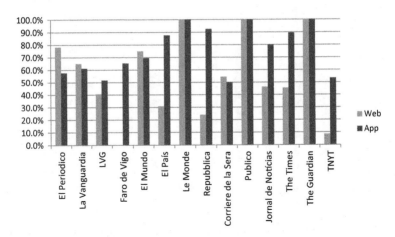

Image 63.2 Reportage on the mobile app and the desktop website of *El País*. *Source* Prepared by the authors from a screenshot of *El País*

Chart 63.3 Percentage of news items with pictures on websites and mobile apps. *Source* Prepared by the authors

predominance to pictures. This comes as no great surprise, given that pictures (Chart 63.3), together with videos, are one of the main attractions for mobile users. So much so that pictures can push headlines to the background.

Furthermore, it should also be noted that on the app, *tags* are brought to the bottom of the screen, whereas on the desktop version this element is placed just at the beginning of the reportage.

Although it is true that these changes seem insignificant they somehow point to the emergence of a new structuration of journalistic text on the small screen.

However, it is remarkable that the drafting remains immutable in the new communicative support.

63.4   Discussion and Conclusions

Are we facing the slow disappearance of reportages in the digital press for mobile devices? There are sceptics who believe that the small screen is not appropriate for long text and the pace of life of today's society. Nevertheless, the data on reading through mobile phone invite us to consider the appropriateness of leading the editorial efforts towards the scene of mobile devices.

Users have changed and the consumption patterns with them. They require digital contents that can be consumed on mobile phones and tablets on which they can encircle the main content of enriched texts, links, pictures, audios and videos, social interaction and sharing, access to supplementary data, etc.

Nowadays, the CEOs of communication companies have concerns about the profitability of mobile technology. Hence, the main cybermedia have adopted a passive attitude and are not able to innovate and thus develop a new model of sustainable business. This scepticism is the main cause to the lack of progress and use of the opportunities offered by geolocation and mobile interaction.

Tablets and smartphones have conquered the field of journalistic reading. The predominance of news items on the specific app versions of the main Spanish digital journals proves that reportages and the rest of interpretative genres are far from being predominant on the small screen.

The successful initiatives in this area show that if the information is not synthesized, transforming the reading experience into a dynamic fluid interaction between the text, the pictures, the hypertext and/or the geolocation, it is highly likely that success will never be achieved.

In conclusion, and taking up the abovementioned Salaverría's classification, we can state that the current phase of genre evolution on mobile devices is stagnated in "repetition". The patterns from the desktop website are repeated and, even if there is a perceived existence of a new structure of design and browsing, there is still a long way to go.

Acknowledgments This article is part of the activities developed within the framework of the research project on mobile communication "Study on journalism through mobile devices in Latin America and development of a research network in Ibero-America and the Nordic Countries of Europe specialised in the analysis of mobile communication", coordinated by Alba Silva Rodríguez from the Pontifical Catholic University of Ibarra in Ecuador. This text has also been written within the framework of the research projects from the group Novos Medios, funded by the "Consolidation and Structure Program of Competitive Research Units" (reference: GPC2014/049) from the Department of Culture, Education and University Management from Xunta de Galicia. Moreover, it is part of the research project Innovation and development of cyber media in Spain. Architecture of the journalistic interactivity in multiple devices: information, conversation and service formats" (Reference: CSO2012-38467-C03-03), funded by the Ministry of Economy and Competitiveness.

References

AIMC. (2015). *Encuesta AIMC a usuarios de Internet.* Available on http://www.aimc.es/-Navegantes-en-la-Red-.html

Canavilhas, J. (2010). Los retos del webperiodismo: lenguaje, recursos humanos y modelos económicos. In N. IVARS (Ed.), *I Congreso Internacional de Comunicación Audiovisual y Publicidad: Internet y la Información* (pp. 50–65). Alicante: Limencop. Retrieved from http://www.in2web.es/cicap/publicaciones/internetylainformacion.pdf

Cantalapiedra, M. J. (2009). ¿Una mera transposición?. Los géneros periodísticos en la Red. *Telos, Cuadernos de Comunicación e Innovación, 59.*

Casasús, J. M., & Ladeveze, L. (1991). *Estilo y géneros periodísticos.* Barcelona: Ariel

Conecta. (2014). Our changing world. La lectura en dispositivos móviles. *Conecta.* Retrieved from http://es.slideshare.net/conectarc/la-lectura-en-dispositivos-mviles

Díaz Noci, J., & Salaverría, R. (2003). *Manual de redacción ciberperiodística.* Barcelona: Ariel.

Edo, C. (2009). La noticia en Internet: cibermedios, blogs y entornos comunicativos emergentes. *Estudios de Periodística, 15.* Madrid: SEP.

Facebook. (2015). Introducing instant articles. In blog de *Facebook.* Retrieved from http://media.fb.com/2015/05/12/instantarticles/

FGE. (2012). *Barómetro de hábitos de lectura y compra de libros 2012.* Retrieved from http://www.federacioneditores.org/0_Resources/Documentos/130207NPR-FGEE-BarometroHabitosdeLectura2012.pdf

Fidalgo, A., & canavilhas. (2009). O celular de Heidegger – comunicaçao ubíqua e distância existencial. *Revista Matrizes.* Retrieved from http://www.matrizes.usp.br/index.php/matrizes/article/view/121/197

Gago, M. (2011). *Deseño editorial e arquitectura da información nos cibermedios internacionais de referencia.* PhD thesis, Santiago de Compostela, University of Santiago de Compostela.

Jenkins, H. (2008). *Convergence Culture. La cultura de la convergencia de los medios de comunicación.* Barcelona: Paidós.

Larrondo, A. (2008). *Los géneros en la redacción ciberperiodística. Contexto, teoría y práctica actual.* País Vasco: Servicio Editorial de la Universidad del País Vasco.

Larrondo, A. (2009). La metamorfosis del reportaje en el ciberperiodismo: concepto y caracterización de un nuevo modelo narrativo. *Comunicación y Sociedad, 12*(2), 59–88.

López, M. (2007). *Cómo se fabrican las noticias: fuentes, selección y planificación.* Barcelona: Paidós.

López, X. (2005). El ciberperiodismo cultiva sus señas de identidad. *Ámbitos, 13–14,* 45–58.

Salaverría, R. (2003). Convergencia de los medios. *Revista Latinoamericana CHASQUI, 081,* 32–39.

Salaverría, R. (2005). *Cibermedios. El impacto de Internet en los medios de comunicación en España.* Sevilla: Comunicación Social.

Salaverría, R., & Cores, R. (2005). Géneros periodísticos en los cibermedios hispanos. In R. SALAVERRÍA (Coord.), *Cibermedios. El impacto de Internet en los medios de comunicación en España* (pp. 145–185). Sevilla: Comunicación Social.

Salesforce. (2014). Mobile behaviour report (Inform). Retrieved from https://www.exacttarget.com/sites/exacttarget/files/deliverables/etmc-2014mobilebehaviorreport.pdf

Silva, A. (2013). Los cibermedios y los móviles: una relación de desconfianza. *Icono 14, 11*(2), 183–207.

Part IX
Education, Science and Cultural Identity

Chapter 64
Crowdfunding: An Alternative for Collaborative Creation and Production in the Spanish Cultural Sector

Mónica López-Golán

Abstract In a context of continuous technological development, the culture sector has undergone many transformations thorough its history. However, in the last years a significant change has been produced. Apart from the digital revolution and the media convergence, there have been other destabilizing factors in the industry, such as new forms of consumption and the economic downturn. This process of change has caused a new ecosystem in which new tendencies constantly arise, based on the Internet and directly associated with the changes in the behavior of the final consumer of the cultural product. One of the most remarkable trends is the emergence of collaborative funding systems (crowdfunding). With this strategy, the creator opens the production phase to users by raising funds. Crowdfunding has become an alternative for the Spanish cultural industry, particularly since 2010 when the first websites specialized in crowdfunding were created. These platforms include categories addressed to all the cultural sectors and offer additional services such as distribution through direct selling or the possibility of taking part in the product as a creator.

Keywords Crowdfunding · Internet · Cultural field · Consumer

64.1 Theoretical Framework

The interest for the studies on the economic impact of culture and media began in the United States in the 70s with pioneers such as Schiller and W. Smythe. The latter starts his theorization trying to hide what he considered to be a blindspot in the study on the role media play for the capital, left by both Marxist theorist and critic economists.

Smythe defines media audiences as goods. Therefore, the main objective of the media is to create blocks of goods to sell them to the capitalism's advertisers, who have to create a profitable market for each good (Quirós and Sierra 2001: 26–27).

M. López-Golán (✉)
Pontificia Universidad Católica del Ecuador Sede Ibarra, Ibarra, Ecuador
e-mail: molopez@pucesi.edu.ec

© Springer International Publishing Switzerland 2017   493
F.C. Freire et al. (eds.), *Media and Metamedia Management*,
Advances in Intelligent Systems and Computing 503,
DOI 10.1007/978-3-319-46068-0_64

In Spain, some years later culture has a remarkable role as an engine for economic growth, being Ramón Zallo and Bustamante the main promoters. Zallo states that the legitimacy of an economy of culture has been proved not only by the significance of its weight on GDP and employment and but also by its social value (Zallo in Bustamante 2011). This author alerts about the need of an economy that covers all the areas and he clarifies that the economic form of culture is not uniform. "Culture can be capital, product or service" (Zallo 2011: 161).

In this study we consider media to be industrial and commercial organizers that produce and deliver goods with use value in a global and digital era, and whose business models has been affected by different factors. Before analyzing these factors, the main economic traces in the cultural sector should be pinpointed:

1. The economic structure in this sector presents the particular feature of being based on a product with a symbolic nature that can neither be depleted, nor destroyed in the act of consumption.
2. Cultural goods are not produced with a sole purpose, but there is a segmentation in the ways of making profits depend on the types of consumption, the sort of product and the private or public nature of industries.
3. The cultural sector has the inherent need of a constant production, characterized by its constant renovation, which will be different depending on the type of product (Zallo 2011: 164).
4. The cultural worker is not usually a wage-earner. He gets money from different formulas such as copyright-based remunerations or price tickets (Tremblay 2011).
5. It is an economy of multiple supply (Zallo 2011: 163).
6. Addressing those risks inherent to the unpredictable character of demand is difficult.
7. It has a social functionality that leads to the generalized demand of culture as knowledge, information, symbolic consumption... (Zallo 2011: 165) and that goes beyond the economic value.
8. Globalization and digitalization have changed the cultural sector, leading to a growing international competence and to the need of opening new markets and business models related to new technologies.
9. With the arrival of new technologies users can copy or reproduce a product and enjoy it as many times as they want.

64.1.1   Culture's Behaviour Amid the Economic Crisis and the Process of Digization

Cultural industry has been one of the most affected by the global economic crisis. Before the crisis, there was a period of growth—its highest peak being in 2008—when the sector benefited from different factors. On the one hand, there was an increase in household expenditures in equipment, access to audiovisual products and a higher demand of cultural products. On the other hand, public expense was constantly increasing as well as the size of the digital contents industry. After that, the downward trend began.

In a sector characterized by its fragmentation, those big groups capitalized by few companies are being the less hit by the crisis. On the contrary, small business have been the biggest contributors to the increase of the unemployment rate in the cultural industry. Moreover, after a cycle of opulence we have found ourselves surrounded by a wide range of cultural products that far exceed demand. As a consequence, the financial crisis has been exacerbated by this over-production.

On the other hand, the period of technological change the cultural sector is going through has not settled productivity models, due to different problems (Moragas I Spa 2011: 20):

- *Reduction of advertising income from traditional media*
- *Difficulties in obtaining credits*
- *Questioning of one of the key aspects of the traditional business model for media*: copyrights and direct payment for information.

Finally, there are other destabilizing elements for digital products in cultural sectors, such as piracy, new ways of use and a tendency to consume free products.

64.1.2 Internet as the Perfect Instrument for Creators and Users of Cultural Products

The Internet has entered the scene in practically every area of culture, forging new ways of developing and distributing products and thus developing into what Manovich (2005) calls "computerized culture". This author highlights the importance of the association between the computer and the new cultural industries, whose evolution has favoured their change from a mere storage device to a tool for creating, delivering and exhibiting.

Given the objectives of this study, we want to highlight the opportunity the Internet offers to creators, giving them the chance to produce, distribute and exhibit their own work. Apart from this, by reviewing the concept of *crowdfunding,* we can state that the net reformulates the traditional value chain established by cultural industries, creating a more reduced one. As a result, cultural creatives have never had a better opportunity to get audience, seek its collaboration, produce a product and distribute it through the Internet.

Likewise, in this time of media convergence and "participatory culture" (Jenkins 2008), the net has changed the way of consuming culture. The Internet turns the user into the main beneficiary of the new uses, mainly instrumental, related to daily life and leisure time:

1. *Social networks, chats and blogs* allow a new type of communication that had not been provided by any other media before. From many to many.
2. *Individuality,* the user has a certain tendency to use the Internet individually.
3. *Immediacy and mobility,* the use of the Internet through mobile devices has led to an immediate use no matter where they are used.
4. *Availability and accessibility,* this feature makes the Internet the main source of information.
5. *A tendency to create avatars in the digital worlds.*
6. *The Internet has created a multitasking user.*

On the other hand, moving the contents of cultural industries to the net has fostered a higher degree of interactivity, giving rise to new consumption habits:

1. *Migration towards consuming contents from traditional media on the Internet.*
2. *A tendency to personalize these contents.*
3. *The high availability of contents and file exchange has led to the storage of thousands of cultural products.*
4. *Depending on the device, the new consumer makes use of the Internet to perform different activities.*

This context of convergence generates the already mentioned "participatory culture". At this point, and to finish this section, we can distinguish four types of participation:

- *Discursive participation* through forums, comments in pieces of news and blogs and chats that the industry makes available for the user.
- *Creative participation,* channelled through contests that require an active commitment from the users and that leads to the *User Generated Content* (UGC).
- *Financial participation (crowdfunding).*
- *Participation for leisure.*

64.2 Methodology

This communication gathers a conglomerate of theoretical reflections obtained from different empirical researching processes from a qualitative perspective based on the analysis of paradigmatic cases. These cases include national referents of funding through crowdfunding, through the website of the project and platforms specialized in this strategy.

64.3   Plataforms and Crowdfunding Projects in Spain

Crowdfunding is a strategy of collective funding which main objective is to gather a quantity of money enough to support the development of a business, cultural or social project. Depending on the project, this formula distinguishes three modalities: *equity crowdfunding (startups); reward-based crowdfunding* (donations and microcredits) and *creative crowdfunding* (artistic and cultural projects), which is the modality to which we refer to in this document.

The most modern version of creative crowdfunding was born in the United States. The British Rock-Band Marillion (1997) has been the pioneer in the phenomenon of online collective funding. In Spain, the first collective-funded project was *El Cosmonauta* (2009), a science fiction film project.

By the end of the last decade the first platforms specialized in crowdfunding were born. Its function is to manage the funding from their different projects. There are general platforms, with projects from different types or specialized in specific fields. Furthermore, there are different fundraising possibilities for promoters (Baeck et al. 2012):

- *All or nothing:* the creator will only get the funds if the fundraising target is met within the time stablished by the platform.
- *Keep it all:* with this option creators get all the proceeds raised within the time established for the campaign.

Concerning platforms specialized in creative crowdfunding in Spain, there are different spreading possibilities both at a national and international level. Hence, *Kickstarter,* a North American platform funded in 2009, is one of the best options for our creatives to fund projects with potential for global dissemination. This platform has the biggest virtual community in the market. Almost 10 million people have contributed with more than 2 billion euros to 95,000 projects (Kickstarter Stats 2015). Cinema and video is the category with the highest percentage of successfully funded projects (Kickstarter 2015).

Spanish platforms can boast of positive data in the cultural sector. *Verkami,* one of the pioneering platforms in this country, has been able to fund almost 3500 projects with the help of 400,000 people. Music and cinema are the most successful categories in this platform, and the most supported project to date has been the documentary film L′Endemà (Verkami 2014). Other successful platform in Spain is *Lánzanos,* born by the end of 2010. *Lánzanos* has 300,000 registered users and has raised more than 5 million euros (Lánzanos 2014).

Finally, *Goteo* is a platform which helps not only to fund creative projects, but also others from different categories, such as educational, social or scientific among others. It is a social network of collective funding and distributed collaboration for projects that offer both individual rewards and collective return.

64.3.1 Some Paradigmatic Cases

In the brief history of cultural crowdfunding in Spain, there are some achievements that are worth mentioning. The first one is *El Cosmonauta,* which raised a considerable part of its funding through crowdfunding thanks to the constant activity of its team, both in digital and face-to-face environments. Besides, *El Cosmonauta* introduced the category micro-investors, i.e. it offered the community the opportunity to donate money in exchange for a percentage of the profits obtained from commercializing and exploiting the film. This project obtained the rest of the budged from private funding and public funding from the Institute of Films and Audiovisual Arts (Instituto de la Cinematografía y de las Artes Audiovisuales, ICAA).

Moreover, *El Cosmonauta* is a transmedia product whose main axis, the film, comes with other products. The film's crew decided their viewers would choose where, when and how they wanted to see the film. This is why *El Cosmonauta* was simultaneously premiered online, on DVD, on TV and in movie theatres.

In June 2014, the system designed to watch the film for free stopped working due to the impossibility of keeping it open, owing to its lack of profitability. Furthermore, the production company had to return the grant given by ICAA. The confluence of all these difficulties has resulted in the company's shutdown (El Cosmonauta 2015).

Other remarkable case is *L'Endemá,* a documentary film about the Catalan independence. It was launched in February 2013 through the platform Verkami and it met its 150,000 economic goal in only 11 days. Its campaign raised 350,000 € from more than 8100 patrons.

The most recent case is *El Español,* a digital newspaper that has just started publishing. This project finished its collective funding phase in February 2015. It raised 3,606,600 €, more than twice the crowdfunding world record for journalism. This new media, chaired by the former director of *El Mundo,* Pedro J. Ramírez, has 5624 stakeholders, responsible for raising the millions. *El Español* thus joins the list of the multiple media projects that opted for fundraising to be created through the Internet and that proves the effectiveness of crowdfunding to fund alternative journalism.

64.4 Discussion and Conclusions

The cultural industry is not alien of the public's practices. Crowdfunding enables a direct relationship between creatives and consumers, allowing the Exchange of ideas or the access to exclusive experiences. Therefore, it is a formula that creates a special union with a public that is potentially interested in a project.

From the community's point of view, the fan phenomenon—fandom—has great importance to generate enthusiasm, implication, dissemination and money. Despite the fact that fans already had their own networks before the Internet came to exist, according to Jenkins, the technological convergence has increased their ability to be promoting agents, and this has not gone unnoticed for those cultural creators using crowdfunding.

Collective funding gives creators a control over the management of the process in all phases. Moreover, it provides them with autonomy and independence to develop a cultural project without business intermediaries. Campaign's promoters and platforms' managers are compelled to follow rules of principles of openness and to render account to patrons in order to generate credibility among the members of their community.

Cases such as *El Cosmonauta,* prove that *crowdfunding* is an alternative funding option. However, this may not be enough to ensure the public acceptance of the project once the process has been concluded.

References

Baeck, P., Collins, L., & Westlake, S. (2012). *How the UK's businesses, charities, government, and financial system can make the most of crowdfunding.* Retrieved on November 8, 2015 from http://www.nesta.org.uk/publications/crowding

Bustamante, E. (2011). *Industria creativas, amenazas sobre la cultura digital.* Barcelona: Gedisa.

El Cosmonauta. (2015). Retrieved on November 8, 2015 from http://www.elcosmonauta.es/blog/?p=267

Jenkins, H. (2008). *Convergence Culture. La cultura de la convergencia de los medios de comunicación.* Barcelona: Paidós.

Kickstarter Stats. (2015). Retrieved on November 8, 2015 from https://www.kickstarter.com/help/stats?ref=footer

Kickstarter. (2015). Retrieved on November 8, 2015 from https://www.kickstarter.com/projects/ryangrepper/coolest-cooler-21st-century-cooler-thats-actually?ref=most_funded

Lánzanos. (2014). Retrieved on November 8, 2015 from http://www.lanzanos.com/

Manovich, L. (2005 [2001]). *El lenguaje de los nuevos medios: la imagen en la era digital.* Barcelona: Paidós.

Moragas I Spa, M. (2011). La comunicación de nuevo escenario mediático. El papel de la investigación. In F. Campos (Coord.), *El nuevo escenario mediático.* Zamora: Comunicación Social.

Quirós, F., & Sierra, C. (2001). *Crítica de la Economía Política de la Comunicación y la Cultura.* Sevilla: Comunicación Social Ediciones y Publicaciones.

Tremblay, G. (2011). Desde la teoría de las industrias culturales. In E. Bustamante (Coord.), *Industrias creativas; amenazas sobre la cultura digital.* Barcelona: Gedisa.

Verkami. (2014). Retrieved on November 8, 2015 from http://www.verkami.com/year/2014/es

Zallo, R. (2011). *Estructuras de la comunicación y la cultura. Políticas para la era digital.* Barcelona: Gedisa.

Chapter 65
Architecture Communication in Online Magazines

Jesús Ángel Coronado Martín, Julia Fontenla Pedreira
and Darío Flores Medina

Abstract Communicate technical data to a non-specialized audience in areas such as architecture is something practically non-existent in the current reality, although the electronic media have facilitated the arrival of certain concepts to a non-specialist public. In the information age, mass media are decisive in disseminating rigorous information in the architecture sector, and nowadays, society dominates key concepts in other fields such as automotive and informatics, getting to know elements such as cylinder capacity, power or operating system. Why is not the same reality in the field of architecture? Data relating to thermal and acoustic coefficients and energy ratings are key for people to obtain the maximum performance of the place where they aspire to build their home and working place. The poor dissemination of technical data is analyzed so that the architecture communication is not rooted in a simple design for discussing the beauty of the sketches made by the architect.

Keywords Communication · Architecture · Specialized information · Technical communication

J.Á. Coronado Martín (✉) · D. Flores Medina
Facultad de Educación, Ciencia y Tecnología,
Universidad Técnica del Norte (UTN), Ibarra, Ecuador
e-mail: jcoronado71@gmail.com

D. Flores Medina
e-mail: dariofloresmedina@gmail.com

J. Fontenla Pedreira
Pontificia Universidad Católica del Ecuador - Sede Ibarra (PUCESI),
Ibarra, Ecuador
e-mail: julia.fontenla@gmail.com

© Springer International Publishing Switzerland 2017
F.C. Freire et al. (eds.), *Media and Metamedia Management*,
Advances in Intelligent Systems and Computing 503,
DOI 10.1007/978-3-319-46068-0_65

65.1 Mass Media and Journalistic Information Specialized in the Field of Architecture

Referring to journalistic communication specialized in the field of architecture implies a type of information that involves the deepening of a series of items linked to both the design and technical elements, always approaching the current reality through specialization. Nowadays there is a large flow of information in society, spurred by the rise of the Internet, and new media are aware of the impact of new ICTs and the multiscreen and multiplatform ecosystem, which is revolutionizing the way media are consumed. All of this requires skilled professionals—journalists, architects, designers and engineers—in this multiscreen area capable of offering a broad and specialized vision of all the data processed in the field of architecture.

It is necessary to start by stressing the fact that specialized journalism is the discipline responsible for establishing a possible management between the informative content providing a globalized synthesis (Estévez Ramírez 1999) to highlight that a specialization in this type of information such as in the field of architecture is vital at deepening in the sector, and even more when we refer to digital media with the ability to transfer specialized knowledge in the area of architecture.

Internet requires the development of specific publishing procedures for digital publications as well as the renewal of the editorial procedures employed by the classic means (Salaverría 2009). It is a fact that affects in exactly the same way to all publications of a specialized nature. But what is specialized journalism? Martinez De Sousa (1981) defines it as that discipline whose "main function is to treat and disseminate news and information of various kinds, echoing the interests of a class (working, capitalist), a sport (soccer, boxing, etc.), a social function (Military, Marine, motorists, artists, financial, religious, etc.) or a science (medicine, biology, astronautics, electronics, etc.)". Martinez Albertos (1984) distinguishes between specialized press and specialist journalism. The specialist press would be all those publications with fixed frequency or without it, always aimed to specific professionals, specialists in a given scientific activity; while the specialist journalism would be aimed at a broad audience without a specific target. All these publications can be extrapolated widely to the universe Web setting out the requirements that the cyber-media demand according to their intrinsic characteristics.

The cyber-media that emphasize journalistic information in the branch of architecture, function as an instrument of mediation between the specialists in the area and the audiences to which they are intended, and for the same reason, both the content and the form and the language of the message are conditioned by the audience it addresses. Thus, the architecture takes place in the online media as it did since its inception in all areas of knowledge.

65.1.1 *Architectural Journals Online as a Means of Data Dissemination*

It is a fact that the online magazines specialized in the field work are a means of information entry in different styles that develop and evolve in different countries, being the scope of professional architects and construction, as well as of the consumer. We can define the architecture online magazines as periodicals devoted to the architecture or to their attached domains: the edifice, the building, the construction, the real estate aspects, public works and, in the same way, they incorporate urbanism issues, monuments or archaeological buildings and even architectural decoration, or by resorting the history of architecture and heritage aspects (Leniaud and Bouvier 2001).

The online magazines specialized in architecture have taken a turn to the subjects treated with respect to the publications printed (bibliography of architects or spokespersons of ideological currents of society). Nowadays, the theme of the online publications is based more on aesthetics and are limited to a conceptual information. These publications work as a commercial and advertising catalog whose main function is to explain another architecture history, made of active ideological filtration, teaching and conviction, but also of intense guardianship toward the consumers and in this way the architecture is placed in the media "in a kind of limbo where the aspirations of the creators and consumers expect converge, where the borders are diluted between the ideal and the real, and where you create a continued expectation toward what is always up to approach" causing a change in the patterns of updated architecture consumption, through these specialized publications as modern styles, groundbreaking and innovative from what society needs (Rodríguez Pedret 2012).

Little by little, the online magazines specialized in architecture have evolved becoming referents for professionals who require an update of knowledge in styles and trends, at failing to find specific sections of the area within the general press. Thus, the online architecture magazines get transformed into an interesting means both for those who like or enjoy this area of knowledge, professionals from the field, as well as for other related professionals as designers, builders, engineers, professionals of the real estate sector, and even students in training.

These digital specialized publications allow to access works that would otherwise be ignored because its absence would imply an exhaustive search in specialized books, to collect data relating to plans, sections, photos, constructive details, etc., also taking into account that its costs are usually high precisely because the sales are more limited and restricted to a highly segmented and scarce audience. Also, the digitization has revolutionized the ways to access of the population in general (and more particularly of architecture and design professionals), to the information target of study. The digital magazines of architecture have increased the number of queries and are in the online age a reference that enables to collect information to advance or acquire a new perspective on new trends and styles (it should not be forgotten that they propose contents related to constructions, allow to publicize buildings, contests, interviews and general information that are published regularly in any geographical area).

This is an information not only interesting for people who belong to the world of architecture but also for all potential buyers of a future dwelling, with more or less passion for the design or technical expertise. All additional data on buildings are key for users (not only technicians working in the field), due to the fact it facilitates consumers the decision at the time of acquiring a dwelling without having any idea of construction. Features such as thermal conductivity, types of materials, ecological or not, acoustic insulation to the airborne noise, etc., are essential elements and specific demanding conditions to determine the quality of a space, not only for the purchase or sale price, but also for the expenses that they generate, a result of the opinions chosen (heating, air conditioning, etc.).

65.2 Methodology

This work has applied a quantitative and qualitative mixed methodology to establish the problem faced by the specialized architecture publications to get access to different types of audiences.

65.2.1 Quantitative

Six digital magazines are taken as reference in the field of architecture (Alive architecture, Architecture Platform, Passages, Digest Architecture, Axxis and Room Digital) from which data is collected to determine what the contents of higher incidence are in the period between October 6th and 20th 2015, and from these contents the proposed elements are analyzed.

Subsequently the technical language used in each of the selected digital media is analyzed, for a later comparison with other disciplines whose technical data are known (automobile field, cellphone and computing devices). In each area three digital magazines are reviewed to analyze the contents exposed.

65.2.2 Qualitative

A survey is conducted to an audience consisting of one hundred people with middle-high educational level and between 25 and 60 years old as maximum, (age range of potential buyers in 2015 according to Planner Exhibitions and VS Valuation Society) who were asked if they used media to obtain information about architecture or data related to this, affecting mainly at the time of acquiring a dwelling. In addition, the justification of its response was set, concepts of the area, and confrontation with other products that do possess specific technical features of the information that is required for an adequate understanding. The survey is presented in point 3. Results, including the percentage of the data obtained.

65.3 Results

The results obtained from the analysis of the architecture magazines are those shown in Table 65.1, showing in "Data", the topics discussed in each one of the news in which it appears and the common words in the architectural jargon that appear in the news analyzed and the frequency of occurrence in percentage.

The results of the technical language analysis are shown in Table 65.2, indicating in the *data* the exposed topics in each one of the news in the areas studied: automotive, cellphone and computing, and the corresponding percentage (%).

The results obtained from the survey are presented in Table 65.3.

Table 65.1 Data obtained from the digital magazines of architecture concerning the analyzed topics and specific words of architecture incorporated in the news

Analyzed topics		Words incorporated in the news			
Data	%	Data	%	Data	%
Architect	100	Architectural language	72	Urban context	52
Project data	20	Construction techniques	48	Architectural program	36
		Energy rating	0	Harmony/balance	32
Situation	96	Color	24	Climate zone	0
Conceptual idea	92	Texture	16	Materials (stone, wood, concrete, brick, etc.)	100
Climate	20	Composition	52	Natural light	36
Land	28	Acoustic insulation	0	Thermal transmittance	0
Program	64	Orientation	56	Core	28
Typology	32	Energy consumption	0	Durability	36
Structure	8	Movement	32	Volume	44
Materials	76	Impermeability level	0	Thermal isolation	0
Recycling/ecology	4	Power generation	12	Symmetry	20

Source self-elaboration

Table 65.2 Data obtained from the digital magazines in the automobile area, of the cellphone and computing devices and their percentage

Automotive industry		Cellphone devices		Computing devices	
Data	%	Data	%	Data	%
Engine	80	Screen	100	Processor	100
Power	70	Resolution	100	RAM memory	100
No. speeds	10	Dimensions	20		
Technical system	100	RAM	100	Hard disk	33
Measures	30	Image stabilizer	40	Operating system	66
Load volumen	30				
Consumption	40	Camera	100	Ports	100
Type of headlights	40	Operating system	60	Card reader	100
Sensors	10	Processor	80		
Acceleration	20	Networks	20	Bluetooth	100
Speed	20	Storage	60	Dimensions	33
Fuel	30	Battery	40		
Type of tires	30	Fingerprint reader	20	Wifi	33
Color	40	Measures	40	Keyboard	33
Interior design	10	Weight	40	Resolution	33
Materials	30	Material	40	Color	33
Finishing	40	Color	40		

Source self-elaboration

65.4 Discussion and Conclusions

The digital magazines of architecture analyzed generally contain only basic information: *Room digital and Passages* have poor data, only specifying initial information and easy to find the different items: architects/designers, the construction site, typology, conceptual idea, printed magazine in which the item is located, and year it was built. *Alive architecture* is the magazine that values more the dissemination of current trends from the cover page, as well as across all of its sections, works, projects, saving a section dedicated to culture. *Architecture Digest*, does not detail any input, contributing in the text with data such as interior and exterior design, materials used, vegetation, customer, functionality or photographs, incorporating audiovisual material with indoor and outdoor tours, even integrating recordings of the development of the construction.

Platform for Architecture and *Axxis* detail the head of construction, landscape architect, project manager, customer, collaborators are and, if they are ecological, engineering, Constructor, area of building and climate of the zone, etc., but it continues without specifying data that involves information more valuable, or data that involve the enlargement of the concept about constructions, which would bring as a direct result that consumers had a more extensive and detailed vision.

Table 65.3 Results of the survey

Survey						
1	Do you use magazines of architecture for information concerning to architecture/design? Yes/No					
	Yes			No		
	19 %			81 %		
	Is the search of the publication in digital or printed format?		What are the purposes? (professionals, hobby, keep me informed, others)	Point out the reasons why you do not use these publications		
	73 % digital	27 % printed	88 % professional	12 % keep informed	87 % technical language, lack of knowledge in the subject		13 % lack of interest
2	At the time of acquiring a dwelling, local, or land for your home or business, would you consider the possibility of searching in a specialized magazine of architecture in order to get more information about the possibilities of purchase? Yes/No					
	Yes			No		
	19 %			81 %		
3	Why?					
	Those who answer yes, are people who belong to the architecture professional sector/related professions or with domain of art or design knowledge, so that they emphasize the relevance of the content at the time of acquiring a new vision on the matter or decision making			Those who answer no, highlight they do not believe they could find guidelines that will assist them in the decision or they did not previously think on it as a method of information aimed to others who are not beyond the professionals of the area		

Source self-elaboration

From the results obtained in the surveys, the majority of the surveyed population (81 %) has never considered the publications specialized in architecture as a means to learn, at considering they are aimed to professionals, when in fact it is possible to obtain data that can be valuable at the moment of acquiring a dwelling and take advantage of the benefits in their homes and places of work, etc.

At posing the query option to these specialized publications, the surveyed people insert it within its range of possibilities but still reluctant to unfounded beliefs of ignorance about language, a language as checked in the comparison tables regarding to other areas (automotive, cellphone or computing) users do present broad notions of terminology that is really very specialized and technical. It is due to globalization that makes it a basic need to possess the most modern device on the market in terms of technology, or the latest model of car with the most innovative features. Data such as energy consumption, energy rating, climatic zone, thermal conductivity, acoustic absorption or thermal isolation are almost as important as memory RAM capacity, image resolution, speed processor, operating system, load volume or type of tire.

The non-dissemination of these technical data responds mainly for two reasons:

1. The buildings are not industrialized products in worldwide sales and with the same features, such as a car or other article. They are not manufactured in series, but in a general way, they are products that are sold on flat, differently to the usual commercial, through spots that bomb daily to potential customers around the world, as it happens with other articles of the sectors analyzed.

2. They do not compete with other similar items (all are different designs, surfaces, etc.), and therefore it is not needed to know each and every one of its technical features, to understand which entails greater profitability for the client. Only the properties of the components that are part of the construction are known and if they are commercial and industrialized materials.

References

Estévez Ramírez, F. (1999). *Áreas de especialización periodística*. Madrid: Fragua.

Leniaud, J. M., & Bouvier, B. (2001). *Les périodiques d'architecture. XVIIIe- XXIe siècle. Recherche d'une méthode critique d'analyse*. Paris: Études et rencontres de l'École des Chartres.

Martinez Albertos, J. L. (1984). *Curso general de redacción periodística*. Barcelona: Mitre.

Martinez De Sousa, J. (1981). *Diccionario general de periodismo*. Madrid: Paraninfo.

Rodríguez Pedret, C. (2012). *Arquitectura en el limbo. Los medios de masas y la difusión de la cultura moderna*. 4IAU 4ª Jornadas Internacionales sobre Investigación en Arquitectura y Urbanismo.

Salaverría, R. (2009). *Cibermedios: El impacto de Internet en los medios de comunicación en España*. Sevilla: Comunicación social.

Chapter 66
A Study of Student and University Teaching Staff Presence on ResearchGate and Academia.edu in Spain

Mar Iglesias-García, Cristina González-Díaz and Lluís Codina

Abstract The main function of a Scientific Social Network (SSN) is to provide researchers and academics with a channel through which to promote their work and make their findings available to others. However, as of today, the majority of studies published on SSN sites (which have been assumed to be used almost exclusively by university teaching staff) provide only a partial picture of the overall SSN user community. Research reported in this article was conducted to test the hypothesis that although the majority of SSN profiles belong to university teaching staff, an ever-growing number of undergraduate, masters and doctoral students are actively using these platforms. An analysis of tallies of ResearchGate and Academia.edu profiles attributable to two distinct groups (university teaching staff on the one hand and undergraduate, graduate and doctoral students on the other) within five different Spanish academic communities indicates that university students at all levels now account for a significant percentage of the total ResearchGate and Academia.edu profiles created and maintained in Spain.

Keywords Scientific social networks (SSN) · Spanish universities · Researchgate · Academia.edu

M. Iglesias-García (✉) · C. González-Díaz
University of Alicante (UA), San Vicente Del Raspeig, Spain
e-mail: mar.iglesias@ua.es

C. González-Díaz
e-mail: cristina.gdiaz@ua.es

L. Codina
University Pompeu Fabra (UPF), Barcelona, Spain
e-mail: lluis.codina@upf.edu

© Springer International Publishing Switzerland 2017
F.C. Freire et al. (eds.), *Media and Metamedia Management*,
Advances in Intelligent Systems and Computing 503,
DOI 10.1007/978-3-319-46068-0_66

66.1 Theoretical Framework

66.1.1 Scientific Social Networking Platforms

Ever since they were created in 2007, Scientific Social Networks (SSN) have been aiming to enhance the visibility, impact and spreading of research findings (Campos Freire and Direito Rebollal 2015). These platforms have a similar structure to that of multi-purpose social networks such as Facebook: they allow users to create a profile, publish their scientific findings, search for and download reference materials and articles uploaded by other scholars. They also facilitate document sharing, encourage collaborations and foster research involving multiple users who work on the same or different disciplines. SSN users can monitor profile traffic, check how many times their posts have been downloaded, see how many followers they have, and measure the impact of their work (González-Díaz et al. 2015). Given the novelty of Scientific Social Networking, most studies conducted on this topic so far have been merely devoted to description and characterisation (Campos et al. 2014; González-Díaz et al. 2015; Martorell et al. 2014).

Most of the studies conducted on SSN sites strongly suggest that the majority of user communities are made up of university teaching staff. This is quite a curious assumption given the fact that no categorical distinctions are made on these sites when it comes to the users' profession or professional status. The starting hypothesis of our research is that although SSN user bases are mainly comprised of academics, more and more undergraduate, graduate and doctoral students are becoming active members of these online communities.

66.1.2 ResearchGate and Academia.edu

Out of the almost fifty SSNs available today, it is Academia.edu (2008) and Researchgate.net (2008) that have achieved the greatest degree of penetration within the academic community: they have more than 35 million users altogether, and they get millions of occasional visitors every month. Each of these sites offers customised features and tools. Academia.edu, which is currently the most popular of the two, features a statistics tool that allows researchers to measure the impact of their work and also provides statistics on page visits, downloads, keywords used in searches and other useful data. A large number of the materials posted and exchanged via this network are related to social sciences. ResearchGate is the second most popular SSN site, and it features a proprietary academic reputation index that is referred to as RG Score. This metric reflects how a user's work is received by other registered network members by taking various factors into account, including online sharing activities, interactions, and peer-to-peer evalua-tions (ResearchGate 2015a, b).

Even if the user bases of Scientific Social Networking sites are growing exponentially, there are no clear indications as to how and for what purposes they are being used or who is using them. As stated above, while it is largely assumed that university teaching staff account for the bulk of activity that takes place on these sites, we have observed that an increasing number of users with different academic profiles are joining the ranks of these communities.

66.2   Methodology

A small-scale exploratory analysis of SSN profiles created by university teaching staff and students was conducted for the purposes of clarifying the size and nature of the current SSN community in Spain. Our study sample comprised teachers and staff from the five largest universities in Spain that have Audiovisual Communication and Advertising departments, namely the Complutense University of Madrid, the University of Seville, the University of the Basque Country, the Autonomous University of Barcelona and the University of Malaga.

66.3   Results

With the sole exception of the University of the Basque Country's teaching staff and students, a greater number of SSN users from the Spanish universities analysed in this study had chosen to create a profile on Academia.edu rather than on ResearchGate. Out of the 5976 schools represented in the sample, the Autonomous University of Barcelona had the highest number of Academia.edu users, even if the size of its academic community (40,771) is half that of the Complutense University of Madrid (800,027). Interestingly enough, the situation for ResearchGate was the opposite: here; it was the Complutense University of Madrid that had the highest number of users (5202).

The presence of teaching staff and students in Scientific Social Networking sites was analysed in percentage terms and compared with the total size of the university's teaching staff and student body. This analysis showed that the largest universities in the sample were not necessarily the ones with the highest percentage of SSN users. For instance, the Autonomous University of Barcelona ranks fourth amongst the studied universities in terms of teaching staff and student body size. However, the percentage of teaching staff and students who use ResearchGate at each school shows that this university actually has the highest SSN user/total teaching staff ratio out of all the student bodies and schools in the sample: 91.9 % of its teaching staff, 10.4 % of its students and 9.3 % of its academic community members have a ResearchGate account. Despite having the largest teaching staff and student body in the sample, the Complutense University of Madrid only comes in second in terms of teaching staff and student engagement on this site. The case for the University of Seville is similar: notwithstanding the overall size of its academic community (68,029 people) and the fact that its teaching staff and student body is the second largest among the schools examined, this university ranks lowest in terms of teaching staff and student interest in ResearchGate: only 69 % of its teachers, 4.8 % of its students and 4.5 % of its academic community members have an account on this site.

The breakdown for Academia.edu is somewhat similar in the sense that the Autonomous University of Barcelona ranks first in terms of across-the-board engagement. These findings are nonetheless conditioned by an interesting statistical anomaly: the percentage of teachers and researchers from the Autonomous University of Barcelona that had a profile on this site was over 100 %.

As discussed below, this aberration is due to the fact that certain types of users often create multiple profiles, a phenomenon that is more common on Academia. edu but can also be seen on ResearchGate. This makes it very difficult to accurately calculate percentages for each user category. When the scope of our examination was narrowed down to the departments of Audiovisual Communication and Advertising of the five universities in question, we saw that the Audiovisual Communication and Advertising departments at the Complutense University of Madrid were ahead of the others in terms of total ResearchGate profiles (48), while the University of Seville's Department of Audiovisual Communication and Advertising came in first in terms of total Academia.edu profiles.

This leadership pattern can be seen in both Scientific Social Networking sites examined. The Departments of Audiovisual Communication, Advertising I and Advertising II of the Complutense University of Madrid altogether scored highest in terms of total teaching staff (21) and student (27) profiles registered on ResearchGate. The fact that student profiles outnumbered teaching staff profiles in this case would appear to support our hypothesis that these sites are not being exclusively used by university teachers and researchers. Although student profiles indeed outnumbered teaching staff profiles at each studied university (with the sole exception of the University of the Basque Country), high incidence of empty and incomplete profiles detected within the sample prevents us from making broad assumptions on this phenomenon.

As previously mentioned, the Department of Audiovisual Communication and Advertising of the University of Seville scored highest in terms of the total number of teaching staff and student profiles registered on Academia.edu. Student profiles connected with this university's audiovisual communication and advertising department outnumbered teaching staff profiles both here and at ResearchGate, although many of them had never been properly completed.

Another metric for determining how strong the presence of a given university or department is on a certain SSN site is the number of documents uploaded by users. Teachers and students from the Audiovisual Communication and Advertising departments at the Complutense University of Madrid uploaded more documents (625) to a given site (ResearchGate) than any other department in the sample. Teachers and students from the University of Seville's Department of Audiovisual

Communication and Advertising ranked second in this category with 386 documents uploaded to the Academia.edu website.

A tally of the number of profiles created by audiovisual communication and advertising teachers and researchers from these five universities indicates that a relatively low percentage have profiles on either site. For example, the Audiovisual Communication and Advertising Department of the Complutense University of Madrid ranks first when it comes to the number of teaching staff profiles registered on ResearchGate, but only 32 % (21 out of 64) of the teachers and researchers in said department actually have a profile on the site. A notable exception is the teaching staff at the Audiovisual Communication and Advertising department of the University of Seville: 83.6 % of its members (41 out of 49) have a profile on Academia.edu. Search results for teaching staff profiles on Academic.edu filtered by university or department indicate that the high number of duplicate and empty profiles on that site is largely due to the confusing array of options one must choose from when creating an account.

These sites in general and Academia.edu in particular have another characteristic that makes it difficult to come up with final figures: the possibility to create several accounts under different names and in multiple languages. In order to make sure the statistics they publish are reliable, SSN sites need to develop a means of factoring in the existence of duplicate profiles and filtering out empty profiles.

66.4   Discussion

The results of this study are in line with the findings of previous studies conducted by González-Díaz et al. (2015) and Iglesias-García et al. (2015). Said research established that teaching staff and student presence on Scientific Social Networking sites is not necessarily proportional to the size of the university that users were affiliated with when they created their SSN profiles. However, a different pattern was revealed when departmental-level analysis was carried out at the same five universities. On the one hand, the Audiovisual Communication and Advertising departments of the Complutense University of Madrid (which has the largest student and teacher populations) accounted for the highest number of ResearchGate profiles. On the other hand, the Department of Audiovisual Communication and Advertising of the University of Seville (which has the second largest student and teacher populations) had the highest number of Academia.edu profiles.

A comparison between the number of university teaching staff profiles registered on each Scientific Social Networking site was also carried out within the framework of this research. Results show that penetration rate in Spain is higher for Academia.edu (over 80 % of teachers from the universities and departments included in this study sample) than for ResearchGate (slightly over 30 % of the same population). Furthermore, two of this study's findings also stand out for their relevance: first, the fact that student profiles outnumbered university teaching staff profiles on the two sites examined and second, the fact that both sites had a high amount of empty profiles.

These two phenomena seem to be clearly related: most empty profiles might be the result of university teachers trying to register on these sites unsuccessfully, or just giving up before completing their profiles. Such a scenario would at least partially explain why there are more student than teacher profiles on these sites. The existence of so many empty and duplicate profiles points to the need for further studies aimed at clarifying two issues: what type of users tend to create empty profiles, and how many of the already existing profiles can be flagged as duplicates. If left unresolved, these issues stemming from design and management weaknesses will eventually undermine the sites' credibility. This is especially true in the case of Academia.edu, where the highest number of empty and duplicate profiles was detected during our study. The fact that some of the profiles lack information about the owner or their scientific work is in clear contradiction of both sites' core purpose, which is disseminating scientific findings.

Although Scientific Social Networks are still in the process of sorting out classic startout problems, they have quickly proven their value as vehicles for collaboration and knowledge exchange. Given the new windows of opportunity that SSNs have opened up for the circulation of scientific knowledge and interaction between scholars, it is well worth ensuring that they will not devolve into mediocre repositories of little use to the academic community. Further studies that go beyond description and characterisation of the subject and rather approach it from a wide range of angles and perspectives will be necessary in order to completely understand the potential, the problems and the ramifications of this groundbreaking, growingly popular concept.

Acknowledgments This research was carried out as a part of project "Active Audiences and Journalism: Interactivity, Web Integration and Findability of Journalistic Information CSO2012-39518-C04-02", which is funded by the National Plan for Scientific and Technical Research and Innovation (Spanish Ministry of Economy and Competitiveness).

References

Campos Freire, F., & Direito Rebollal, S. (2015). La gestión de la visibilidad de la ciencia en las redes sociales digitales. In J. Rom, J. Cuenca & K. Zilles (Eds.), *Negotiating (in)visibility: managing attention in the digital sphere*. VIII International Conference on Communication and Reality, 117–124.

Campos Freire, F., Rivera Rogel, D., & Rodríguez, C. (2014). La presencia e impacto de las universidades de los países andinos en las redes sociales digitales. *Revista Latina de Comunicación Social, 69*, 571–592. doi:10.4185/RLCS-2014-1025

González-Díaz, C., Iglesias-García, M., & Codina, L. (2015). Presencia de las universidades españolas en las redes sociales digitales científicas: caso de los estudios de comunicación. *El profesional de la información, 24*(5), 640–647. doi:10.3145/epi.2015.sep.12

Iglesias-García, M., González-Díaz, C., & Codina, L. (2015). Estudio de los perfiles de ResearGate y Academia.edu en las universidades españolas: el caso de los Departamentos de Comunicación Audiovisual y Publicidad. XESCOM I Simposio Internacional sobre Gestión de la Comunicación. Pontevedra (España).

Martorell, S., Canet, F., & Codina, L. (2014). Canalizar audiencias académicas: Propuesta de una red social para investigadores en estudios fílmicos. *Hipertext. net* 12.

ResearchGate (2015a). Celebrating 6 million members. Retrieved December 18, 2015 from https://www.researchgate.net

ResearchGate (2015b). 8 out of 8 million. Retrieved December 18, 2015 from https://www.researchgate.net/blog/post/8-out-of-8-million

Printed in the United States
By Bookmasters